Recycled Materials in Civil Engineering Application

Recycled Materials in Civil Engineering Application

Editor

Malgorzata Ulewicz

Basel • Beijing • Wuhan • Barcelona • Belgrade • Novi Sad • Cluj • Manchester

Editor
Malgorzata Ulewicz
Czestochowa University of Technology
Czestochowa, Poland

Editorial Office
MDPI
St. Alban-Anlage 66
4052 Basel, Switzerland

This is a reprint of articles from the Special Issue published online in the open access journal *Materials* (ISSN 1996-1944) (available at: https://www.mdpi.com/journal/materials/special_issues/recycled_mat_civil).

For citation purposes, cite each article independently as indicated on the article page online and as indicated below:

Lastname, A.A.; Lastname, B.B. Article Title. *Journal Name* **Year**, *Volume Number*, Page Range.

ISBN 978-3-0365-9540-5 (Hbk)
ISBN 978-3-0365-9541-2 (PDF)
doi.org/10.3390/books978-3-0365-9541-2

© 2023 by the authors. Articles in this book are Open Access and distributed under the Creative Commons Attribution (CC BY) license. The book as a whole is distributed by MDPI under the terms and conditions of the Creative Commons Attribution-NonCommercial-NoDerivs (CC BY-NC-ND) license.

Contents

About the Editor vii

Preface ix

Malgorzata Ulewicz
Recycled Materials in Civil Engineering Application
Reprinted from: *Materials* **2023**, *16*, 7057, doi:10.3390/ma16227075 1

Tianhao Zhao, Yong Lv, Jianzhong Chen, Pengfei Song, Mingqing Sun, Xiaoyu Zhang and Li Huang
Effect of Glass Fiber-Reinforced Plastic Waste on the Mechanical Properties of Concrete and Evaluation of Its Feasibility for Reuse in Concrete Applications
Reprinted from: *Materials* **2023**, *16*, 6772, doi:10.3390/ma16206772 5

Nelli G. Muradyan, Avetik A. Arzumanyan, Marine A. Kalantaryan, Yeghiazar V. Vardanyan, Mkrtich Yeranosyan, Malgorzata Ulewicz, et al.
The Use of Biosilica to Increase the Compressive Strength of Cement Mortar: The Effect of the Mixing Method
Reprinted from: *Materials* **2023**, *16*, 5516, doi:10.3390/ma16165516 23

Yao Xiao, Zhengguang Wu and Yongfan Gong
Study on Alkali-Activated Prefabricated Building Recycled Concrete Powder for Foamed Lightweight Soils
Reprinted from: *Materials* **2023**, *16*, 4167, doi:10.3390/ma16114167 33

Tomasz Kalak, Patrycja Szypura, Ryszard Cierpiszewski and Malgorzata Ulewicz
Modification of Concrete Composition Doped by Sewage Sludge Fly Ash and Its Effect on Compressive Strength
Reprinted from: *Materials* **2023**, *16*, 4043, doi:10.3390/ma16114043 45

Alina Pietrzak and Malgorzata Ulewicz
Influence of Post-Consumer Waste Thermoplastic Elastomers Obtained from Used Car Floor Mats on Concrete Properties
Reprinted from: *Materials* **2023**, *16*, 2231, doi:10.3390/ma16062231 63

Fouad Boukhelf, Daniel Lira Lopes Targino, Mohammed Hichem Benzaama, Lucas Feitosa de Albuquerque Lima Babadopulos and Yassine El Mendili
Insight into the Behavior of Mortars Containing Glass Powder: An Artificial Neural Network Analysis Approach to Classify the Hydration Modes
Reprinted from: *Materials* **2023**, *16*, 943, doi:10.3390/ma16030943 77

Paulina Kostrzewa-Demczuk, Anna Stepien, Ryszard Dachowski and Rogério Barbosa da Silva
Influence of Waste Basalt Powder Addition on the Microstructure and Mechanical Properties of Autoclave Brick
Reprinted from: *Materials* **2023**, *16*, 870, doi:10.3390/ma16020870 101

Katarzyna Borek, Przemysław Czapik and Ryszard Dachowski
Cement Bypass Dust as an Ecological Binder Substitute in Autoclaved Silica–Lime Products
Reprinted from: *Materials* **2023**, *16*, 316, doi:10.3390/ma16010316 125

Sanja Dimter, Martina Zagvozda, Tea Tonc and Miroslav Šimun
Evaluation of Strength Properties of Sand Stabilized with Wood Fly Ash (WFA) and Cement
Reprinted from: *Materials* **2022**, *15*, 3090, doi:10.3390/ma15093090 139

Jakub Jura and Malgorzata Ulewicz
Assessment of the Possibility of Using Fly Ash from Biomass Combustion for Concrete
Reprinted from: *Materials* **2021**, *14*, 6708, doi:10.3390/ma14216708 165

Muhammad Sufian, Safi Ullah, Krzysztof Adam Ostrowski, Ayaz Ahmad, Asad Zia, Klaudia Śliwa-Wieczorek, et al.
An Experimental and Empirical Study on the Use of Waste Marble Powder in Construction Material
Reprinted from: *Materials* **2021**, *14*, 3829, doi:10.3390/ma14143829 . . . . **181**

Zinoviy Blikharskyy, Khrystyna Sobol, Taras Markiv and Jacek Selejdak
Properties of Concretes Incorporating Recycling Waste and Corrosion Susceptibility of Reinforcing Steel Bars
Reprinted from: *Materials* **2021**, *14*, 2638, doi:10.3390/ma14102638 . . . . **199**

Małgorzata Olejarczyk, Iwona Rykowska and Włodzimierz Urbaniak
Management of Solid Waste Containing Fluoride—A Review
Reprinted from: *Materials* **2022**, *15*, 3461, doi:10.3390/ma15103461 . . . . **215**

Magdalena Dobiszewska, Orlando Bagcal, Ahmet Beycioğlu, Dimitrios Goulias, Fuat Köksal, Maciej Niedostatkiewicz and Hüsamettin Ürünveren
Influence of Rock Dust Additives as Fine Aggregate Replacement on Properties of Cement Composites—A Review
Reprinted from: *Materials* **2022**, *15*, 2947, doi:10.3390/ma15082947 . . . . **239**

About the Editor

Malgorzata Ulewicz

Malgorzata Ulewicz graduated from the Czestochowa University of Technology and the Higher Pedagogical School in Czestochowa. She completed her PhD degree and training at the Faculty of Engineering, Process, Materials and Applied Physics at the Czestochowa University of Technology. She is employed as a professor at the Czestochowa University of Technology and is the vice-dean of the Faculty of Civil Engineering.

Her research interests are related to materials engineering, in particular, the process of recycling materials, the use of waste materials in the synthesis of new materials, metal separation techniques, and environmental protection. Her didactic interests include the processes of the utilization and recycling of materials, sustainable construction, management, and social ecology.

Malgorzata Ulewicz is the author or co-author of 250 articles, including 9 books and monographs, 100 articles in journals, 15 chapters in monographs, and a number of papers based on original material research results presented at national and international conferences. She is the Editor-in-chief of the *Construction of Optimized Energy Potential* journal and a member of the Environmental Protection and Waste Management Committee of the Polish Academy of Sciences and the Building Engineering Committee of the Polish Academy of Sciences. She is also a member of several scientific committees for conferences and journals.

Preface

This Special Issue "Recycled Materials in Civil Engineering Application" covers an important aspect of the latest research in the field of materials produced with recycled materials that may be used in the construction sector. The production of materials, especially construction materials, using recycled materials is a key element in the development of a circular economy. Various types of recycled materials are used for their production, including plastics, metal alloys, glass, and ceramics, as well as industrial waste. Due to the diverse physico-chemical properties of the waste and recycled materials used, the synthesis of building materials using them requires in-depth research, which creates challenges for scientists and engineers. Therefore, this Special Issue presents the results of the latest research in the field of the synthesis of building materials, their properties, and their impact on the environment. Other issues are also presented, including a model for predicting the hydration process and technologies for removing fluorine from various wastes. This Special Issue contains a total of twelve original scientific papers and two review articles. Both laboratory research results and interesting and practical solutions are presented. The published research shows that the use of recycled materials to produce materials for the construction sector is a key element in reducing the consumption of natural resources and leads to circular material management.

Malgorzata Ulewicz
Editor

Editorial

Recycled Materials in Civil Engineering Application

Malgorzata Ulewicz

Faculty of Civil Engineering, Czestochowa University of Technology, Dabrowskiego 69 Street, 42-201 Czestochowa, Poland; malgorzata.ulewicz@pcz.pl; Tel.:+48-343250935

Citation: Ulewicz, M. Recycled Materials in Civil Engineering Application. *Materials* **2023**, *16*, 7057. https://doi.org/10.3390/ma16227075

Received: 2 November 2023
Accepted: 3 November 2023
Published: 8 November 2023

Copyright: © 2023 by the author. Licensee MDPI, Basel, Switzerland. This article is an open access article distributed under the terms and conditions of the Creative Commons Attribution (CC BY) license (https:// creativecommons.org/licenses/by/ 4.0/).

In recent years, the construction sector has shown great interest in the use of various by-products and industrial waste, as well as the consumer products used. The use of various waste and recycled materials by the construction sector fits well into the idea of sustainable construction. Replacing at least partial natural resources with waste materials allows them to be preserved for future generations and is also the first step toward a circular economy. It should be remembered that the design and production of building materials should take into account appropriate standards and environmental protection requirements. Moreover, the use of waste from various industries and recycling materials in building materials must be preceded by a series of laboratory tests to make them harmless to the environment and people.

This Special Issue of "Recycled Materials in Civil Engineering Application" publishes the latest results of scientific research on the use of waste and recycled material for the production of mortar, concrete, ceramic and other building material. Research in this area was conducted in many centers and universities around the world, including the USA, France, Pakistan, Turkey, Poland, Ukraine, Croatia, Brazil, China, Armenia and Chile. In this issue, twelve original scientific papers and two review articles were published. The topics of these articles cover a number of aspects that are important for construction and the environment, including determining the physicochemical and mechanical properties of mortar, concrete and ceramic materials produced from waste and recycled materials. A model for predicting the hydration process of a new alternative binder is also proposed, and methods and technologies for removing excess fluorine from various wastes, which are beneficial to the environment, are presented. Both laboratory research results and interesting and practical solutions are presented.

Zhao et al. [1] present the influence of adding glass fiber-reinforced plastic (GFRP) waste (in the range of 1–5%) on the properties of concrete. It has been shown that the addition of GFRP powder does not have a significant effect on the mechanical properties of concrete, while the addition of a small amount of GFRP cluster slightly improves the compressive and tensile strength of concrete. Pietrzak and Ulewicz [2] determined the influence of this addition (in the range of 2.5–10%) of post-consumer thermoplastic elastomer (TPE) derived from used car mats on the physical and mechanical properties of concrete. It has been shown that only the addition of post-consumer elastomer waste in an amount of up to 2.5% of the cement mass can be used as a replacement for sand and gravel aggregate in concrete without reducing its mechanical strength and microstructure. In turn, Kalak et al. [3] determined the mechanical properties of concrete produced with the addition of fly ash from the thermal conversion of municipal sewage sludge (SSFA). A correlation was demonstrated between the mechanical strength of waste-modified concrete and various parameters, including the composition of concrete mixtures (amount of sand and gravel, cement, SSFA), water-to-cement ratio (w/c) and sand point. In the scope examined by the authors, the addition of SSFA did not reduce the compressive strength of the concrete produced, and despite the lack of legal regulations regarding the physicochemical properties of SSFA, this material could be used for concrete.

In the production of concrete, fly ash, slag and biosilica were also used. Jura and Ulewicz [4] determined the properties of concrete produced with the addition of fly ash from the combustion of wood-sunflower biomass in a fluidized bed boiler. It was shown that fly ash used in an amount of 10–30% can be added as a substitute for sand in the production of concrete without impairing its mechanical properties (compressive strength and resistance to low temperatures) compared to the control concrete as the heavy metal ions present in the ashes do not have a negative impact on the environment. By contrast, Blikharskyy et al. [5] determine the mechanical properties of concrete produced with the addition of ground-granulated blast furnace slag (GGBFS) and fly ash (FA) and the corrosion susceptibility of reinforcing bars. It was found that the degree of fragmentation of GGBFS and FA and their simultaneous combination impacted the kinetics of increasing the strength of concrete. The highest compressive strength (6.5% higher) was obtained from concrete containing 10% GGBFS and 10% FA. In these types of concrete, the protection of the steel reinforcement was not compromised due to the low degree of substitution of Portland cement and its tighter microstructure. In turn, Muradyan et al. [6] determined the effect of adding biosilica and two different methods of mixing it with Portland cement (mixing the dry ingredients directly and mixing after dissolving biosilica in water) on the compressive strength of cement mortars. It has been shown that both mixing methods give positive results, and the highest compressive strength is achieved by mortars with the addition of 10% biosilica.

Waste materials were also used to produce other building materials, including autoclaved bricks and ballast mixtures. Bork et al. [7] used cement bypass dust (CBPD) to produce autoclaved bricks. It has been shown that the amount of CBPD dust used in autoclaved products depends on the chemical composition of the dust and, in particular, on the content of free CaO. The modification of the traditional silica–calcium mixture with bypass dust changes its phase composition and introduces new phases into the system in the form of portlandite and sylvine. The use of CBPD as a lime substitute does not inhibit the formation of hydrated lime silicates, characteristic of autoclaved products. On the contrary, an increase in the share of waste dust in products results in the formation of denser structures with a greater degree of crystallization, i.e., tobermorite. The complete replacement of quicklime with CBPD contributes to increasing the compressive strength of the manufactured bricks (by 5%) while reducing their bulk density (by 5.3%). On the other hand, Kostrzewa-Demczuk et al. [8] present the influence of adding basalt powder waste on the physical and mechanical properties of manufactured lime–sand products (silicates). It has been shown that the use of basalt flour at an amount of 10% as a component of lime–sand product brings positive effects (a double increase in compressive strength compared to traditional silicate). Unfortunately, with the increase in the addition of waste basalt flour, a decrease in the performance parameters of the manufactured products was recorded, although these products still met the conditions for traditional silicate bricks. Sufian et al. [9] determined the influence of adding marble flour (in the range of 5–30%) as a clay substitute on the properties of the produced bricks. It was shown that, with the addition of marble flour, the compressive strength and bulk density of the bricks decreased, but their water absorption capacity and porosity improved. Bricks made with the addition of 5–20% marble flour have adequate compressive strength in relation to the values required by standards, and the addition of marble waste does not significantly affect the amount of salt effloresce occurring in the bricks. In turn, Dimter et al. [10] determined the mechanical properties (California Bearing Ratio, compressive and intermediate tensile strength and resistance to freeze/thaw cycles) of wood fly ash (WFA) sand mixtures produced with different amounts of cement. It has been shown that WFA has a significant stabilizing effect on the sand mixture and improves its load-bearing capacity. Adding a small amount of cement causes the hydraulic reaction in the stabilized mixture to be more intense, resulting in greater strength and better resistance to freezing. The test results have shown that by replacing part of the sand with the addition of WFA (30%), a mixture with greater strength (by 90.7%) was obtained compared to the mixture of sand and cement alone, which contributes to savings during the construction of the pavement structure.

This Special Issue also presents an innovative solution that is useful for cement producers. Boukhelf et al. [11] present an artificial neural network (ANN) model to predict the hydration process of a new alternative binder. This model overcomes the lack of input parameters of physical models, providing a realistic explanation with fewer inputs and fast calculations. The verification of this model was carried out for mortar produced on the basis of CEM I and CEM III cement with added glass powder as the cement substitution. The proposed model could be useful for cement producers as it facilitates the quick identification of different hydration modes of new binders, using only the heat of the hydration test as an input parameter. An innovative approach to demolition waste was presented by Xiao et al. [12]. The authors proposed a process for preparing foamed lightweight soil using alkaline-activated concrete flour derived from prefabricated waste. It is best to use concrete flour powder, fly ash and slag at the amount of 60, 20 and 20%, respectively, to prepare light soil. The lightweight soil proposed by the authors is convenient to build without compaction, and its production costs are lower compared to filling soil or reinforced soil.

This Special Issue also includes two interesting review articles. Dobiszewska et al. [13] present a comprehensive review of literature reports on the use of rock dust from various geological origins for the production of mortars and concrete. The influence of rock dust as a substitute for fine aggregate on the properties of cement composites, such as workability, segregation and bleeding, mechanical properties and the durability of hardened concrete and mortar, was analyzed and assessed. The use of environmentally friendly rock dust in a specified amount to replace fine aggregate in cement-based composites has been shown to improve many properties in both the fresh and hardened state. The authors draw attention to the lack of information on the influence of particle size distribution in stone dust itself and on the properties of concrete and mortar. A detailed analysis of the particle size distribution can help in the decision to use stone flour because it affects the internal structure and many related mechanical properties of the manufactured materials. On the other hand, reports by Olejarczyk et al. [14] show that the production of building materials from various wastes containing hazardous substances, including fluorine, is still very limited. However, this type of waste can be used as additives or admixtures for the production of new construction materials after undergoing "solidification/stabilization" processes. The authors proposed several methods for reducing the concentration of fluoride ions in waste and several materials that can be used as fluorine adsorbents while also taking into account the low costs of processing these wastes.

All articles published in this Special Issue have been peer-reviewed by renowned experts. As a guest editor, I would like to thank all authors for their valuable contribution to the development of building materials engineering and the reviewers for their comments and suggestions that significantly improved the quality of the published articles. I would also like to express my sincere thanks to the Section Managing Editor, Ms. Freda Zhang, for her kind assistance in preparing this Special Issue of the Journal.

Conflicts of Interest: The author declares no conflict of interest.

References

1. Zhao, T.; Lv, Y.; Chen, J.; Song, P.; Sun, M.; Zhang, X.; Huang, L. Effect of Glass Fiber-Reinforced Plastic Waste on the Mechanical Properties of Concrete and Evaluation of Its Feasibility for Reuse in Concrete Applications. *Materials* **2023**, *16*, 6772. [CrossRef] [PubMed]
2. Pietrzak, A.; Ulewicz, M. Influence of Post-Consumer Waste Thermoplastic Elastomers Obtained from Used Car Floor Mats on Concrete Properties. *Materials* **2023**, *16*, 2231. [CrossRef] [PubMed]
3. Kalak, T.; Szypura, P.; Cierpiszewski, R.; Ulewicz, M. Modification of Concrete Composition Doped by Sewage Sludge Fly Ash and Its Effect on Compressive Strength. *Materials* **2023**, *16*, 4043. [CrossRef] [PubMed]
4. Jura, J.; Ulewicz, M. Assessment of the Possibility of Using Fly Ash from Biomass Combustion for Concrete. *Materials* **2021**, *14*, 6708. [CrossRef] [PubMed]
5. Blikharskyy, Z.; Sobol, K.; Markiv, T.; Selejdak, J. Properties of Concretes Incorporating Recycling Waste and Corrosion Susceptibility of Reinforcing Steel Bars. *Materials* **2021**, *14*, 2638. [CrossRef] [PubMed]

6. Muradyan, N.G.; Arzumanyan, A.A.; Kalantaryan, M.A.; Vardanyan, Y.V.; Yeranosyan, M.; Ulewicz, M.; Laroze, D.; Barseghyan, M.G. The Use of Biosilica to Increase the Compressive Strength of Cement Mortar: The Effect of the Mixing Method. *Materials* **2023**, *16*, 5516. [CrossRef] [PubMed]
7. Borek, K.; Czapik, P.; Dachowski, R. Cement Bypass Dust as an Ecological Binder Substitute in Autoclaved Silica–Lime Products. *Materials* **2023**, *16*, 316. [CrossRef] [PubMed]
8. Kostrzewa-Demczuk, P.; Stepien, A.; Dachowski, R.; Silva, R.B.d. Influence of Waste Basalt Powder Addition on the Microstructure and Mechanical Properties of Autoclave Brick. *Materials* **2023**, *16*, 870. [CrossRef] [PubMed]
9. Sufian, M.; Ullah, S.; Ostrowski, K.A.; Ahmad, A.; Zia, A.; Śliwa-Wieczorek, K.; Siddiq, M.; Awan, A.A. An Experimental and Empirical Study on the Use of Waste Marble Powder in Construction Material. *Materials* **2021**, *14*, 3829. [CrossRef] [PubMed]
10. Dimter, S.; Zagvozda, M.; Tonc, T.; Šimun, M. Evaluation of Strength Properties of Sand Stabilized with Wood Fly Ash (WFA) and Cement. *Materials* **2022**, *15*, 3090. [CrossRef] [PubMed]
11. Boukhelf, F.; Targino, D.L.L.; Benzaama, M.H.; Lima Babadopulos, L.F.d.A.; El Mendili, Y. Insight into the Behavior of Mortars Containing Glass Powder: An Artificial Neural Network Analysis Approach to Classify the Hydration Modes. *Materials* **2023**, *16*, 943. [CrossRef] [PubMed]
12. Xiao, Y.; Wu, Z.; Gong, Y. Study on Alkali-Activated Prefabricated Building Recycled Concrete Powder for Foamed Lightweight Soils. *Materials* **2023**, *16*, 4167. [CrossRef] [PubMed]
13. Dobiszewska, M.; Bagcal, O.; Beycioğlu, A.; Goulias, D.; Köksal, F.; Niedostatkiewicz, M.; Ürünveren, H. Influence of Rock Dust Additives as Fine Aggregate Replacement on Properties of Cement Composites—A Review. *Materials* **2022**, *15*, 2947. [CrossRef] [PubMed]
14. Olejarczyk, M.; Rykowska, I.; Urbaniak, W. Management of Solid Waste Containing Fluoride—A Review. *Materials* **2022**, *15*, 3461. [CrossRef] [PubMed]

Disclaimer/Publisher's Note: The statements, opinions and data contained in all publications are solely those of the individual author(s) and contributor(s) and not of MDPI and/or the editor(s). MDPI and/or the editor(s) disclaim responsibility for any injury to people or property resulting from any ideas, methods, instructions or products referred to in the content.

Article

Effect of Glass Fiber-Reinforced Plastic Waste on the Mechanical Properties of Concrete and Evaluation of Its Feasibility for Reuse in Concrete Applications

Tianhao Zhao [1], Yong Lv [1], Jianzhong Chen [1,*], Pengfei Song [2], Mingqing Sun [1], Xiaoyu Zhang [1] and Li Huang [1]

[1] Hubei Key Laboratory of Theory and Application of Advanced Material Mechanics, School of Science, Wuhan University of Technology, Wuhan 430070, China; thob06@163.com (T.Z.)
[2] Hengrun Group New Materials Co., Ltd., Hengshui 053100, China
* Correspondence: cjzwhut@163.com

Abstract: The disposal of glass fiber-reinforced plastic (GFRP) waste has become an urgent issue in both the engineering and environmental fields. In this study, the feasibility of reusing mechanically recycled GFRP in concrete was evaluated. Secondary screening of the recycled material was conducted to obtain different types of products, and the recycled GFRP (rGFRP) was characterized. Subsequently, the effect of rGFRP on concrete performance was evaluated using different dosages (0%, 1%, 3%, 5%) of rGFRP powder and rGFRP cluster (with different sizes and fiber contents) to replace fine aggregate in concrete preparation. The experimental results indicated that the addition of rGFRP powder has no significant impact on the mechanical properties of concrete, while the addition of a small amount of rGFRP cluster slightly improves the compressive strength and splitting tensile strength of concrete. Additionally, the short fibers in rGFRP improve the failure mode of concrete, and increased fiber content and longer fiber length demonstrate a more pronounced reinforcing effect. The challenges and potential directions for future research in the realm of reusing rGFRP in concrete are discussed at the end. A systematic process for reusing GFRP waste in concrete is proposed to address the primary challenges and provide guidance for its practical engineering application.

Keywords: glass fiber-reinforced plastic (GFRP); concrete; mechanical properties; mechanical recycling; optimized reuse process

1. Introduction

In recent years, the use of glass fiber-reinforced plastic (GFRP) has been widely adopted in the aerospace, construction, and automotive industries [1–6]. However, as the production and usage of GFRP products continue to rise, a significant amount of GFRP waste is expected to reach the end of its service life in the future [7,8]. Additionally, a considerable amount of GFRP waste is generated during the production and processing of GFRP products, posing a significant waste challenge [9]. Therefore, the efficient and cost-effective management of GFRP waste has gained increasing attention.

Many countries have imposed restrictions and prohibitions on traditional landfilling and incineration methods [10]. Therefore, it is crucial to develop innovative GFRP waste management approaches. Previous studies have explored the chemical and thermal recycling of GFRP. However, due to the high recycling costs and the complexity of recycling processes, thermal and chemical recycling may not be the most suitable options for GFRP [11–16].

At present, mechanical recycling has emerged as the most practical and feasible method for managing GFRP waste due to its low cost, minimal environmental impact, and capability to efficiently process large quantities of GFRP waste [17]. The mechanical recycling process involves a series of steps, including sorting, cutting, grinding, and screening, aimed at reducing the size of GFRP waste [18,19].

The construction industry has made significant efforts to reduce waste and promote recycling to mitigate adverse environmental impacts [20–23]. Using recycled GFRP (rGFRP) obtained through mechanical recycling to produce concrete seems to be a promising solution. This approach not only reduces the necessity for landfilling such materials, but also preserves acceptable concrete quality, in some cases even higher [24–27].

Asokan et al. [28] found that the addition of GFRP waste powder improved the mechanical properties of concrete, despite a slightly higher w/c ratio. Correia et al. [29] used GFRP waste particles to partially replace the fine aggregate in concrete. They found that when the recycled GFRP content was high, there was a significant decrease in all mechanical properties. García et al. [30] optimized the recycling process and used the obtained rGFRP material to prepare microconcrete specimens. They found that adding 1% rGFRP improved the compressive strength of the microconcrete specimens. Tittarelli et al. [31] used GFRP powder from a shipyard as an industrial by-product and replaced 5% and 10% of the natural sand. They found that the compressive strength of the concrete significantly decreased by about 40% and 50%, respectively. Dehghan et al. [32] investigated the effect of rGFRP on the mechanical properties of accelerated mortar and found that there was no significant improvement in compressive strength, but an increase in splitting tensile strength.

The results from the previously mentioned studies reveal significant differences. Even when the amount of GFRP waste added is similar, the impact on concrete's mechanical properties varies distinctly. This variation can be attributed to (1) differences in GFRP waste size, including particle size and fiber length, resulting from various recycling processes, and (2) differences in the composition of GFRP waste from different sources, particularly in terms of fiber content. Therefore, it is necessary to systematically evaluate the effect of these factors on concrete performance.

In this study, an experimental study was conducted to investigate the effects of mechanically recycled GFRP with different sizes, compositions, and contents on the performance of concrete, and to evaluate its feasibility of reuse in concrete. Subsequently, the challenges faced by reusing GFRP waste in concrete at the current stage were analyzed, and existing reuse methods were summarized and optimized, enabling us to propose a systematic reuse program.

2. Materials and Methods

2.1. Material Origin and Components

The rGFRP used in this study was mechanically recycled from decommissioned wind turbine blades. GFRP material is composed of E-glass fiber and unsaturated polyester resin. After the GFRP waste was crushed and sieved, secondary screening using a 1.18 mm sieve was conducted to obtain two types of recycled materials, namely, rGFRP powder and rGFRP cluster. As shown in Figure 1, the rGFRP powder had a granular form with a considerable amount of resin dust, while the rGFRP cluster took on a fluffy cluster form. Concrete specimens were prepared using P.O. 42.5 cement. Crushed limestone with a maximum particle size of 20 mm was used as a coarse aggregate. The crushing index of gravel is 16%, meeting the requirements of GB/T 14685-2022 [33]. Natural river sand with a fineness modulus of 2.68 was used as the fine aggregate. Figure 2 shows the grading curves of the rGFRP and sand. To improve the concrete properties, polycarboxylate superplasticizer (1.5% of cementitious materials content), fly ash, and mineral powder were added to all concrete. The polycarboxylate superplasticizer has a water-reducing rate of 25% and a solid content of 19%. The chemical compositions of cement, fly ash, mineral powder, and rGFRP were determined using X-ray fluorescence spectroscopy (Malvern Panalytical Zetium, Almoro, The Netherlands) via the pellet method and are listed in Table 1. All the raw materials used complied with GB 50164 [34].

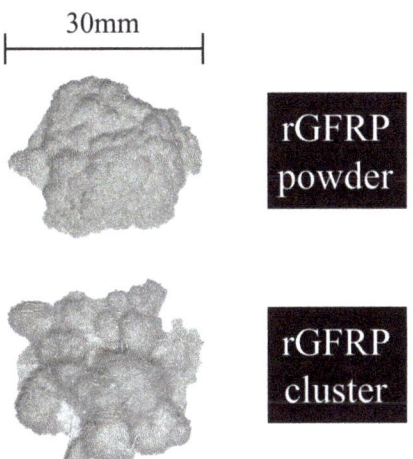

Figure 1. rGFRP powder and rGFRP cluster.

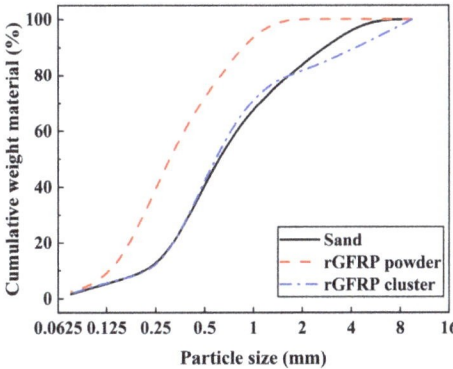

Figure 2. Particle size distribution of rGFRP and sand.

Table 1. The chemical composition of raw materials according to XRF analysis (wt%).

Components	Na_2O	MgO	Al_2O_3	SiO_2	P_2O_5	SO_3	K_2O	CaO	Fe_2O_3	LOI
Cement	0.2	2.1	5.3	19.9	0.1	2.4	0.8	61.4	2.8	4.4
Fly ash	0.9	1.0	31.1	54.0	0.7	0.7	2.0	4.0	4.2	-
Mineral powder	-	6.0	17.7	34.5	-	1.6	-	34.0	1.0	0.8
rGFRP	0.4	1.9	10.0	40.4	0.1	0.1	0.4	19.9	0.8	25.4

2.2. Concrete Mix Proportions

Seven different concrete mixes were used in this study, as outlined in Table 2. The first mix, labeled as CC, was a control mix containing no rGFRP. The remaining six mixes, labeled as P01, P03, P05, C01, C03, and C05, contained varying amounts of rGFRP powder or cluster, replacing 0%, 1%, 3%, and 5% of the fine aggregate. The only variable across all the mixes was the substitution rate of rGFRP to investigate the effect of rGFRP on concrete properties.

Table 2. Mix proportion of rGFRP concrete (kg/m^3).

Concrete Mixture	rGFRP Type	CA	FA	Cement	Superplasticizer	Fly Ash	Mineral Powder	Water	rGFRP
CC	--	1011	732	333	7.1	71.4	71.4	182	0
P01	powder	1011	725	333	7.1	71.4	71.4	182	7
P03	powder	1011	710	333	7.1	71.4	71.4	182	22
P05	powder	1011	695	333	7.1	71.4	71.4	182	37
C01	cluster	1011	725	333	7.1	71.4	71.4	182	7
C03	cluster	1011	710	333	7.1	71.4	71.4	182	22
C05	cluster	1011	695	333	7.1	71.4	71.4	182	37

2.3. Specimen Preparation and Testing Procedures

The process of specimen preparation and testing is outlined in Figure 3. To prepare the concrete specimens, a single-shaft horizontal mixer was used. The first step involved the sequential addition of the coarse aggregate, fine aggregate, and cementitious material into the mixer, followed by dry mixing for one minute. Subsequently, 50% of the required water and superplasticizer were added, followed by mixing for 1.5 min. Finally, the remaining water and superplasticizer were added, and mixing continued for an additional 2 min. During this mixing process, the rGFRP was gradually added to ensure its thorough dispersion throughout the concrete.

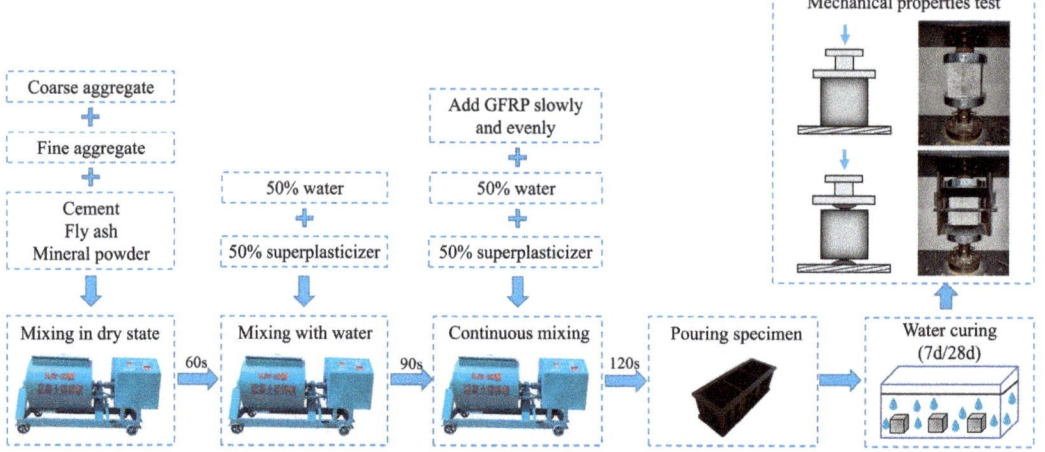

Figure 3. Preparation and testing process of rGFRP concrete specimen.

Following the mixing process, the fresh concrete was tested for slump according to the standards outlined in GB/T 50080 [35]. The concrete was then poured into molds, and all specimens were demolded after 24 h of casting. To ensure consistency, all specimens were cast from the same concrete batch. They were then cured by being immersed in a saturated Ca(OH)$_2$ solution at room temperature until the testing day. For each type of concrete, a total of twelve cubes, measuring 100 mm × 100 mm × 100 mm, were prepared.

After the concrete attained the appropriate curing age, a series of performance tests were conducted, including mechanical property and density tests, on the hardened concrete. A summary of the test specifications and specimen sizes can be found in Table 3.

The composition of rGFRP was analyzed using a simultaneous thermal analyzer (NETZSCH STA449F3 Jupiter, Selb, Germany). Additionally, the microscopic morphology of rGFRP was characterized using scanning electron microscopy (JEOL JSM-IT300, Tokyo, Japan).

Table 3. Summary of tests performed, specifications, and specimen sizes.

	Test	Specification	Specimen Size
Fresh state	Slump	GB/T 50080 [35]	--
Hardened state	Compressive strength	GB/T 50081 [36]	$100 \times 100 \times 100$ mm^3
	Tensile splitting strength	GB/T 50081 [36]	$100 \times 100 \times 100$ mm^3
	Density	BS EN 12390-7 [37]	$100 \times 100 \times 100$ mm^3

3. Results

3.1. Test on the rGFRP

The mass loss curves for the rGFRP are depicted in Figure 4. The trend of mass change with temperature was similar for both recycled materials until the temperature reached 400 °C. The initial mass loss from room temperature to 260 °C was primarily attributed to water evaporation. From 260 °C to 400 °C, rapid mass loss occurred due to the decomposition and combustion of the resin. However, a notable difference between the two materials was observed: the rGFRP cluster stabilized in mass after 400 °C, whereas the mass loss rate of the rGFRP powder significantly decreased from 400 °C to 600 °C, attributed to the burning of residual carbon. Beyond 600 °C, the mass loss rate decelerated, and the curve plateaued. The rGFRP powder experienced a mass loss of approximately 38%, while the rGFRP cluster experienced a mass loss of around 20%. The thermogravimetric analysis revealed that the rGFRP powder consisted of approximately 72% glass fiber and 38% organic materials, while the rGFRP cluster contained roughly 80% glass fiber and 20% organic materials.

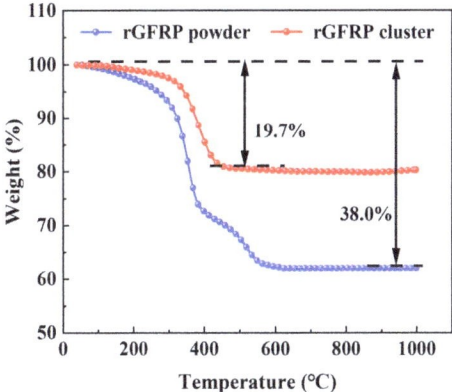

Figure 4. Comparison of mass loss of two types of rGFRP.

The microscopic morphology of rGFRP is shown in Figure 5. The analysis indicated the presence of a significant number of short fibers in both recycled materials, primarily in the monofilament form, with lengths ranging from 50 μm to 1000 μm. A few fibers appeared in bundles consisting of several fiber monofilaments. Some fibers had damaged resin layers, and resin shedding was evident, leading to the observation of a substantial amount of flaky and granular resin. Notably, the average fiber length in the rGFRP cluster was longer than that in the rGFRP powder, owing to the disparity between the two recycled materials after sieving. Additionally, more fibers were detected in the SEM images of the rGFRP cluster, which is consistent with the findings from the thermogravimetric analysis.

Based on the characterization, it is evident that the two recycled materials exhibit varying proportions of glass fibers and organic resins. Additionally, distinctions are observed in particle size (powdered component) and fiber length (fibrous component).

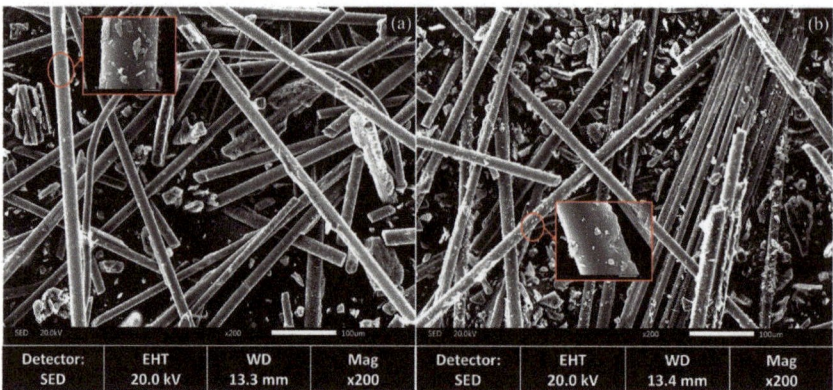

Figure 5. Microstructure comparison of different types of rGFRP: (**a**) rGFRP powder, (**b**) rGFRP cluster.

3.2. Test on the Concrete

3.2.1. Effect of rGFRP on the Workability of Fresh Concrete

The workability of concrete is a crucial performance indicator that depends on the amount of water added during concrete mixing. This study compared the slump of different test groups to examine the influence of rGFRP type and content on concrete workability. All test groups had the same w/c ratio, ensuring uniform water content during concrete mixing.

Before casting the test specimens, slump tests were conducted on each mixture. The slump was measured for six different batches of concrete for each mixture to reduce errors caused by manipulation or other factors. The test results are shown in Table 4 and Figure 6.

Table 4. Slump of fresh concrete (mm).

Concrete Mixture	Test No.						Mean	SD
	1	2	3	4	5	6		
CC	220	229	210	214	215	207	216	7.83
P01	200	210	196	182	183	193	194	10.60
P03	170	159	150	173	180	155	165	11.60
P05	68	51	80	54	39	72	61	15.25
C01	180	166	156	176	193	183	176	13.06
C03	114	107	143	142	126	122	126	14.60
C05	28	43	36	69	35	67	46	17.45

The slump of the concrete decreased linearly with an increase in rGFRP content. This outcome aligns with findings by other researchers, who have found that the high specific surface area of rGFRP, compared to natural sand, necessitates a larger surface to be covered by a water film during mixing. This leads to a decrease in free water in fresh concrete and an increase in the water demand for concrete [38]. From a rheological perspective, the presence of short fibers in rGFRP elevates the yield stress and plastic viscosity of fresh concrete [39], contributing to a reduction in slump. Furthermore, the SEM images of rGFRP also unveiled the presence of numerous irregularly shaped and sized resin particles, which could have an adverse impact on workability.

Additionally, the negative effect of rGFRP clusters on the concrete slump was more pronounced due to their longer fiber length and higher fiber content, leading to higher yield stress and plastic viscosity of fresh concrete.

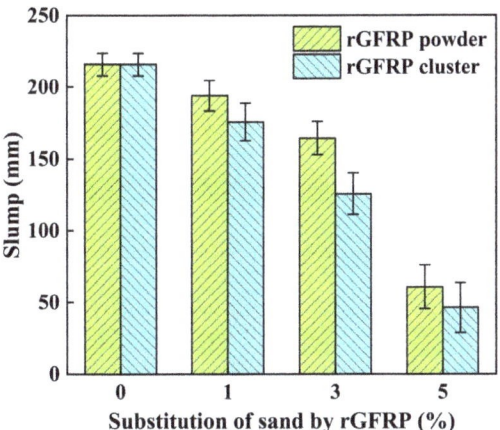

Figure 6. Effect of rGFRP on the slump of fresh concrete.

3.2.2. Effect of rGFRP on the Compressive Strength of Concrete

The compressive strength of CC and rGFRP concrete cube specimens was tested after 7 and 28 days of curing using a universal testing machine (MTS CMT5106, Shanghai, China), following the procedures specified in GB/T 50081 [36]. The specimens were subjected to a displacement loading rate of 0.5 mm/min and loaded until the concrete failed. Each set of experiments included three concrete samples to ensure the reproducibility of the results.

The compressive failure mode of the concrete is shown in Figure 7. The control concrete (CC) failure mode is characterized by large pieces of concrete peeling off and wide cracks penetrating the entire cross-section of the concrete, resulting in a conical failure pattern. In contrast, the failure mode of rGFRP concrete is characterized by multiple cracks, fewer large pieces of concrete peeling off, and a complete failure pattern. The short fibers in rGFRP restricted the initiation and propagation of some cracks, constrained the relative slip on the crack surface, and made the failure process smoother. This positive effect was more noticeable with an increase in the content of rGFRP. Due to its higher fiber content and longer fiber length, the rGFRP cluster exhibited stronger crack resistance compared to rGFRP powder.

The results of the compressive strength tests are shown in Table 5 and Figure 8. The 7-day compressive strength of rGFRP powder concrete initially increased, and then, decreased with increasing rGFRP content. Among all the groups, the 7-day compressive strength of P01 was the highest, measuring 37.6 MPa, a 7.7% increase compared to the control group (34.9 MPa). On the other hand, the 7-day compressive strengths of the C01, C03, and C05 groups were slightly lower than that of the control group, with C01 having the lowest value of 31.6 MPa, representing a decrease of 9.5%.

The 28-day compressive strength values of concrete containing rGFRP powder were similar to those of control concrete without rGFRP (43.2 MPa), indicating that rGFRP powder had no significant effect on the 28-day compressive strength of concrete. Conversely, the influence of the rGFRP cluster on the 28-day compressive strength of concrete was more noticeable. With its content of 1%, the 28-day compressive strength of concrete was 46.1 MPa, representing a 6.8% increase. However, when the content was further increased, the compressive strength decreased. Specifically, the compressive strength decreased from 41.7 MPa to 38.5 MPa as the content was increased from 3% to 5%.

These phenomena can be attributed to several factors: (1) The presence of CaO, Al_2O_3, SiO_2, and other polymers in rGFRP can strengthen the compressive strength of concrete, and the short fibers in rGFRP can inhibit the development of concrete cracks [40,41]. (2) The irregular size and shape of rGFRP may affect its bonding with other concrete components. (3) During the crushing of GFRP, some of the organic resin attached to the fibers is stripped off, as evident in SEM images. These resin particles have a lower specific modulus and transmit and resist significantly lower stresses in the concrete compared to other components [42], leading to more severe stress concentrations. When the first effect predominates, the compressive strength increases; otherwise, it decreases significantly.

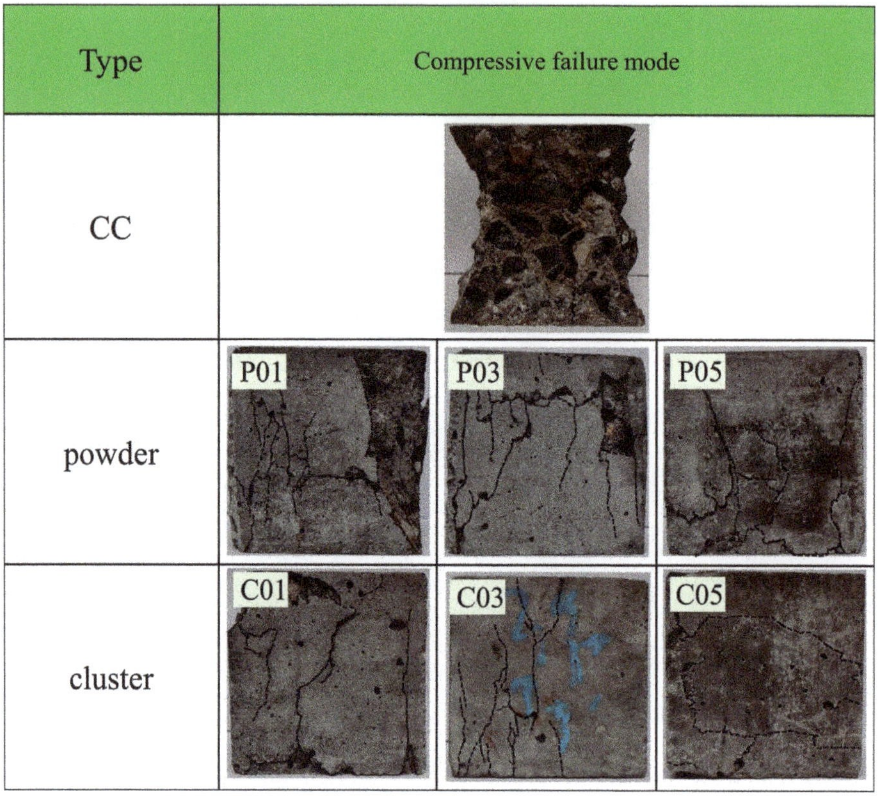

Type	Compressive failure mode		
CC			
powder	P01	P03	P05
cluster	C01	C03	C05

Figure 7. Compressive failure modes of rGFRP concrete specimens.

Table 5. Compressive strength of rGFRP concrete (MPa).

Concrete Mixture	rGFRP Content (%)	7-Day Compressive Strength (MPa)					28-Day Compressive Strength (MPa)				
		Specimen No.			Mean	SD	Specimen No.			Mean	SD
		1	2	3			1	2	3		
CC	0	33.45	34.15	37.17	34.92	1.97	45.17	40.43	43.92	43.17	2.45
P01	1	38.15	37.37	37.15	37.55	0.52	43.63	41.58	42.95	42.72	1.04
P03	3	32.67	32.23	30.55	31.82	1.12	45.40	44.61	40.33	43.45	2.73
P05	5	31.93	30.65	30.43	31.01	0.81	46.31	44.61	40.27	43.73	3.12
C01	1	30.30	30.16	34.39	31.62	2.40	49.61	47.55	41.16	46.11	4.41
C03	3	32.23	36.22	33.70	34.05	2.02	43.02	44.24	37.76	41.67	3.45
C05	5	34.89	32.93	32.11	33.31	1.43	39.38	37.58	38.61	38.52	0.91

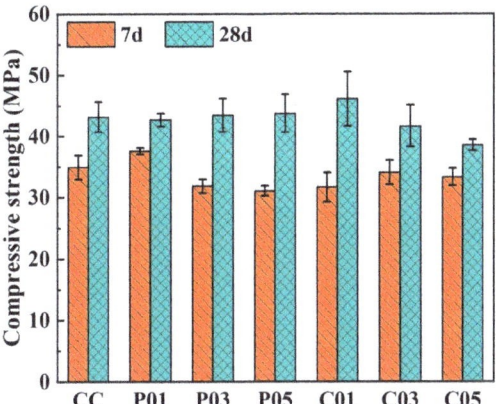

Figure 8. Effect of rGFRP on the compressive strength of concrete.

3.2.3. Effect of rGFRP on the Splitting Tensile Strength of Concrete

Splitting tensile tests were conducted on concrete cube specimens according to GB/T 50081 [36]. A strip was positioned between the upper and lower circular spacers and the specimen to evenly distribute the applied load. Uniformly distributed pressure was applied to the central plane of the specimen to generate approximately uniform tensile stress in the vertical plane of the external force. The equivalent splitting tensile strength f_{ts} was calculated using the following equation:

$$f_{ts} = \frac{2F}{\pi A} = 0.637\frac{F}{A} \quad (1)$$

where A is the area of the splitting surface of the specimen, and F is the maximum load recorded at the time of concrete failure. The concrete splitting tensile failure mode is shown in Figure 9.

The control group specimen exhibited single crack initiation under tensile load. As the load increased, the crack initially initiated at the center of the cross-section of the specimen, and then, propagated towards the edge strip. However, concrete specimens containing rGFRP powder and rGFRP cluster exhibited multi-crack initiation with similar failure modes. In addition to the main through-crack located at the center of the section, numerous branch cracks also existed. These branch cracks were shorter in length and did not traverse the entire surface of the concrete. The addition of more rGFRP resulted in a more pronounced multi-crack morphology. This indicated that the anisotropic microstructure of material properties, caused by the short fibers in rGFRP, changed the stress distribution and crack expansion direction. This dispersed crack expansion plays a restraining role in the main crack expansion.

As shown in Table 6 and Figure 10, the incorporation of rGFRP did not lead to a significant improvement in the splitting tensile strength of the concrete, contrary to expectations. Even when the concrete was incorporated with the rGFRP cluster containing more and longer fibers, the 7d and 28d splitting tensile strength of the concrete did not show a significant increase. The 28d splitting tensile strength of concrete reached its maximum value when the content of both rGFRPs was 3%, which was 4.0% and 3.1% higher than that of the control group (3.8 MPa), respectively. As the rGFRP content increased, the splitting tensile strength of rGFRP concrete tended to increase first, and then, decrease.

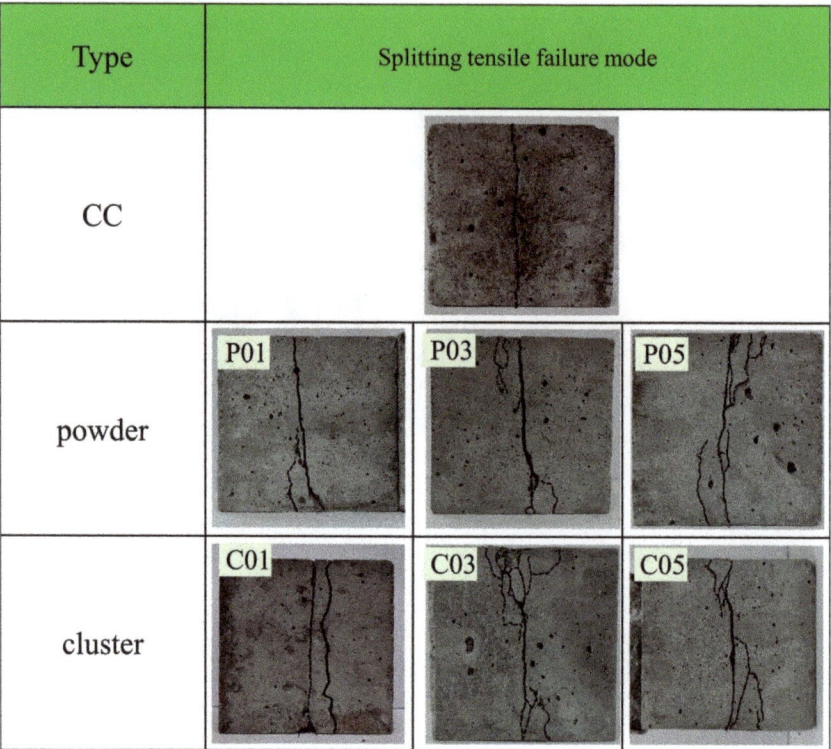

Figure 9. Splitting tensile failure modes of rGFRP concrete specimens.

Table 6. Splitting tensile strength of rGFRP concrete (MPa).

Concrete Mixture	rGFRP Content (%)	7-Day Splitting Tensile Strength (MPa)					28-Day Splitting Tensile Strength (MPa)				
		Specimen No.			Mean	SD	Specimen No.			Mean	SD
		1	2	3			1	2	3		
CC	0	2.86	2.92	3.30	3.03	0.24	3.63	4.23	3.60	3.82	0.36
P01	1	3.05	3.12	2.99	3.05	0.07	3.35	4.21	4.02	3.86	0.45
P03	3	2.90	2.86	2.92	2.89	0.03	4.30	3.86	3.78	3.98	0.28
P05	5	2.73	2.62	2.93	2.76	0.16	3.83	4.09	3.93	3.95	0.13
C01	1	3.20	2.99	2.98	3.06	0.12	3.34	4.10	3.92	3.79	0.40
C03	3	3.37	3.04	3.62	3.34	0.29	3.77	4.29	3.76	3.94	0.30
C05	5	2.73	3.11	2.89	2.91	0.19	3.65	3.26	3.54	3.48	0.20

During the splitting tensile test of the rGFRP concrete specimens, the load–displacement curves exhibited two types of curve, as shown in Figure 11. The first type of curve corresponded to the load–displacement curves of all sample groups except C05. It was characterized by a rapid increase in load with an increase in displacement in the early loading stage. After reaching the peak load, the specimens cracked and were damaged rapidly, losing their load-bearing capacity instantaneously, followed by a rapid decrease in load until failure.

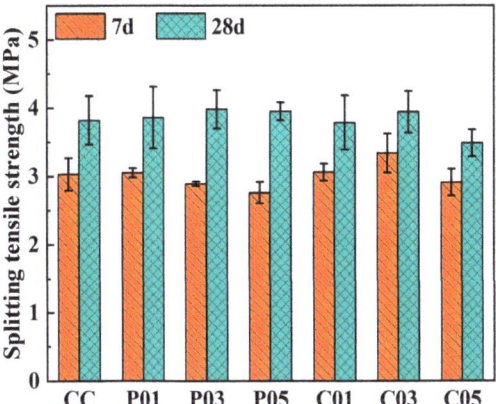

Figure 10. Effect of rGFRP on the splitting tensile strength of concrete.

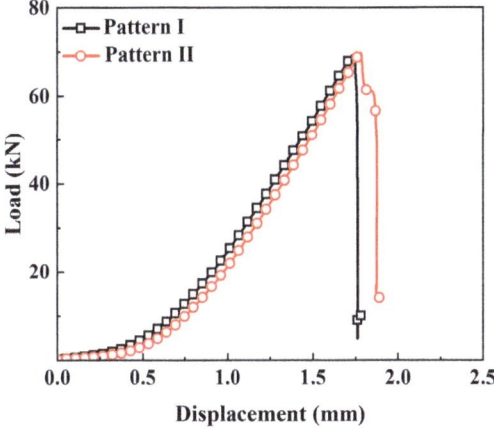

Figure 11. Two types of concrete splitting tensile load–displacement curves.

Furthermore, it was observed that the load–displacement curves exhibited by the concrete specimens of group C05 displayed distinctive behavior characterized by a second type of curve. The ascending portion of this type of curve shared similarities with that of the first type of curve. However, following the attainment of its peak, the load curve demonstrated a secondary peak after a period of decline. This phenomenon is often referred to as strain hardening. Subsequently, the load rapidly decreased until the concrete completely failed due to damage.

The underlying mechanism is as follows: When the applied load peaks and the concrete develops cracks, the stress is transferred to the fibers. Consequently, more fibers participate in the load-bearing process following cracking. The bond between these fibers and the matrix material creates an anchoring effect, thereby increasing the concrete's load-bearing capacity. Simultaneously, some of the fibers break under the influence of the load, absorbing a certain amount of energy. However, when the fibers are pulled out or broken, they lose their load-bearing capacity, and the curve experiences a rapid descent.

Due to the limited length and number of fibers, the descending portion of the curve after reaching the peak is not gradual. This indicates that the reinforcing effect of these short fibers on the matrix is not optimal.

3.2.4. Effect of rGFRP on the Microstructure of Concrete

To examine the mechanism of short fibers in rGFRP on the concrete microstructure, the small fragments that broke off during the compression and splitting tensile tests on rGFRP concrete were analyzed. The observation results are shown in Figure 12.

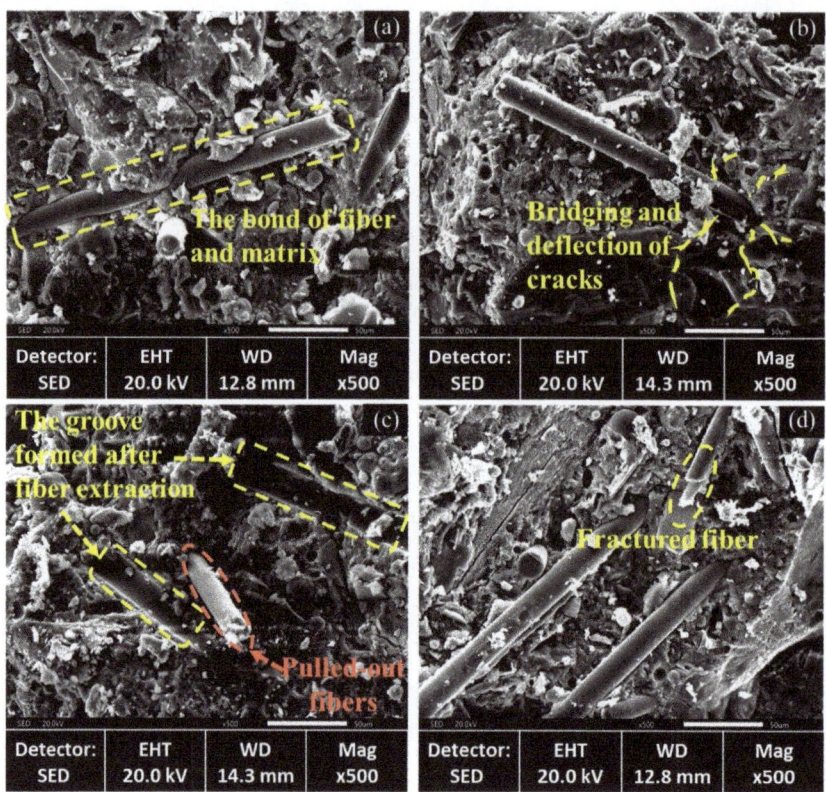

Figure 12. Enhancement mechanism of rGFRP in the base material.

Upon examining the SEM images, it was observed that more fibers were pulled out than broken, suggesting that the bond between the fibers and the matrix was not robust, as shown in Figure 12a. As the fibers began to bear loads, a significant proportion of them slid without reaching their failure load, thereby failing to improve the strength of the concrete. The presence of fibers had a bridging and deflection effect on cracks, as shown in Figure 12b. As micro cracks propagated towards the fibers, these bridging and deflection effects prevented local stress concentration, making the concrete failure mode more flexible. As shown in Figure 12c, the fibers were pulled out, leaving visible grooves, and the fractured fiber is observed in Figure 12d.

3.2.5. Effect of rGFRP on the Density of Concrete

To investigate the effect of rGFRP on the density of concrete, the dry density of the concrete was calculated after 28 days of curing. To minimize errors, 12 concrete specimens were selected for density measurement in each group. The results, shown in Figure 13, indicate that the density of the control concrete was 2.42 g/cm^3. As the amount of rGFRP increased, the density of the concrete decreased linearly. Since the density of rGFRP is lower than that of fine aggregate, increasing the content of rGFRP in concrete will lead to a reduction in the density of concrete.

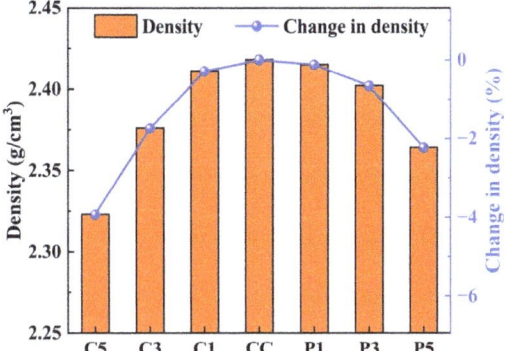

Figure 13. Effect of rGFRP on the density of concrete.

When the rGFRP content was 3%, the compressive strength and splitting tensile strength of concrete improved, and the density of concrete reduced by 0.66% and 1.74%, respectively. Replacing 3% of fine aggregate with rGFRP in concrete appears to be a favorable option since it slightly increases strength while reducing density. When the rGFRP cluster content was 5%, the strength of the concrete decreased slightly, but the density of the concrete reduced by 3.93%, significantly decreasing the weight of the concrete structure. Therefore, in practical applications, determining the optimal dosage of rGFRP should be based on its intended use.

4. Optimization of Reusing rGFRP in Concrete

While some progress has been made in researching the mechanical recycling of rGFRP, its reuse in concrete has been primarily limited to laboratory research. Different scholars have focused on various aspects of this research, thus resulting in conclusions that may not be universally applicable. As a result, there appears to be a gap between the current state of research and its practical implementation in the industry. Yazdanbakhsh and Bank [42] believed that the main challenge in reusing rGFRP in concrete lies in the change in performance after its addition, rather than the uncertainty of recycled raw materials' performance. The properties of rGFRP may significantly vary depending on its application and type. To achieve the industrialization of rGFRP concrete, it is essential to develop a comprehensive process to ensure its feasibility.

There is a lack of a widely applicable program for reusing GFRP. Building upon this study and other pertinent research, we propose an optimized rGFRP reuse program to address this gap. As shown in Figure 14, this program aims to establish a correlation between the properties of rGFRP concrete and the primary properties of rGFRP, defining the feasible range of rGFRP, and determine the optimal mix proportions for each rGFRP concrete through experimentation.

Figure 14. Optimization procedure for rGFRP reuse.

4.1. Selection of Recycling Technology

Currently, there are three ways to reuse rGFRP obtained through mechanical recycling in concrete. One approach involves processing GFRP waste into granules, which can serve as a fine aggregate replacement in concrete. Another method involves cutting GFRP waste into large pieces, suitable for use as a coarse aggregate replacement in concrete. Furthermore, GFRP waste can also be employed as fiber-shaped reinforcing bars by modifying the mechanical recycling processes. All three options for using rGFRP in concrete are attractive. Replacing aggregate comprising the largest proportion of concrete with rGFRP can maximize the reuse of GFRP recycled materials. Fiber-shaped rGFRP, retaining its relatively intact structure and performance, can be regarded as weakened fibers and has a significant enhancing effect on the mechanical properties of concrete.

Yazdanbakhsh et al. [43] conducted an experiment using FRP-RA made from waste GFRP rebar in concrete, partially replacing coarse aggregate. The results showed that the addition of FRP-RA reduces the strength of concrete.

Fu et al. [38] developed a new method for the mechanical recycling of GFRP waste by machining it into macro-fiber to reinforce concrete. The research demonstrated that adding macro-fibers improved the flexural strength and toughness of the concrete.

The results of prior studies in this research indicate that the impact of rGFRP on concrete properties varied according to the particle size levels. Likewise, Mastali et al. [44] observed that recycled GFRP fibers of varying lengths had diverse effects on the workability and mechanical properties of concrete. Various mechanical recycling processes can modify the shape and properties of rGFRP, affecting different aspects of concrete. Additionally, it is crucial to recognize that the cost of recycling GFRP waste is associated with the mechanical recycling process, highlighting the need to consider multiple factors when selecting a recycling method. Therefore, characterizing rGFRP accurately is crucial to identifying the most appropriate recycling process.

4.2. Characterization of rGFRP

Recycling plants that process waste from various sources of GFRP produce waste materials with different compositions. When incorporated into concrete, these waste materials can significantly affect the properties of the concrete, as shown in a study by García et al. [30], where significant differences in mechanical properties were observed in microconcrete prepared using rGFRP from four different sources. The unknown properties of rGFRP greatly constrain its potential for reuse. Therefore, it is essential to develop criteria and procedures for characterizing recycled GFRP waste from various sources.

The recycled waste should be tested to (a) measure its physical properties, (b) determine the type and content of its components, and (c) characterize its microstructure. Testing methods may include sieve analysis or laser particle size determination to determine particle size distribution, thermal analysis, and spectroscopic techniques such as XRD, XRF, and FT-IR to determine the type and content of fibers, resins, and other components in rGFRP, and scanning electron microscopy to determine the distribution of fibers and other components. Based on the data obtained from the tests and the properties of concrete containing GFRP waste, the usable range of the material can be determined.

Currently, there are limited studies in this area, and the performance of the raw materials used varies greatly. There is a lack of qualitative and quantitative understanding of the effects of the chemical composition and microstructure parameters of rGFRP on concrete performance.

4.3. Optimization of Mix Proportions

Mix proportion optimization aims to improve the performance of concrete and maximize the consumption of rGFRP while specifying performance requirements. Yazdanbakhsh and Bank [42] acknowledged that various forms of GFRP waste might negatively impact the performance of concrete. However, they believed that adjusting concrete mix proportions could resolve this issue. Previous research has shown that increasing the rGFRP content can significantly decrease concrete workability. However, increasing the w/c ratio to improve workability can result in decreased concrete strength. Therefore, multiple orthogonal mix proportion designs are required for optimizing mix proportions.

Varying mix proportions should be used for rGFRP concrete with different rGFRP content to ensure the desired workability and mechanical properties of the concrete, as well as to determine the optimal dosage of different types of rGFRP and their corresponding mix ratio. Optimization methods may include the use of superplasticizer, the addition of fly ash, adjustment of the w/c ratio, and the optimization of aggregate grading, among others.

With the development of computer technology, machine learning has been widely applied in the field of concrete mix design [45–48]. Compared to traditional design methods, machine learning can more accurately simulate the relationship between material variables and concrete performance to formalize the problem [49,50]. Therefore, using machine learning to design and optimize the mix proportion of rGFRP concrete is undoubtedly attractive, as it can greatly save time and reduce costs, all while achieving more accurate predictions of concrete performance [51].

4.4. Evaluation of Concrete Quality

The performance of rGFRP concrete exhibits significant variation due to the inherent variability in concrete's internal structure and the uneven distribution of rGFRP within it. So, it is essential to establish the quantitative relationship between factors such as the size and dosage of GFRP waste and the mechanical properties of concrete, which will determine their application in practical engineering projects. Currently, there is limited research in this direction, yet it is crucial because it can provide guidance for the mechanical recycling process of GFRP and the overall reuse of GFRP waste in concrete.

Apart from evaluating the workability and mechanical properties of concrete, it is crucial to emphasize its durability, especially in alkali–aggregate reaction testing. The alkali in cement can react with the silica in GFRP waste, leading to an alkali–silica reaction (ASR) that triggers destructive expansion and shortens the lifespan of concrete. Hence, thorough testing and evaluation of the alkali–aggregate reaction is necessary to ensure the durability and long-term performance of the rGFRP concrete. Moreover, feasible methods and technologies, such as adding mineral admixtures to suppress the alkali–aggregate reaction, need to be explored and applied to maximize the reliability and durability of rGFRP concrete.

5. Conclusions

GFRP products are increasingly being used in various industrial fields. However, the speed of GFRP waste accumulation is continuously increasing, which necessitates the development of more sustainable solutions within the GFRP industry. In light of this, this study conducted experiments to assess the reuse potential of recycled GFRP in concrete. The following conclusions were drawn from the experiments:

The addition of rGFRP reduced the slump of concrete. However, at lower content levels, the effect was negligible.

The influence of a low content (1–5%) of rGFRP powder on the strength of concrete was minimal. The addition of a small amount of rGFRP cluster slightly improved the compressive strength and splitting tensile strength of concrete. The presence of short fibers in rGFRP improved the compressive and splitting tensile failure mode of concrete. Moreover, higher fiber content and longer fiber length exhibited a stronger reinforcing effect.

The addition of rGFRP decreased the density of concrete, indicating that it can somewhat facilitate the lightweight design of concrete structures.

To guarantee the quality and performance of concrete when using various GFRP waste materials, this study discussed and optimized existing reuse methods. It proposes an optimized procedure for GFRP reuse to maximize the technical and economic benefits. This procedure aims to facilitate the reuse of GFRP waste in engineering applications and advance its industrial-scale implementation.

Currently, there is a lack of understanding regarding the microstructural mechanisms underlying the impact of rGFRP on the mechanical properties of concrete. Further research is required to delve into this aspect, which will help advance the practical application of GFRP waste in engineering projects.

Author Contributions: Conceptualization, T.Z. and J.C.; methodology, T.Z., J.C. and Y.L.; formal analysis, X.Z. and L.H.; investigation, P.S. and M.S.; resources, J.C. and P.S.; data curation, T.Z. and Y.L.; writing—original draft preparation, T.Z.; writing—review and editing, T.Z., J.C., Y.L. and M.S.; visualization, T.Z. and Y.L.; supervision, J.C. and Y.L.; funding acquisition, J.C. All authors have read and agreed to the published version of the manuscript.

Funding: This research was funded by the Fundamental Research Funds for the Central Universities (WUT:2021III059JC).

Institutional Review Board Statement: Not applicable.

Informed Consent Statement: Not applicable.

Data Availability Statement: The data are contained within this article.

Conflicts of Interest: The authors declare no conflict of interest.

References

1. Berardi, V.P.; Perrella, M.; Feo, L.; Cricri, G. Creep behavior of GFRP laminates and their phases: Experimental investigation and analytical modeling. *Compos. Part B-Eng.* **2017**, *122*, 136–144.
2. Rafiee, R.; Ghorbanhosseini, A. Developing a micro-macromechanical approach for evaluating long-term creep in composite cylinders. *Thin-Walled Struct.* **2020**, *151*, 106714. [CrossRef]
3. Zhu, Y.X.; Chen, Z.Y.; Jiang, Y.B.; Zhao, T.; Li, Y.; Chen, J.Z.; Fang, D.N. Design, fabrication and stiffness analysis of a novel GFRP sandwiched pipe with stiffened core. *Thin-Walled Struct.* **2020**, *156*, 106982. [CrossRef]
4. Farinha, C.B.; de Brito, J.; Veiga, R. Assessment of glass fibre reinforced polymer waste reuse as filler in mortars. *J. Clean. Prod.* **2019**, *210*, 1579–1594. [CrossRef]
5. Imjai, T.; Garcia, R.; Guadagnini, M.; Pilakoutas, K. Strength Degradation in Curved Fiber-reinforced Polymer (FRP) Bars Used as Concrete Reinforcement. *Polymers* **2020**, *12*, 1653. [CrossRef]
6. Imjai, T.; Guadagnini, M.; Garcia, R.; Pilakoutas, K. A practical method for determining shear crack induced deformation in FRP RC beams. *Eng. Struct.* **2016**, *126*, 353–364. [CrossRef]
7. Tittarelli, F.; Shah, S.P. Effect of low dosages of waste GRP dust on fresh and hardened properties of mortars: Part 1. *Constr. Build. Mater.* **2013**, *47*, 1532–1538. [CrossRef]
8. Zabihi, O.; Ahmadi, M.; Liu, C.; Mahmoodi, R.; Li, Q.; Ferdowsi, M.; Naebe, M. A Sustainable Approach to the Low-Cost Recycling of Waste Glass Fibres Composites towards Circular Economy. *Sustainability* **2020**, *12*, 641. [CrossRef]
9. Castro, A.C.M.; Carvalho, J.P.; Ribeiro, M.C.S.; Meixedo, J.P.; Silva, F.J.G.; Fiuza, A.; Dinis, M.L. An integrated recycling approach for GFRP pultrusion wastes: Recycling and reuse assessment into new composite materials using Fuzzy Boolean Nets. *J. Clean. Prod.* **2014**, *66*, 420–430. [CrossRef]
10. Conroy, A.; Halliwell, S.; Reynolds, T. Composite recycling in the construction industry. *Compos. Part A Appl. Sci. Manuf.* **2006**, *37*, 1216–1222. [CrossRef]
11. Åkesson, D.; Foltynowicz, Z.; Christéen, J.; Skrifvars, M. Microwave pyrolysis as a method of recycling glass fibre from used blades of wind turbines. *J. Reinf. Plast. Compos.* **2012**, *31*, 1136–1142. [CrossRef]
12. Oliveux, G.; Dandy, L.O.; Leeke, G.A. Current status of recycling of fibre reinforced polymers: Review of technologies, reuse and resulting properties. *Prog. Mater. Sci.* **2015**, *72*, 61–99. [CrossRef]
13. Job, S. Recycling composites commercially. *Reinf. Plast.* **2014**, *58*, 32–38. [CrossRef]
14. Lin, J.; Guo, Z.X.; Hong, B.; Xu, J.Q.; Fan, Z.P.; Lu, G.Y.; Wang, D.W.; Oeser, M. Using recycled waste glass fiber reinforced polymer (GFRP) as filler to improve the performance of asphalt mastics. *J. Clean. Prod.* **2022**, *336*, 130357. [CrossRef]
15. Shima, H.; Takahashi, H.; Mizuguchi, J. Recovery of Glass Fibers from Fiber Reinforced Plastics. *Mater. Trans.* **2011**, *52*, 1327–1329. [CrossRef]
16. Guido, G.; Tomohito, K.; Tomoyuki, M.; Toshiaki, Y. Recovery of glass fibers from glass fiber reinforced plastics by pyrolysis. *J. Mater. Cycles Waste Manag.* **2013**, *15*, 122–128.
17. Meira Castro, A.C.; Ribeiro, M.C.S.; Santos, J.; Meixedo, J.P.; Silva, F.J.G.; Fiúza, A.; Dinis, M.L.; Alvim, M.R. Sustainable waste recycling solution for the glass fibre reinforced polymer composite materials industry. *Constr. Build. Mater.* **2013**, *45*, 87–94. [CrossRef]
18. Krauklis, A.; Karl, C.; Gagani, A.; Jørgensen, J. Composite Material Recycling Technology—State-of-the-Art and Sustainable Development for the 2020s. *J. Compos. Sci.* **2021**, *5*, 28. [CrossRef]

19. Pickering, S.J. Recycling technologies for thermoset composite materials—Current status. *Compos. Part A Appl. Sci. Manuf.* **2006**, *37*, 1206–1215. [CrossRef]
20. Akbar, A.; Liew, K.M. Assessing recycling potential of carbon fiber reinforced plastic waste in production of eco-efficient cement-based materials. *J. Clean. Prod.* **2020**, *274*, 123001. [CrossRef]
21. Ahmed, S.; Ali, M. Use of agriculture waste as short discrete fibers and glass-fiber-reinforced-polymer rebars in concrete walls for enhancing impact resistance. *J. Clean. Prod.* **2020**, *268*, 122211. [CrossRef]
22. Pietrzak, A.; Ulewicz, M. Influence of Post-Consumer Waste Thermoplastic Elastomers Obtained from Used Car Floor Mats on Concrete Properties. *Materials* **2023**, *16*, 2231. [CrossRef] [PubMed]
23. Ulewicz, M. Recycled Materials for Concrete and Other Composites. *Materials* **2021**, *14*, 2279.
24. Hendriks, C.F.; Janssen, G.M.T. Use of recycled materials in constructions. *Mater. Struct.* **2003**, *36*, 604–608.
25. Zhou, B.; Zhang, M.; Wang, L.; Ma, G. Experimental study on mechanical property and microstructure of cement mortar reinforced with elaborately recycled GFRP fiber. *Cem. Concr. Compos.* **2021**, *117*, 103908. [CrossRef]
26. Sebaibi, N.; Benzerzour, M.; Abriak, N.E.; Binetruy, C. Mechanical properties of concrete-reinforced fibres and powders with crushed thermoset composites: The influence of fibre/matrix interaction. *Constr. Build. Mater.* **2012**, *29*, 332–338. [CrossRef]
27. Li, Y.-F.; Hsu, Y.-W.; Syu, J.-Y.; Chen, B.-Y.; Song, B. Study on the Utilization of Waste Thermoset Glass Fiber-Reinforced Polymer in Normal Strength Concrete and Controlled Low Strength Material. *Materials* **2023**, *16*, 3552. [CrossRef] [PubMed]
28. Asokan, P.; Osmani, M.; Price, A.D.F. Improvement of the mechanical properties of glass fibre reinforced plastic waste powder filled concrete. *Constr. Build. Mater.* **2010**, *24*, 448–460. [CrossRef]
29. Correia, J.R.; Almeida, N.M.; Figueira, J.R. Recycling of FRP composites: Reusing fine GFRP waste in concrete mixtures. *J. Clean. Prod.* **2011**, *19*, 1745–1753.
30. García, D.; Vegas, I.; Cacho, I. Mechanical recycling of GFRP waste as short-fiber reinforcements in microconcrete. *Constr. Build. Mater.* **2014**, *64*, 293–300. [CrossRef]
31. Tittarelli, F.; Kawashima, S.; Tregger, N.; Moriconi, G.; Shah, S.P. Effect of GRP by-product addition on plastic and hardened properties of cement mortars. In Proceedings of the Second International Conference on Sustainable Construction Materials and Technologies, Ancona, Italy, 28–30 June 2010; pp. 28–30.
32. Dehghan, A.; Peterson, K.; Shvarzman, A. Recycled Glass Fiber Reinforced Polymer Additions to Portland Cement Concrete. *Constr. Build. Mater.* **2017**, *146*, 238–250. [CrossRef]
33. GB/T 14685-2022; SAC (Standardization Administration of the People's Republic of China). Pebble and Crushed Stone for Construction. Standards Press of China: Beijing, China, 2022.
34. GB 50164-2011; Ministry of Housing and Urban-Rural Development of the People's Republic of China. Standard for Quality Control of Concrete. China Architecture & Building Press: Beijing, China, 2011.
35. GB/T 50080-2016; Ministry of Housing and Urban-Rural Development of the People's Republic of China. Standard for Test Method of Performemance on Ordinary Fresh Concrete. China Architecture & Building Press: Beijing, China, 2016.
36. GB/T 50081-2019; Ministry of Housing and Urban-Rural Development of the People's Republic of China. Standard for Test Methods of Concrete Physical and Mechanical Properties. China Architecture & Building Press: Beijing, China, 2019.
37. BS EN 12390-7:2009; British Standards Institution. Testing Hardened Concrete. Part 7. Density of Hardened Concrete. British Standards Institution: London, UK, 2009.
38. Fu, B.; Liu, K.C.; Chen, J.F.; Teng, J.G. Concrete reinforced with macro fibres recycled from waste GFRP. *Constr. Build. Mater.* **2021**, *310*, 125063. [CrossRef]
39. Alberti, M.G.; Enfedaque, A.; Galvez, J.C. The effect of fibres in the rheology of self-compacting concrete. *Constr. Build. Mater.* **2019**, *219*, 144–153. [CrossRef]
40. Kazmi, S.M.S.; Munir, M.J.; Wu, Y.F.; Patnaikuni, I. Effect of macro-synthetic fibers on the fracture energy and mechanical behavior of recycled aggregate concrete. *Constr. Build. Mater.* **2018**, *189*, 857–868. [CrossRef]
41. Yang, Q.; Xian, G.; Karbhari, V.M. Hygrothermal ageing of an epoxy adhesive used in FRP strengthening of concrete. *J. Appl. Polym. Sci.* **2008**, *107*, 2607–2617. [CrossRef]
42. Yazdanbakhsh, A.; Bank, L.C. A Critical Review of Research on Reuse of Mechanically Recycled FRP Production and End-of-Life Waste for Construction. *Polymers* **2014**, *6*, 1810–1826. [CrossRef]
43. Yazdanhakhsh, A.; Bank, L.C.; Chen, C. Use of recycled FRP reinforcing bar in concrete as coarse aggregate and its impact on the mechanical properties of concrete. *Constr. Build. Mater.* **2016**, *121*, 278–284. [CrossRef]
44. Mastali, M.; Abdollahnejad, Z.; Dalvand, A.; Sattarifard, A.; Illikainen, M. 19-Comparative effects of using recycled CFRP and GFRP fibers on fresh- and hardened-state properties of self-compacting concretes: A review. *New Mater. Civ. Eng.* **2020**, 643–655. [CrossRef]
45. Lee, H.-S.; Lim, S.-M.; Wang, X.-Y. Optimal Mixture Design of Low-CO_2 High-Volume Slag Concrete Considering Climate Change and CO_2 Uptake. *Int. J. Concr. Struct. Mater.* **2019**, *13*, 56.
46. Ziolkowski, P.; Niedostatkiewicz, M.; Kang, S.-B. Model-Based Adaptive Machine Learning Approach in Concrete Mix Design. *Materials* **2021**, *14*, 1661. [CrossRef]
47. Ziolkowski, P.; Niedostatkiewicz, M. Machine Learning Techniques in Concrete Mix Design. *Materials* **2019**, *12*, 1256. [CrossRef]
48. Song, Y.; Wang, X.; Li, H.; He, Y.; Zhang, Z.; Huang, J. Mixture Optimization of Cementitious Materials Using Machine Learning and Metaheuristic Algorithms: State of the Art and Future Prospects. *Materials* **2022**, *15*, 7830. [CrossRef] [PubMed]

49. Chou, J.-S.; Pham, A.-D. Enhanced artificial intelligence for ensemble approach to predicting high performance concrete compressive strength. *Constr. Build. Mater.* **2013**, *49*, 554–563. [CrossRef]
50. Mashhadban, H.; Kutanaei, S.S.; Sayarinejad, M.A. Prediction and modeling of mechanical properties in fiber reinforced self-compacting concrete using particle swarm optimization algorithm and artificial neural network. *Constr. Build. Mater.* **2016**, *119*, 277–287. [CrossRef]
51. Rosa, A.C.; Hammad, A.W.A.; Boer, D.; Haddad, A. Use of operational research techniques for concrete mix design: A systematic review. *Heliyon* **2023**, *9*, e15362. [CrossRef]

Disclaimer/Publisher's Note: The statements, opinions and data contained in all publications are solely those of the individual author(s) and contributor(s) and not of MDPI and/or the editor(s). MDPI and/or the editor(s) disclaim responsibility for any injury to people or property resulting from any ideas, methods, instructions or products referred to in the content.

Article

The Use of Biosilica to Increase the Compressive Strength of Cement Mortar: The Effect of the Mixing Method

Nelli G. Muradyan [1], Avetik A. Arzumanyan [1], Marine A. Kalantaryan [1], Yeghiazar V. Vardanyan [1], Mkrtich Yeranosyan [2], Malgorzata Ulewicz [3], David Laroze [4] and Manuk G. Barseghyan [1,*]

[1] Faculty of Construction, National University of Architecture and Construction of Armenia, 105 Teryan Street, Yerevan 0009, Armenia
[2] Innovation Center for Nanoscience and Technologies, A.B. Nalbandyan Institute of Chemical Physics NAS RA, 5/2 P. Sevak Street, Yerevan 0014, Armenia; myeranos@ysu.am
[3] Faculty of Civil Engineering, Czestochowa University of Technology, Dabrowskiego 69 Street, PL 42-201 Czestochowa, Poland
[4] Instituto de Alta Investigación, Universidad de Tarapacá, Casilla 7D, Arica 1000000, Chile; dlarozen@academicos.uta.cl
* Correspondence: manuk.barseghyan@nuaca.am or manuk.g.barseghyan@gmail.com

Abstract: In this work, the effect of biosilica concentration and two different mixing methods with Portland cement on the compressive strength of cement-based mortars were investigated. The following values of the biosilica concentration of cement weight were investigated: 2.5, 5, 7.5, and 10 wt.%. The mortar was prepared using the following two biosilica mixing methods: First, biosilica was mixed with cement and appropriate samples were prepared. For the other mixing method, samples were prepared by dissolving biosilica in water using a magnetic stirrer. Compressive tests were carried out on an automatic compression machine with a loading rate of 2.4 kN/s at the age of 7 and 28 days. It is shown that, for all cases, the compressive strength has the maximum value of 10% biosilica concentration. In particular, in the case of the first mixing method, the compressive strength of the specimen over 7 days of curing increased by 30.5%, and by 36.5% for a curing period of 28 days. In the case of the second mixing method, the compressive strength of the specimen over 7 days of curing increased by 23.4%, and by 47.3% for a curing period of 28 days. Additionally, using the first and second mixing methods, the water absorption parameters were reduced by 22% and 34%, respectively. Finally, it is worth noting that the obtained results were intend to provide valuable insights into optimizing biosilica incorporation in cement mortar. With the aim of contributing to the advancement of construction materials, this research delves into the intriguing application of biosilica in cement mortar, emphasizing the significant impact of mixing techniques on the resultant compressive strength.

Keywords: biosilica; compressive strength; water absorption; cement mortar; mixing method

1. Introduction

Currently, cementitious materials are the most widely used construction materials. However, they have low tensile strength and are highly susceptible to cracking. To address this issue, several research efforts have focused on enhancing the cement structure with micro- or nano-level reinforcements. The cement mortars were produced using microsilica [1–7] as well as nanosilica [8–21]. Both materials were also used together to modify the properties of the mortars, which improved the mechanical properties of the tested mortars [2–4,7]. The results of the researched cement composites with nanosilica confirmed its acceleration action on C_3S hydration and formation gel C-S-H, and the modification of composite viscosity improving the tightness of the cement matrix. However, the addition of nanosilica, whose particles are much finer than microsilica, causes a rapid decrease in the consistency of the fresh mix of cement mortars. The microsilica and nanosilica were often

Citation: Muradyan, N.G.; Arzumanyan, A.A.; Kalantaryan, M.A.; Vardanyan, Y.V.; Yeranosyan, M.; Ulewicz, M.; Laroze, D.; Barseghyan, M.G. The Use of Biosilica to Increase the Compressive Strength of Cement Mortar: The Effect of the Mixing Method. *Materials* **2023**, *16*, 5516. https://doi.org/10.3390/ma16165516

Academic Editor: Dolores Eliche Quesada

Received: 8 June 2023
Revised: 24 July 2023
Accepted: 1 August 2023
Published: 8 August 2023

Copyright: © 2023 by the authors. Licensee MDPI, Basel, Switzerland. This article is an open access article distributed under the terms and conditions of the Creative Commons Attribution (CC BY) license (https:// creativecommons.org/licenses/by/ 4.0/).

used together with other waste materials, i.e., broken glass [8–12], waste ceramics [1], or fly ash [5,13,14,19]. The use of waste is particularly important if we want building materials to be produced in accordance with the idea of a circular economy.

Various approaches have been explored to enhance the mechanical properties of cementitious materials, including the incorporation of additives, that can modify the microstructure and hydration characteristics of cement-based systems, thereby influencing their mechanical behavior. A new trend in research is the use of not only microsilica, which is a waste from the metallurgical industry (created during the processing of metallic silicon, ferrosilicon, or other silicon alloys), but also the use of biosilica of biogenic origin or extracted from rice husk ash using the thermochemical method [22–28]. Using diatomite (DE) as a reinforcement in cement-based materials has shown promising results [25]. These materials exhibit higher resistance to leaching, improved Cl impermeability, and reduced drying shrinkage, making them suitable for use in reinforced structures located in freshwater, seawater, or humid environments. Alternatively, DE-containing cement-based composites can serve non-reinforced structures (e.g., concrete blocks) [26,27]. It is well known that biosilica, a naturally occurring silica derived from diatoms, has been proposed as a potential additive to improve the mechanical properties of cement-based materials. Biosilica has several advantages over traditional silica-based additives, including its biodegradability and renewable nature [22]. The application of biosilica as an additive in cement mortar holds significant importance in the field of construction materials. Its unique properties, including high surface area, pozzolanic activity, and micro-sized particles, present a promising avenue to address the challenges of improving the mechanical properties of cement-based materials.

In a recent study [28], it was demonstrated that DE with a biosilica content of 83+ wt.% can effectively replace up to 40 wt.% of Portland cement in mortar and concrete, resulting in significant improvements in compressive strength. By partially replacing Portland cement with high-quality DE, the study found that the global warming potential could be reduced by 50% and emissions of PM10 and lead from mortar/concrete production could also be minimized, in addition to the strength enhancements [29]. In [22], it was shown that the compression strength of biocement mortar was greatly enhanced with the addition of biosilica. Nevertheless, the inclusion of biosilica particles resulted in a reduction in both the water vapor transmission rate and the setting time of the mortar. In [23], the researchers investigated the durability performance and underlying methods of DE-containing cementitious composites by partially replacing Portland cement with 30 wt.% diatomite. The results showed that the cementitious materials containing DE biosilica exhibited remarkable impermeability and leaching resistance after 28 days.

The effectiveness of an additive in cement mortar depends not only on its inherent properties but also on the method of incorporation into the mortar mix. Mixing methods play a crucial role in achieving a homogeneous dispersion of additives, ensuring their proper interaction with the cementitious matrix, and influencing the overall performance of the resulting composite. Therefore, understanding the effect of mixing methods on the utilization of biosilica in cement mortar is vital for optimizing its effectiveness and harnessing its full potential.

With the aim of contributing to the advancement of construction materials, this research delves into the intriguing application of biosilica in cement mortar, emphasizing the significant impact of mixing techniques on the resultant compressive strength. In this work, the effect of biosilica concentration and two different mixing methods with Portland cement on the compressive strength of cement-based mortar were investigated.

2. Materials and Methods

2.1. Materials

As a binder for the mortars in this study, ordinary Portland cement 52.5 (GOST 31108-2020) was used, which is available at Ararat cement Factory in Yerevan, Armenia. Table 1 displays the chemical composition and physical properties of the cement used in the study [30,31], in accordance with GOST EN 196-1-2002 [32], 196-2-2002 [33], and 196-3-2002 [34]. Meanwhile, Tables 2 and 3 provide the physical and chemical properties of the sand and biosilica, respectively, that were used in the study. The biosilica (Effect group, Yerevan, Armenia) used as an additive in these mortars, which is amorphous silica of biogenic origin that resulted in the combined activation of the natural diatomite. The FTIR spectra and SEM image of the above-mentioned biosilica have been presented in Figure 1a,b, respectively. Several characteristic absorption bands can be observed in the spectra, providing insights into the molecular structure of biosilica. In the FTIR spectrum the general peak observed was around 1050–1100 cm^{-1}, which corresponds to the characteristic Si-O-Si stretching vibrations. This peak confirms the presence of silica (SiO_2) in the biosilica sample, which is essential for its potential pozzolanic activity and contribution to the mechanical properties of cement mortar. On the other hand, from the SEM image, it can be observed that the biosilica particles exhibit a predominantly spherical or irregular shape. As can be seen from Figure 2, the particle size distribution (ranging from nanoscale to micrometer scale) appears to be relatively uniform, which suggests that the biosilica promotes its potential interaction with the cementitious matrix.

Table 1. Physical properties and chemical composition of cement.

Characteristics	Days	Results Obtained
Standard consistency (%)	-	30
Specific gravity (g/cm^3)	-	3.1
Blain's fineness (m^2/kg)	-	354.8
Compressive strength (MPa)	3 days	21
	7 days	37
	28 days	51
Setting time (min)	Initial	50
	Final	310
Chemical composition of cement (wt.%)		

Al_2O_3	SiO_2	Fe_2O_3	CaO	MgO	SO_3	Loss of ignition	Insol. Resid.	Free CaO
3.93	21.9	2.17	62.2	1.1	2.1	3.2	1.9	1.5

Table 2. Physical properties of sand.

Fineness Modulus	Specific Gravity	Zone	Bulk Density in Compact State (kg/m^3)	Bulk Density in Loose State (g/cm^3)
2.35	2.44	II	1739	1.57

Table 3. The metal oxide content in biosilica (wt.%).

SiO_2	Al_2O_3	Fe_2O_3	K_2O	MgO
88.92	6.1	2.8	1.34	0.84

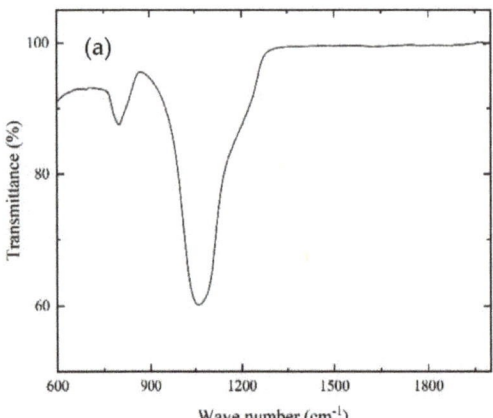

Figure 1. FTIR spectra (**a**) and SEM image (**b**) of biosilica.

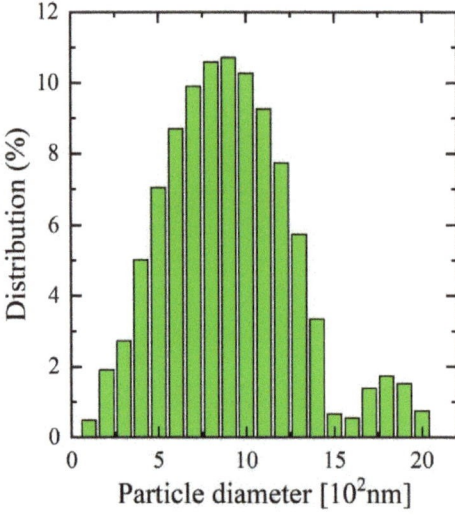

Figure 2. Particle size distribution of biosilica.

2.2. Mixing and Sample Preparation

The w/c ratio used in the present study was 0.47, and the used cement/sand proportion was 1:4 [31]. Two different mixing methods of biosilica were used for sample preparation. First, biosilica and cement were mixed (E095 Mortar mixer, Matest, Treviolo, Italy) for 2.0 min, then this mixture and sand were mixed for 2 min, and the final mixture was mixed with water for 5 min. In the case of second mixing method, the cement and sand were mixed for 2 min, biosilica and water were mixed with magnetic mixer for 5 min, then both obtained mixtures were mixed together for 5 min. The size of the molds were 40 mm × 40 mm × 160 mm. Using a vibrating table (C278, Matest, Treviolo, Italy) the mortar was subjected to 30 s of compaction. Likewise, various mortars were created, including one with no biosilica and others with biosilica content of 2.5, 5.0, 7.5, and 10% of the weight of the cement. After 24 h, the specimens were de-molded, and the mortar samples was immersed in water at 20 ± 0.2 °C (Figure 3).

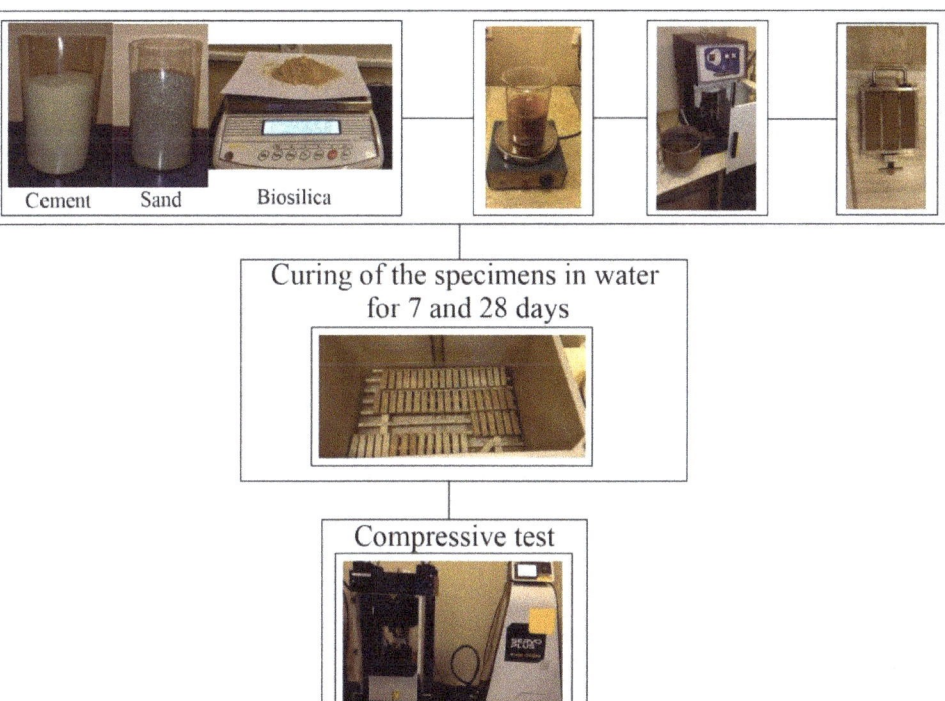

Figure 3. Diagram of the experimental procedure.

2.3. Compressive Strength and Water Absorption Testing

Three specimens were randomly selected from each batch to measure their compressive strength using the Concrete Compression Machine 2000 kN automatic, Servo-Plus Progress, in accordance with standard EN 196-1. The measured specimen dimensions for testing compressive strength were 40 mm × 40 mm. The compressive tests were performed at the ages of 7 and 28 days using an automatic pressure machine (C089) with a loading rate of 2.4 kN/s. In this study, we also paid attention to water absorption characteristics, with tests being conducted in accordance with GOST 12730.3-2020 [35].

Water absorption refers to the capacity of a material to absorb and retain water, and it is assessed by measuring the water saturation of a test sample [GOST 12730.3-2020 Concretes. Method of determination of water absorption]. The test samples were dried at 105 °C until they reached a constant weight and were then weighed under air-dry conditions (m_1). Then, to determine the mass of the saturated test samples, they were submerged in a container of water at a temperature of (20 ± 2) °C, ensuring that the water level was positioned 50 mm above the upper mark of the test samples. At 24 h intervals, the test samples were weighed under air conditions with an accuracy of no more than 0.1% (m_2). Before weighing, the surfaces of the saturated samples were wiped with a damp cloth. The test sample is considered saturated when the difference between consecutive weighings does not exceed 0.1%. After completing the aforementioned steps, the water absorption (W) of the test samples was determined using the following formula:

$$W = \frac{m_2 - m_1}{m_1} \cdot 100\%$$

where W is the water absorption in percentage, m_2 is the mass of the test sample after saturation, and m_1 is the mass of the air-dried test sample. By applying this formula, the water absorption value can be calculated based on the weight difference between the saturated and air-dried samples, expressed as a percentage of the initial dry weight.

3. Results

Figure 4a,b shows the compressive strength of cement mortars with different wt.% of biosilica for 7 and 28 days, respectively. B0, B1, B2, B3, and B4 correspond to the 0%, 2.5%, 5%, 7.5%, and 10% of biosilica, respectively. The presented results were obtained when biosilica was mixed with cement (mixing method 1). The results indicate that the compressive strength of each composition increases with the increase in curing period. This is associated with increased hydration over time. On the other hand, the compressive strength increases with the increase in biosilica concentration. It is shown that for all cases the compressive strength has the maximum value of 10% biosilica concentration. In particular, the strength of the specimens over 7 days of curing increased by 30.5%, and by 36.5% for a curing period of 28 days.

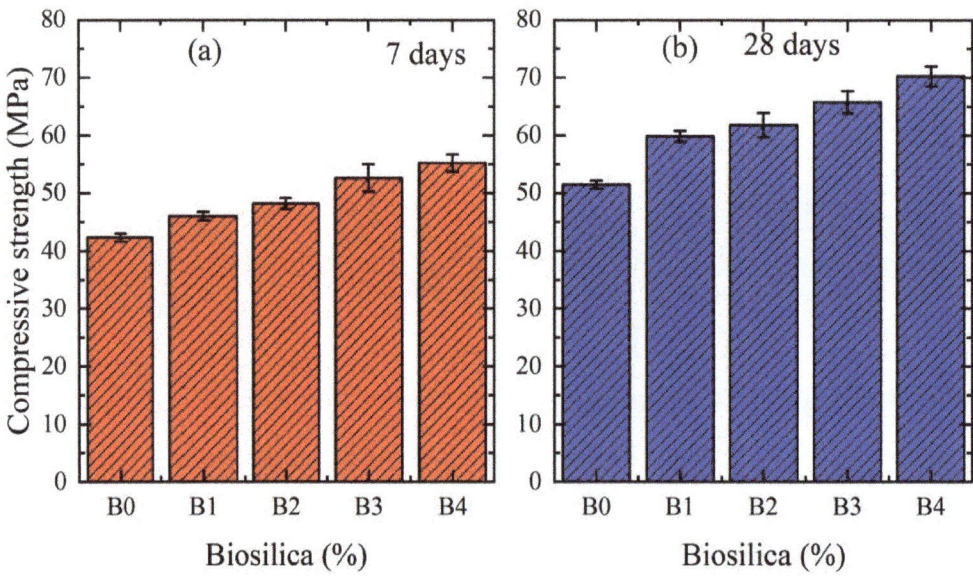

Figure 4. Compressive strength of cement mortars with a different wt.% of biosilica. The results are given for mixing method 1. (**a**) for 7 days, (**b**) for 28 days.

Figure 5a,b shows the compressive strength of cement mortars with a different wt.% of biosilica for 7 and 28 days, respectively. These results were obtained when biosilica was mixed with water (mixing method 2). The same effects of the above parameters (curing period and biosilica concentration) on the compressive strength were also observed for this mixing method. In this case, the compressive strength of the specimens over 7 days of curing increased by 23.4%, and by 47.3% for a curing period of 28 days.

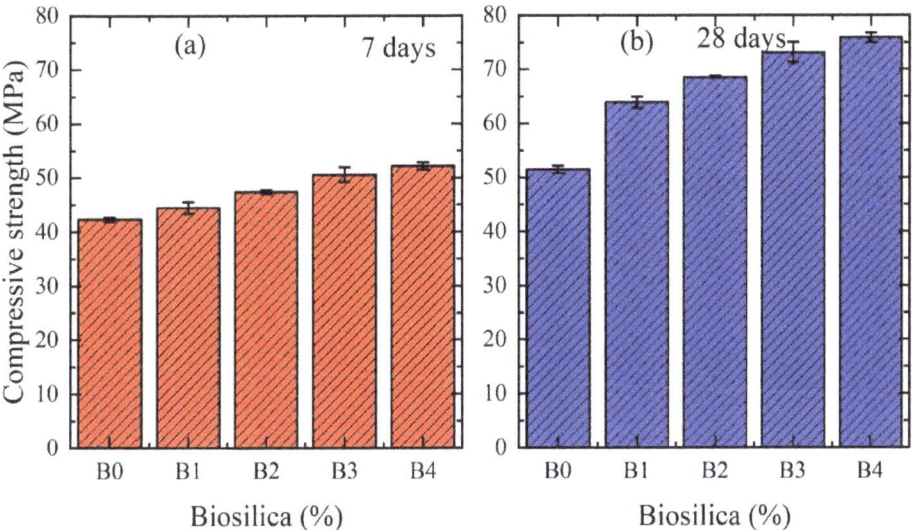

Figure 5. Compressive strength of cement mortars with a different wt.% of biosilica. The results are given for mixing method 2. (**a**) for 7 days, (**b**) for 28 days.

As can be seen from Figures 3 and 4, in all cases, by increasing the content of biosilica in cement mortar, the porosity of the interfacial transition zone significantly diminishes, and that is why the compressive strength always increases (in the Supplementary Materials we present the flexural strength results).

Figure 6a,b shows the water absorption of cement mortars with a different wt.% of biosilica for 28 days in the case of using two different mixing methods (method 1—Figure 5a, method 2—Figure 5b), for which the parameters are already presented above. From the obtained results, the minimum value of water absorption has the same 10% biosilica concentration. In the case of mixing method 1, the water absorption parameters reduced by 22%, and 34% in the case of mixing method 2.

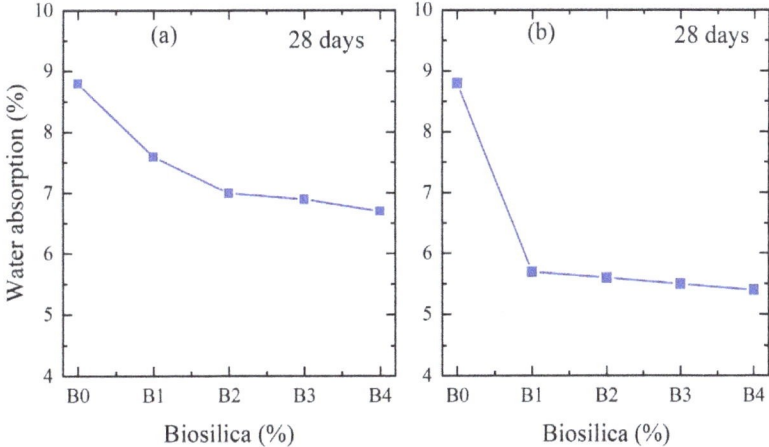

Figure 6. Water absorption of cement mortars with a different wt.% of biosilica. The results are given for mixing method 1 (**a**) and mixing method 2 (**b**), respectively.

The incorporation of a substantial amount of biosilica can result in the filling of gaps and consequently enhance the compressive strength and reduce water absorption. Biosilica particles can absorb compressive loads and thus provide high strength. Due to their small size, biosilica particles can settle into the voids in cement, leading to increased density. The pores become nondeformable due to the presence of biosilica particles, which helps maintain structural integrity. It follows that a homogeneous distribution of biosilica particles can underlie the results obtained due to various mixing methods, and it is obvious that the degree of particle distribution homogeneity is higher in the case of the second mixing method.

4. Conclusions

In this work, the mechanical properties such as compressive strength and water absorption of cement-based mortar with different concentrations of biosilica have been investigated. The mortar was prepared using the following two biosilica mixing methods: First, biosilica was mixed with cement and appropriate samples were prepared. For the other mixing method, samples were prepared by dissolving biosilica in water using a magnetic stirrer. It is shown that, for all cases, the compressive strength has the maximum and water absorption has the minimum value for 10% biosilica concentration. The obtained results show that, in the case of first mixing method, the compressive strength of the specimen over 7 days of hardening increased by 30.5%, and by 36.5% for a curing period of 28 days. In the case of the second mixing method, the compressive strength of the specimen over 7 days of hardening increased by 23.4%, and by 47.3% for a curing period of 28 days. Using mixing methods 1 and 2, the water absorption parameters reduced by 22% and 34%, respectively.

It can also be concluded that using mixing method 2 allows for an increase over 28 days in the mortar's compressive strength by 29.6% and a reduction in water absorption by 35.3% compared with mixing method 1.

Finally, we can conclude that different mixing methods will lead to variations in the dispersion and distribution of biosilica within the mortar matrix, thereby affecting the resulting compressive strength and water absorption. By systematically studying and comparing different mixing methods, this study intends to provide valuable insights into the optimization of biosilica incorporation in cement mortar.

Supplementary Materials: The following supporting information can be downloaded at: https://www.mdpi.com/article/10.3390/ma16165516/s1, Figure S1: Flexural strength of cement mortars with different wt.% of biosilica. The results given for the mixing method 1. Figure S2: Flexural strength of cement mortars with different wt.% of biosilica. The results given for the mixing method 2.

Author Contributions: Conceptualization, M.G.B. and A.A.A.; methodology, N.G.M.; validation, M.G.B., N.G.M. and M.A.K.; formal analysis, Y.V.V.; investigation, N.G.M., M.A.K., A.A.A., D.L. and M.Y.; resources, Y.V.V.; data curation, N.G.M.; writing—original draft preparation, M.G.B. and A.A.A.; writing—review and editing M.U., D.L. and M.G.B.; visualization, M.G.B.; supervision, M.G.B.; project administration, Y.V.V.; funding acquisition, Y.V.V., D.L. and M.G.B. All authors have read and agreed to the published version of the manuscript.

Funding: The authors would like to acknowledge the financial support of the Science Committee of the Republic of Armenia (Project no. 21AG-1C008).

Institutional Review Board Statement: Not applicable.

Informed Consent Statement: Not applicable.

Data Availability Statement: The data presented in this study are available upon request from the corresponding author.

Conflicts of Interest: The authors declare no conflict of interest.

References

1. Nayana, A.M.; Rakesh, P. Strength and durability study on cement mortar with ceramic waste and micro-silica. *Mater. Today Proc.* **2018**, *5*, 24780–24791. [CrossRef]
2. Li, L.G.; Zhu, J.; Huang, Z.H.; Kwan, A.K.H.; Li, L.J. Combined effects of micro-silica and nano-silica on durability of mortar. *Constr. Build. Mater.* **2017**, *157*, 337–347. [CrossRef]
3. Aboul-Nour, L.A.; Mohamed, H.A.; Teama, M. Strength Properties of Nano and Micro Silica Mortar. *Egypt. Int. J. Eng. Sci. Technol.* **2020**, *31*, 35–43.
4. Li, L.G.; Huang, Z.H.; Ng, P.M.; Zhu, J.; Kwan, A.K.H. Effects of Micro-silica and Nano-silica on Fresh Properties of Mortar. *Mater. Sci.* **2017**, *23*, 362–371. [CrossRef]
5. Mahmood, R.A.; Kockal, N.U. Effects of Silica Fume and Micro Silica on the Properties of Mortars Containing Waste PVC Plastic Fibers. *Microplastics* **2022**, *1*, 587–609. [CrossRef]
6. Khan, K.; Ullah, M.F.; Shahzada, K.; Amin, M.N.; Bibi, T.; Wahab, N.; Aljaafari, A. Effective use of micro-silica extracted from rice husk ash for the production of high-performance and sustainable cement mortar. *Constr. Build. Mater.* **2020**, *258*, 119589. [CrossRef]
7. Kooshkaki, A.; Eskandari-Naddaf, H. Effect of porosity on predicting compressive and flexural strength of cement mortar containing micro and nano silica by multi-objective ANN modeling. *Constr. Build. Mater.* **2019**, *212*, 176–191. [CrossRef]
8. Sikora, P.; Augustyniak, A.; Cendrowski, K.; Horszczaruk, E.; Rucinska, T.; Nawrotek, P.; Mijowska, E. Characterization of Mechanical and Bactericidal Properties of Cement Mortars Containing Waste Glass Aggregate and Nanomaterials. *Materials* **2016**, *9*, 701. [CrossRef]
9. Sikora, P.; Horszczaruk, E.; Rucinska, T. The effect of nanosilica and titanium dioxide on the mechanicaland self-cleaning properties of waste-glass cement mortar. *Procedia Eng.* **2015**, *108*, 146–153. [CrossRef]
10. Skoczylas, K.; Rucińska, T. Strength and durability of cement mortars containing nanosilica and waste glass fine aggregate. *Cem. Lime Concr.* **2018**, *21*, 206–215.
11. Cai, Y.; Xuan, D.; Poon, C.S. Effects of nano-SiO_2 and glass powder on mitigating alkali-silica reaction of cement glass mortars. *Constr. Build. Mater.* **2019**, *201*, 295–302. [CrossRef]
12. Lu, J.X.; Poon, C.S. Improvement of early-age properties for glass-cement mortar by adding nanosilica. *Cem. Concr. Compos.* **2018**, *89*, 18–30. [CrossRef]
13. Gupta, S.; Krishnan, P.; Kashani, A.; Kua, H.W. Application of biochar from coconut and wood waste to reduce shrinkage and improve physical properties of silica fume-cement mortar. *Constr. Build. Mater.* **2020**, *262*, 120688. [CrossRef]
14. Hou, P.K.; Kawashima, S.; Wang, K.J.; Corr, D.J.; Qian, J.S.; Shah, S.P. Effects of colloidal nanosilica on rheological and mechanical properties of fly ashe mortar. *Cem. Concr. Compos.* **2013**, *35*, 12–22. [CrossRef]
15. del Bosque, I.E.S.; Martin-Pastor, M.; Martínez-Ramírez, S.; Blanco-Varela, M.T. Effect of temperature on C3S and C3S ţ nanosilica hydration and C-S-H structure. *J. Am. Ceram. Soc.* **2013**, *96*, 957–965. [CrossRef]
16. Ghafari, E.; Costa, H.; Júlio, E. The effect of nanosilica addition on flowability, strength and transport properties of ultra high performance concrete. *Mater. Des.* **2014**, *59*, 1–9. [CrossRef]
17. Singh, L.P.; Karade, S.R.; Bhattacharyya, S.K.; Yousuf, M.M.; Ahalawat, S. Beneficial role of nano silica in cement-based materials—A review. *Constr. Build. Mater.* **2013**, *47*, 1069–1077. [CrossRef]
18. Li, H.; Xiao, H.G.; Yuan, J.; Ou, J. Microstructure of cement mortar with nano-particles. *Compos. Part B Eng.* **2004**, *35*, 185–189. [CrossRef]
19. Joshaghani, A.; Moeini, M.A. Evaluating the effects of sugar cane bagasse ash (SCBA) and nanosilica on the mechanical and durability properties of mortar. *Construct. Build. Mater.* **2017**, *153*, 818–831. [CrossRef]
20. Horszczaruk, E.; Mijowska, E.; Cendrowski, K.; Sikora, P. Influence of the new method of nanosilica addition on the mechanical properties of cement mortars. *Cem. Lime Concr.* **2014**, *5*, 308–315.
21. Flores-Vivian, I.; Pradoto, R.G.K.; Moini, M.; Kozhukhova, M.; Potapov, V.; Sobolev, K. The e_ect of SiO_2 nanoparticles derived from hydrothermal solutions on the performance of portland cement based materials. *Front. Struct. Civ. Eng.* **2017**, *11*, 436–445. [CrossRef]
22. Sahibulla, S.M.M.; Jaisingh, S.J. Pozzolanic biosilica, biochar, and egg shell in setting time, WVTR, and compression strength of biocement mortar: A Taguchi GRA validation. *Biomass Convers. Biorefinery* **2021**. [CrossRef]
23. Li, J.; Jin, Q.; Zhang, W.; Li, C.; Monteiro, P.J.M. Microstructure and durability performance of sustainable cementitious composites containing high-volume regenerative biosilica. *Resour Conserv. Recycl.* **2022**, *178*, 106038. [CrossRef]
24. Hosseini, M.M.; Shao, Y.; Whalen, J.K. Biocement production from silicon-rich plant residues: Perspectives and future potential in Canada. *Biosyst. Eng.* **2011**, *110*, 351–362. [CrossRef]
25. Ahmadi, Z.; Esmaeili, J.; Kasaei, J.; Hajialioghli, R. Properties of Sustainable Cement Mortars Containing High Volume of Raw Diatomite. *Sustain. Mater. Technol.* **2018**, *16*, 47–53.
26. Khodabakhshian, A.; de Brito, J.; Ghalehnovi, M.; Shamsabadi, E.A. Mechanical environmental and economic performance of structural concrete containing silica fume and marble industry waste powder. *Constr. Build. Mater.* **2018**, *169*, 237–251. [CrossRef]
27. Subramanian, N. Introduction to Reinforced Concrete. In *Design of Reinforced Concrete Structures*; Harper & Row: New York, NY, USA, 2013.

28. Li, J.; Zhang, W.; Li, C.; Monteiro, P.J.M. Green concrete containing diatomaceous earth and limestone: Workability, mechanical properties, and life-cycle assessment. *J. Clean. Prod.* **2019**, *223*, 662–679. [CrossRef]
29. Li, J.; Zhang, W.; Li, C.; Monteiro, P.J. Eco-friendly mortar with high-volume diatomite and fly ash: Performance and life-cycle assessment with regional variability. *J. Clean. Prod.* **2020**, *261*, 121224. [CrossRef]
30. Arzumanyan, A.A.; Tadevosyan, V.G.; Muradyan, N.G.; Navasardyan, H.V. Study of "Saralsk" Deposit for Practical Applications in Construction. *J. Arch. Eng. Res.* **2021**, *1*, 3–6. [CrossRef]
31. Muradyan, N.G.; Gyulasaryan, H.; Arzumanyan, A.A.; Badalyan, M.M.; Kalantaryan, M.A.; Vardanyan, Y.V.; Laroze, D.; Manukyan, A.; Barseghyan, M.G. The Effect of Multi-Walled Carbon Nanotubes on the Compressive Strength of Cement Mortars. *Coatings* **2022**, *12*, 1933. [CrossRef]
32. *GOST EN 196-1-2002*; Methods of Testing Cement. Part 1. Determination of Strength. Available online: https://www.armstandard.am/en/standart/5556 (accessed on 31 July 2023).
33. *GOST EN 196-2-2002*; Methods of Testing Cement. Part 2. Chemical Analysis of Cement. Available online: https://www.armstandard.am/en/standart/5557 (accessed on 31 July 2023).
34. *GOST EN 196-3-2002*; Methods of Testing Cement. Part 3. Determination of Setting Time and Soundness. Available online: https://www.armstandard.am/en/standart/5558 (accessed on 31 July 2023).
35. *GOST 12730.3-2020*; Concretes. Method of Determination of Water Absorption. Available online: https://www.armstandard.am/en/standart/5508 (accessed on 31 July 2023).

Disclaimer/Publisher's Note: The statements, opinions and data contained in all publications are solely those of the individual author(s) and contributor(s) and not of MDPI and/or the editor(s). MDPI and/or the editor(s) disclaim responsibility for any injury to people or property resulting from any ideas, methods, instructions or products referred to in the content.

Article

Study on Alkali-Activated Prefabricated Building Recycled Concrete Powder for Foamed Lightweight Soils

Yao Xiao [1], Zhengguang Wu [2] and Yongfan Gong [2,*]

[1] Suzhou Polytechnic Institute of Agriculture, Suzhou 215008, China
[2] College of Civil Science and Engineering, Yangzhou University, Yangzhou 225127, China
* Correspondence: yfgong@yzu.edu.cn; Tel.: +86-15850663317

Abstract: The advantage of a prefabricated building is its ease of construction. Concrete is one of the essential components of prefabricated buildings. A large amount of waste concrete from prefabricated buildings will be produced during the demolition of construction waste. In this paper, foamed lightweight soil is primarily made of concrete waste, a chemical activator, a foaming agent, and a foam stabilizer. The effect of the foam admixture on the wet bulk density, fluidity, dry density, water absorption, and unconfined compressive strength of the material was investigated. Microstructure and composition were measured by SEM and FTIR. The results demonstrated that the wet bulk density is 912.87 kg/m^3, the fluidity is 174 mm, the water absorption is 23.16%, and the strength is 1.53 MPa, which can meet the requirements of light soil for highway embankment. When the foam content ranges from 55% to 70%, the foam proportion is increased and the material's wet bulk density is decreased. Excessive foaming also increases the number of open pores, which reduces water absorption. At a higher foam content, there are fewer slurry components and lower strength. This demonstrates that recycled concrete powder did not participate in the reaction while acting as a skeleton in the cementitious material with a micro-aggregate effect. Slag and fly ash reacted with alkali activators and formed C-N-S(A)-H gels to provide strength. The obtained material is a construction material that can be constructed quickly and reduce post-construction settlement.

Keywords: alkali activated; prefabricated building; foamed lightweight soil; properties; micro-structure

Citation: Xiao, Y.; Wu, Z.; Gong, Y. Study on Alkali-Activated Prefabricated Building Recycled Concrete Powder for Foamed Lightweight Soils. *Materials* 2023, 16, 4167. https://doi.org/10.3390/ma16114167

Academic Editors: Malgorzata Ulewicz and René de Borst

Received: 6 March 2023
Revised: 29 May 2023
Accepted: 30 May 2023
Published: 2 June 2023

Copyright: © 2023 by the authors. Licensee MDPI, Basel, Switzerland. This article is an open access article distributed under the terms and conditions of the Creative Commons Attribution (CC BY) license (https://creativecommons.org/licenses/by/4.0/).

1. Introduction

Prefabricated buildings have the advantages of convenient construction, high construction efficiency, and low costs [1]. It has become a development trend in the construction industry [2–4]. For example, the penetration rate of prefabricated buildings has surpassed 70% in Sweden [5–7]. Additionally, as concrete is a materials for prefabricated buildings, a large amount of waste concrete building will be produced during the demolition of construction waste. So far, recycled aggregates have been widely used in civil engineering in accordance with some codes for their application of them, such as GB/T25176-2010 "Recycled fine aggregate for concrete and mortar", GB/T 25177-2010 "Recycled coarse aggregate for concrete and mortar" [8]. However, due to low activity and high porosity, recycled concrete powder has been stockpiled for the long term. As reviewed by some researchers [9,10], recycled concrete powder has been highlighted as a sustainable approach. However, when the dosage is greater than 15%, it has adverse effects on the strength, flowability, and durability of the formed specimen. Therefore, except for a small amount used for low-value products, recycled concrete powder from building waste has not yet been effectively utilized; therefore, it is of great social significance and economic benefit to solve the low activity and realize resource utilization of RCP. Currently, there are recent studies reported on the preparation of alkali-activated recycled concrete powder [11–13]. The potential activity of a large amount of SiO$_2$ and Al$_2$O$_3$ in the recycled concrete powder can be improved by alkali activator activation. Hence, as granulated blast furnace slag (GBFS) and fly ash (FA), RCP has been selected by researchers as a raw geopolymer material, which is synthesized by regulating compounds, fineness, and the activation technology of RCP [14–16]. The advantages of alkali-activated recycled concrete powder cementitious material include a simple preparation process, no calcination, low energy consumption, low costs, and a large market. It can essentially achieve zero emissions of "three wastes" [17–19].

Gong [20] activated recycled concrete powder with NaOH and mixed sodium silicate and discovered that the compressive strength was 18.5 percent higher than that of recycled concrete without an alkali activator. Some researchers [21] found that the activation effect of the NaOH activator is higher than that of the sodium sulfate base activator. Wang [22] revealed that adding an alkali activator promotes the formation of C-S-H and makes the microstructure of recycled micro-powder concrete more compact. The double mixing of the alkali activator and salt activator further promotes the hydration reaction of recycled concrete powder, improves the strength of recycled concrete powder, and produces hydration products; a dense microstructure occurs. Rovnan [23] demonstrated that building waste powder can have pozzolanic properties after being ground to a certain fineness and can react with alkaline components. This property allows it to be used as an active admixture in concrete or as a raw material for the production of geopolymer. Bassani [24] investigated the alkali activation potential of fine particles of recycled construction and demolition waste (CDW) aggregate when mixed with an appropriate alkaline solution. A waste recycled concrete powder is used to produce alternative paving blocks through the alkali activation process.

Foamed lightweight soil is a kind of light material that is prepared by physical method from foaming agent aqueous solution into foam, mixed and stirred with component cementitious materials, water, bubbles, admixtures, additives, etc. in a certain proportion, and hardened by physical and chemical action [25–27]. In recent years, due to the advantages of light weight, high strength, good fluidity, convenient construction, and so on, it has been widely used in road engineering, geotechnical engineering, and reinforcement engineering in China [28–30]. However, there are several issues with Portland cement as a traditional cementitious in the preparation process of foamed lightweight soil, such as high energy consumption, large pollution, resource waste, high carbon emissions, and so on. Moreover, when cement is used as the raw material of foamed lightweight soil, it is also prone to burst foam and loses its original shape. Therefore, various researchers are focusing on alkali-activated cementitious material as a replacement for cement [31–34]. Through some research [35–37], it was found that the curing system has a significant impact on the properties and microstructure of the recycled concrete powder. In general, an increase in reaction temperature can significantly increase the hydration rate of recycled concrete powder, which has a very beneficial effect on improving the strength of geopolymer.

According to the published literature, alkali-activated recycled concrete powder cementitious materials can achieve similar strength, compatibility, and dry shrinkage properties to cement paste. The production process of alkali-inspired materials consumes less power than cement and is a new type of green material. In this paper, we propose to seek a method for the comprehensive utilization of concrete powder, at the same time, preparing foamed lightweight soil for road engineering. It is suitable for replacement or filling as the main material where conventional materials cannot meet the requirements. It can not only meet the requirements of unit weight but also meet the strength requirements. It can be widely used in fields such as highways, buildings, pipelines, and mine goaf backfilling.

2. Materials and Methods
2.1. Raw Materials

Recycled concrete powder (RCP) is produced from broken waste prefabricated building concrete. It is collected by a dust removal device to manufacture recycled aggregate with a specific surface area of more than 350 m^2/kg. The slag powder was S95 grade, with an activity index of 96%. The fly ash was of grade II with a low lime content. The chemical compositions of recycled concrete powder, slag, and fly ash are shown in Table 1. As can be seen in Figures 1 and 2, the main components of recycled concrete powder are calcite, silicon dioxide, and dolomite. Among them, calcite is the phase formed after carbonization, while silicon dioxide and dolomite are the phases existing in natural stone and mortar. The medium particle size of the recycled concrete powder is 45 μm.

Table 1. Chemical composition of raw materials.

NO.	SiO_2	CaO	Al_2O_3	Fe_2O_3	MgO	K_2O	TiO_2	Na_2O	Others
RCP (%)	34.3	42.1	8.2	6.2	4.5	1.2	1.0	1.0	1.5
Slag (%)	30.0	38.1	13.6	0.6	12.5	0.4	0.6	0.3	3.9
Fly ash (%)	61.9	2.4	28.8	2.5	0.8	1.5	1.04	0.3	0.76

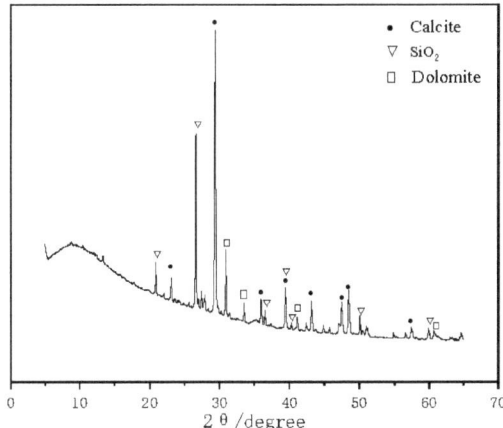

Figure 1. XRD pattern of recycled concrete powder.

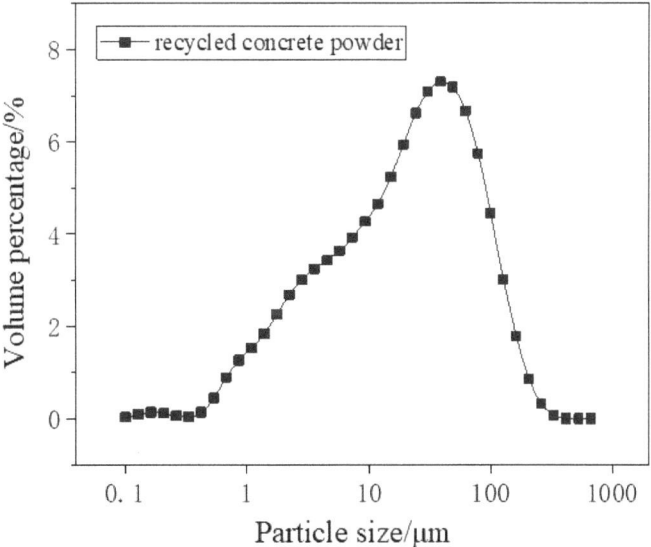

Figure 2. The particle size of recycled concrete powder.

The alkali activator formulated in this paper is based on sodium hydroxide and industrial water glass as raw materials. The sodium hydroxide was used to reduce the modulus of the industrial water glass solution from 3.3 to 1.4. The dosage of water glass (sodium silicate) was 6%, which was formulated according to the ratio of Na_2O to cementitious materials. The foaming agent used in this study was sodium dodecyl sulfate (YS-200), with an analytical purity level (AR), and the foam stabilizer is calcium stearate with an analytical purity level (AR).

2.2. Design of Mix Proportion

The experiments in this study were guided by the specifications JTG D30-2015 and TJG F1001-2011. It is proposed that the construction wet capacity of foam lightweight soil should be 5.0~11.0 kg/m^3, and the flow value is 170~190 mm. This experiment began with a single-factor study to investigate the effect of foam admixture on the fluidity, dry density, compressive strength, and water absorption of foamed lightweight soil, to determine the optimal performance of recycled micronized cementitious material foamed lightweight soil. According to the specification, the ratio design of foam lightweight soil can be calculated according to the following steps. Keeping the total mass of the construction wet weight unchanged, which was measured to be 600~1000 kg/m^3, then gradually adjusting the amount of cementitious material. The components contained 60% recycled concrete powder, 20% slag, and 20% fly ash. Water glass as an activator was measured at 6%. With water to cementitious materials ratio of 0.45 (the total amount of components is 891 kg/m^3, 794 kg/m^3, 694 kg/m^3, 595 kg/m^3, and 476 kg/m^3, respectively). After mixing, five mix ratio test groups with a wet density of 600 kg/m^3–1000 kg/m^3 were obtained. Calcium stearate is mixed at 2% of the powder mass. The specific mix ratio is shown in Table 2.

Table 2. Mix proportions of foamed lightweight soil.

Sample	W/C	Components (kg/m^3)					R_{fw} (kg/m^3)
		RCP	Slag	Fly Ash	Water Glass	Foam	
F1		535	178	178	53.5	1090	1000
F2		476	159	159	47.6	1189	900
F3	0.45	416	139	139	41.6	1288	800
F4		357	119	119	35.7	1387	700
F5		278	99	99	29.8	1986	600

2.3. Preparation

First, the foaming agent is mixed with water according to the dilution ratio, and the foaming machine (FP-5) is started to prepare foam. Second, pour the weight of recycled concrete powder, slag powder, fly ash, and calcium stearate into the mixing pot and mix evenly. Third, according to the design requirements of density grade, add corresponding foam, then pour the prepared foam light soil into the corresponding test mold. Lastly, move the entire test mold into a constant temperature cement standard curing box with a relative humidity of >90% and a temperature of 20 °C ± 0.5 °C. Then, remove the formwork 24 h after sample preparation and place it in the curing box again for curing. Upon arrival, take it out for property testing. The Preparation process of foamed lightweight soil has been shown in Figure 3.

2.4. Methods

Foamed lightweight soil pastes with varying foam contents were prepared, the w/c was set as 0.45, and a cuboid specimen measuring 40 mm × 40 mm × 160 mm was formed; six samples were tested per group. The wet bulk density weight is the weight per unit volume of freshly mixed foamed lightweight soil. According to the relevant criterion of the specification requirements, its wet bulk density should be less than 1100 kg/m^3. The fluidity was tested according to GB/T17671-2021, the dry density was tested according to GB/T 5486-2008. Water absorption was tested in accordance with JG/T 266-2011, and unconfined compressive strength in accordance with CECS 249-2008.

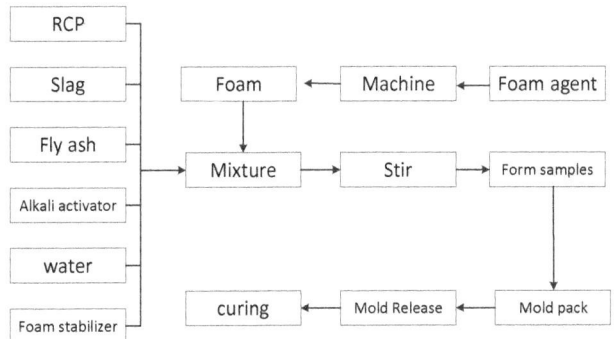

Figure 3. Preparation process of foamed lightweight soil.

The infrared spectrum is primarily used to examine the composition and structure of foamed lightweight soil pastes. Cary 610/670 micro-infrared spectrometer produced by Varian in the United States is used for infrared spectrum testing. Carl Zeiss' Gemini SEM 300 ZEISS field emission scanning electron microscope system was used.

3. Results

3.1. Wet Bulk Density

The wet bulk density of foamed lightweight soil decreases significantly with the increase in foam admixture, as shown in Figure 4. This is due to the use of many bubbles rather than gelling materials, which increases the proportion of generated pores while decreasing the quality of the gelling components. It can be found that the measured wet bulk density is larger compared with the theoretically calculated value. This is because the foam is crushed by the slurry during mixing. The foam soil in a mold was discovered to collapse the phenomenon of F5 to a certain degree, which is due to the addition of excessive foam, resulting in the cementing component being unable to effectively lubricate the foam wall, and the ruptured foam released free water mixed into the slurry, equivalent to improving the water–cement ratio, increasing the wet bulk density of the F5 group.

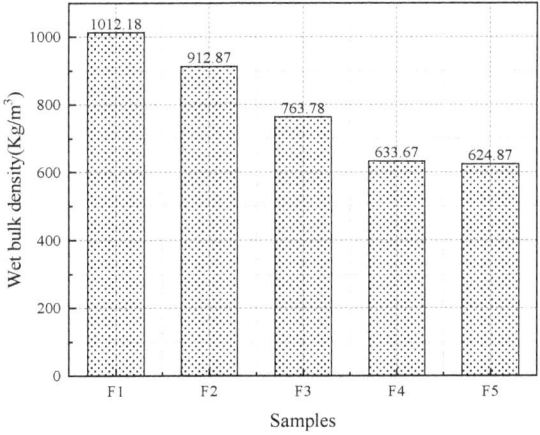

Figure 4. Wet bulk density of foamed lightweight soil.

3.2. Fluidity

Fluidity is an important indicator to measure whether foamed lightweight soil meets the specification requirements. The fluidity should be controlled at 170~190 mm, and the test results are shown in Figure 5. As shown in Figure 5, fluidity decreases significantly with the increase in foam content. This is because the poor fluidity of the foam necessitates slurry wrapping lubrication, and increased production results in insufficient slurry. Calcium stearate has a surface-activating effect and can enhance the liquid wall of the foam. Calcium stearate has hydrophobic properties and can combine with hydroxyl groups to form hydrophobic substances to achieve a bubble retention effect. The overall viscosity of the foam after calcium stearate pacification is large, which reduces the fluidity and requires the cementitious material to play a lubricating role. The alkali-activated slurry has the modifying effect of alkali activation, resulting in an unbalanced charge and a flocculent gel adsorbing a large amount of free water, which will make a low-fluidity slurry.

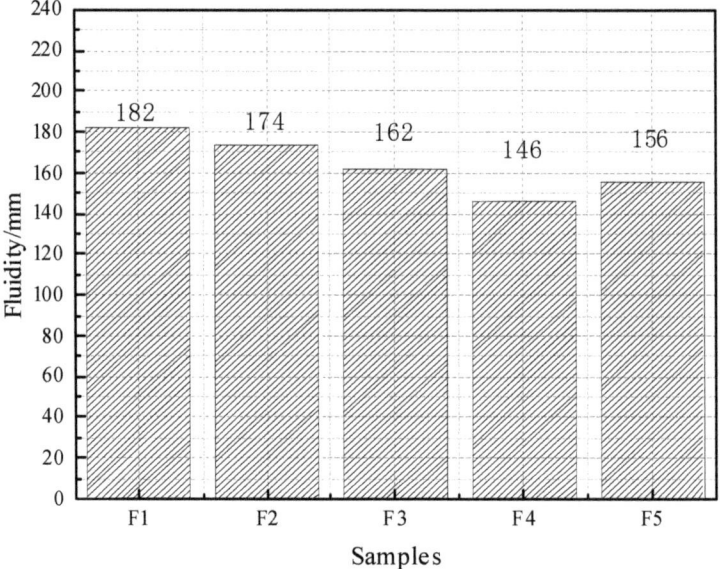

Figure 5. Fluidity of foamed lightweight soil.

Figure 6 shows a freshly mixed recycled powder light soil sample with calcium stearate added to the left and an unadded sample on the right. Many bubbles accumulate on the surface of the foam stabilizer component without the addition of foam stabilizer. This is due to the high flow rate of slurries without foam stabilizers, and the foam is more likely to float to the surface and burst during the stirring process. Despite the addition of calcium stearate to this test sample, excessive foam caused the structural instability of foam in the slurry, the deterioration of bonding performance, and failure to float. In sample F5, many foams burst, the proportion of cementitious materials was increased, and the moisture was released, resulting in a return to fluidity. According to the data, only the F1 and F2 groups meet the criterion requirements.

Figure 6. Effect of calcium stearate on bubbles.

3.3. Dry Density and Water Absorption

According to the test results shown in Figure 7, the water absorption rate of foamed lightweight soil gradually increases with the increase of the foam content. This is because when the volume content of foam increases, the open pores and connecting pores on its surface also increase, and its ability to absorb water is significantly improved. The increase in water absorption rate gradually accelerates as the foam content increases because when the foam content is too high, the stability of the foam decreases, and small foams easily merge into large foams. Because of the capillary effect, an appropriate pore size is more likely to retain adsorbed water, so the water absorption rate increases linearly when the content is high.

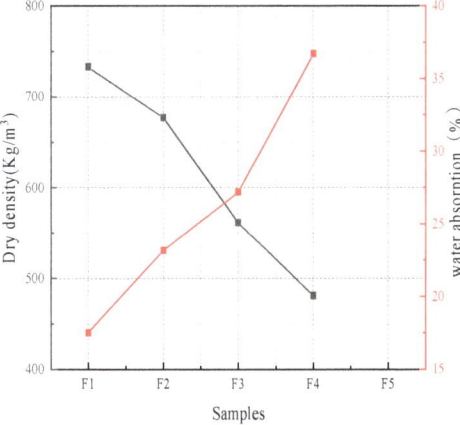

Figure 7. Dry density and water absorption of foamed lightweight soil.

Figure 8 illustrates the surface morphology of samples F4 and F2, respectively. Sample F4 has insufficient slurry, a too-thin foam wall, and a smaller foam burst due to the addition of a large amount of foam. At the same time, some combinations produce foam with larger pore sizes, a more discrete foam distribution, and an uneven structure. The P2 specimen's slurry and foam are in good condition, and it can be seen that the fine foam is evenly distributed throughout the material, and the sample with this ratio is in better condition. This is because the uniformity of pores improves the molding effect, and better pore structure can improve the strength of the specimen.

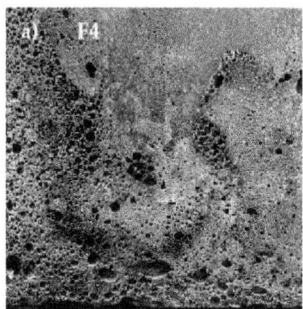 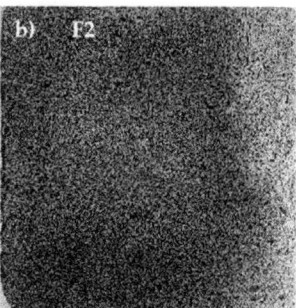

Figure 8. Surface morphology of foamed lightweight soil.

3.4. Unconfined Compressive Strength

According to the specification, cubic test blocks with specimen sizes of 100 mm × 100 mm × 100 mm are loaded under unrestricted conditions. As illustrated in Figure 9, the compressive strength gradually decreases with the increase in foam content. F1, F2, F3, and F4 exceed the standard value of 0.6 MPa. This is due to the fact that as the foam content increases, the amount of cementitious material decreases while the porosity increases. This leads to a decrease in the internal load-bearing skeleton of the specimen and a decrease in the strength support part, resulting in a decline in strength. When the foam content is high, the internal foam is unstable and will burst to form large and connected pores, as shown in the image above. Large pores and connected pores cause uneven stress distribution within the material, resulting in weak parts and deteriorating strength. When the foam content is moderate, its distribution is better, and the structure is uniform. Relatively good structural morphology can improve the strength of the material. After testing, all other components can meet the criterion conditions, with the exception of the F5 group, which shows the collapse phenomenon of foamed lightweight soil in the mold.

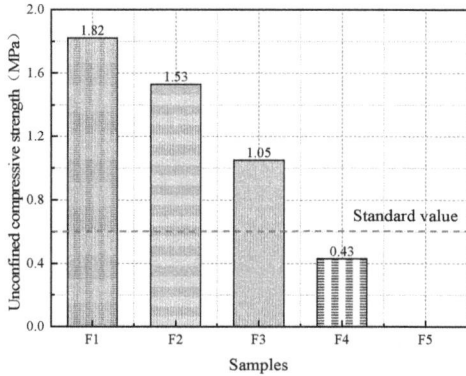

Figure 9. Unconfined compressive strength of foamed lightweight soil.

3.5. TG Analysis

According to the TG image of F2 in the Figure 10, at the stage of 20~100 °C, with the increase in age, the weight loss rate at 28 days show a certain increase compared to 3 days. As the time increases, the C-A-S-H gel components increase and the amount of free water adsorbed increases in the foamed lightweight soil. At the stage of 100~300 °C, the weight loss rate is stable and the residual free water gradually decreases. At the stage of 400~500 °C, there was no significant weight loss. However, due to the secondary hydration reaction between $Ca(OH)_2$ and the active mixture, residual $Ca(OH)_2$ is extremely prone to carbonization into $CaCO_3$. Between 600 and 700 °C, the test sample loses a large amount of weight. It is due to the endothermic decomposition of calcite in the sample, which releases CO_2 and converts to CaO. It can be seen that there is a difference in the decomposition state between the alkali-activated cementitious material and the raw recycled concrete powders. The peak TG value of the cementitious material is around 700 °C, while the peak TG value of the raw recycled concrete powders is around 735 °C. This is becausee the calcite particles formed by carbonization of the cementitious material are very small and can be decomposed at lower temperatures.

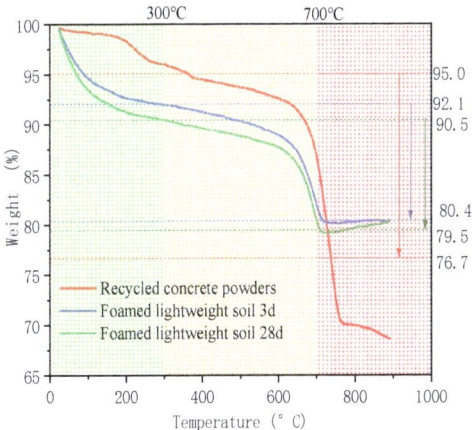

Figure 10. TG analysis of foamed lightweight soil.

3.6. Phase Analysis Results of FTIR

The FTIR spectra of the foamed lightweight soil of F2 after 3 and 28 days of curing are shown in Figure 11. It is clear that the vibrational absorption peak of the positive silicate Si-O bond at around 445 cm^{-1} smooths out. Around 445 cm^{-1} to 625 cm^{-1} there are the flexural vibration absorption peaks of Si-O-Si or Si-O-Al of chain-shaped and layered silicates. This indicates that under the action of the alkali activator, silicate polymerization occurs, resulting in different degrees of product polymerization. The waveforms of these products overlap on the curve, so the spectrum tends to be flat. The four-coordinate aluminum expansion vibration peak at 873 cm^{-1} weakens with time, and Si-O-R at 979 cm^{-1} weakens and shifts to a lower wave number. Similarly, the carbonate expansion vibration peak at 1419 cm^{-1} weakens and disappears, indicating further depolymerization of these groups with age. However, the foamed lightweight soil of the F2 change trend was similar to that of pure recycled concrete powders. This shows that S-O bonds, Al-O bonds, and C-O bonds are the main products in alkali-activated materials systems. Although C-N-S(A)-H gels covered the original gels, the structure and shape of gels are similar, and the bonding energy becomes stronger.

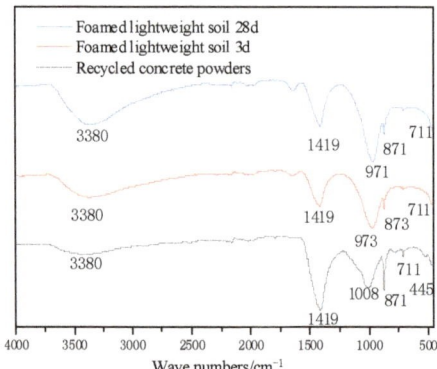

Figure 11. FTIR of foamed lightweight soil.

3.7. Microstructure Results of SEM

SEM images of the foamed lightweight soil of F2 were observed at different ages. As shown in Figure 12a, the microscopic images of the 3-day-old pastes show more cracks in the surface. This might occur because of the low activity of recycled concrete powders. Simultaneously, the alkali activation reaction of recycled concrete powder, slag, and fly ash cannot be completely reacted in the early stages, which leads to the low strength of foamed lightweight soil. As Figure 12b shows, it is difficult to find wide cracks in the surface of 28-day-old paste. This indicated that in the alkali activating system, the hydration of the foamed lightweight soil can effectively fill gaps between early hydration products. Significantly, the prolongation of time not only improves compactness but also improves the strength of foamed lightweight soil. Therefore, the hydration of alkali-activated foamed lightweight soils is a continuous process and cannot fully react in the early stages.

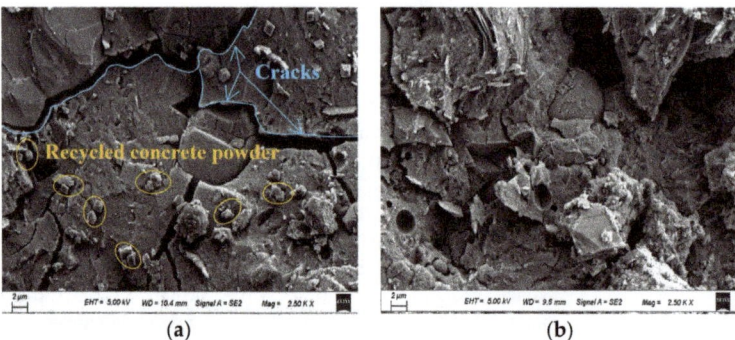

Figure 12. SEM images of foamed lightweight soil: (**a**) 3 d; (**b**) 28 d.

4. Conclusions

The process of foamed lightweight soil preparation using alkali-activated recycled concrete powder has been proposed, and the effect of foam content on the material is revealed in terms of strength, microstructure, and other parameters. Based on our experimental results, the following conclusions and recommendations can be drawn:

1. As alkali-activated raw materials, when the recycled powder, fly ash, and slag occupied 60%, 20%, and 20%, respectively, it can be used to prepare foamed lightweight soil. The wet bulk density is 912.87 kg/m^3, the fluidity is 174 mm, the water absorption is 23.16%, and the strength is 1.53 MPa, all of which can meet the requirements for light soil for highway embankment.

2. The addition of an appropriate amount of foam stabilizer greatly improves the foam retention of the material. When the foam content ranges from 55% to 70%, it results in an increase in the proportion of foam and a decrease in the material's wet bulk density. When the foam content is low, the slurry does not wrap the excess foam well, and the material's fluidity decreases significantly. Too much foam increases the number of open pores and it can easily merge into large foam without stability, which reduces water absorption. At a higher foam content, there are fewer slurry components and lower strength.
3. There are more cracks in the surface of early hydration products, as the reaction progresses, the foamed lightweight soil can effectively fill gaps between early hydration products. However, the hydration of alkali-activated foamed lightweight soils is a continuous process and cannot fully react in the early stages.
4. Compared with the general fill soil or reinforced soil, foamed lightweight soil is convenient for construction without compaction. It is a construction material that can be constructed quickly and reduce post construction settlement, and has higher economic benefits, and is therefore worth spreading and applying.

Author Contributions: Y.X., Z.W. and Y.G. are the main contributors to this research. They conducted the main experimental program, analyzed the experimental results, drafted the research paper, processed experimental data, and proofread the paper. All authors have read and agreed to the published version of the manuscript.

Funding: This work was supported by the policy guidance program (Cooperation of Industry, Education and Academy) of Jiangsu Province (BY2020661), the Yangzhou city science and technology plan for a cooperation project between the city and the university (YZ2021165).

Institutional Review Board Statement: Not applicable.

Informed Consent Statement: Not applicable.

Data Availability Statement: Data are contained within the article.

Conflicts of Interest: The authors declare no conflict of interest.

References

1. Wang, K.; Ren, L.; Yang, L. Excellent carbonation behavior of rankinite prepared by calcining the C-S-H: Potential recycling of waste concrete powders for prefabricated building products. *Materials* **2018**, *11*, 1474. [CrossRef]
2. Migunthanna, J.; Rajeev, P.; Sanjayan, J. Investigation of waste clay brick as partial replacement in geopolymer binder. *Constr. Build. Mater.* **2023**, *365*, 130107. [CrossRef]
3. Zeybek, O.; Ozkilic, Y.O.; Karalar, M.; Celik, A.I.; Qaidi, S.; Ahmad, J.; Burduhos-Nergis, D.D.; Burduhos-Nergis, D.P. Influence of replacing cement with waste glass on mechanical properties of concrete. *Materials* **2022**, *15*, 7513. [CrossRef]
4. Sosa, M.E.; Zega, C.J. Experimental and estimated evaluation of drying shrinkage of concrete made with fine recycled aggregates. *Sustainability* **2023**, *15*, 7666. [CrossRef]
5. Navaratnam, S.; Ngo, T.; Gunawardena, T.; Henderson, D. Performance review of prefabricated building systems and future research in Australia. *Buildings* **2019**, *9*, 38. [CrossRef]
6. Liu, X.; Liu, E.; Fu, Y. Reduction in Drying Shrinkage and Efflorescence of Recycled Brick and Concrete Fine Powder–Slag-Based Geopolymer. *Appl. Sci.* **2023**, *13*, 2997. [CrossRef]
7. Yuan, C.; Chen, Y.; Liu, D.; Lv, W.; Zhang, Z. The basic mechanical properties and shrinkage properties of recycled micropowder UHPC. *Materials* **2023**, *16*, 1570. [CrossRef] [PubMed]
8. Jang, H.; Kim, J.; Sicakova, A. Effect of aggregate size on recycled aggregate concrete under equivalent mortar volume mix design. *Appl. Sci.* **2021**, *11*, 11274. [CrossRef]
9. Shi, Y.N.; Weng, H.X.; Yu, J.Q.; Gong, Y.F. Study on modification and mechanism of construction waste to solidified silt. *Materials* **2023**, *16*, 2780. [CrossRef]
10. Bondar, D.; Nanukuttan, S. External sulphate attack on alkali-activated slag and slag/fly ash concrete. *Buildings* **2022**, *12*, 94. [CrossRef]
11. Gencel, O.; Nodehi, M.; Hekimoglu, G.; Ustaoglu, A.; Sarı, A.; Kaplan, G.; Bayraktar, O.Y.; Sutcu, M.; Ozbakkaloglu, T. Foam concrete produced with recycled concrete powder and phase change materials. *Sustainability* **2022**, *14*, 7458. [CrossRef]
12. Tang, Q.; Ma, Z.M.; Wu, H.X.; Wang, W. The utilization of eco-friendly recycled powder from concrete and brick waste in new concrete: A critical review. *Cem. Concr. Compos.* **2020**, *114*, 103807. [CrossRef]

13. Xu, J.; Kang, A.H.; Wu, Z.G.; Gong, Y.F.; Xiao, P. The effect of mechanical-thermal synergistic activation on the mechanical properties and microstructure of recycled powder geopolymer. *J. Clean. Prod.* **2021**, *327*, 129477. [CrossRef]
14. Bun, P.; Cyr, M.; Laniesse, P.; Idir, R.; Bun, K.N. Concrete made of 100% recycled materials-Feasibility study. *Resour. Conserv. Recycl.* **2022**, *180*, 106199. [CrossRef]
15. Miraldo, S.; Lopes, S.; Pacheco-Torgal, F.; Lopes, A. Advantages and shortcomings of the utilization of recycled wastes as aggregates in structural concretes. *Constr. Build. Mater.* **2021**, *298*, 123729. [CrossRef]
16. Fořt, J.; Vejmelková, E.; Koňáková, D.; Ablblová, N.; Čáchová, M.; Keppert, M.; Rovnaníková, P.; Černý, R. Application of waste brick powder in alkali activated aluminosilicates: Functional and environmental aspects. *J. Clean. Prod.* **2018**, *194*, 714–725. [CrossRef]
17. Ye, T.; Xiao, J.Z.; Duan, Z.H.; Li, S.S. Geopolymers made of recycled brick and concrete powder—A critical review. *Constr. Build. Mater.* **2022**, *330*, 127232. [CrossRef]
18. Liu, M.; Hu, R.H.; Zhang, Y.C.; Wang, C.Q.; Ma, Z.M. Effect of ground concrete waste as green binder on the micro-macro properties of eco-friendly metakaolin-based geopolymer mortar. *J. Build. Eng.* **2023**, *68*, 106191. [CrossRef]
19. Kim, J.; Jang, H. Closed-loop recycling of C&D waste: Mechanical properties of concrete with the repeatedly recycled C&D powderas partial cement replacement. *J. Clean. Prod.* **2022**, *343*, 130977.
20. Gong, Y.F.; Fang, Y.H.; Yan, Y.R.; Chen, L.Q. Investigation on alkali activated recycled cement mortar powder cementitious material. *Mater. Res. Innov.* **2014**, *18*, 784–787. [CrossRef]
21. Robayo-Salazar, R.; Valencia-Saavedra, W.; de Gutiérrez, R.M. Reuse of powders and recycled aggregates from mixed construction and demolition waste in alkali-activated materials and precast concrete units. *Sustainability* **2022**, *14*, 9685. [CrossRef]
22. Wang, X.; Yan, Y.; Tong, X.; Gong, Y. Investigation of mineral admixtures on mechanical properties of alkali-activated recycled concrete powders cement. *Buildings* **2022**, *12*, 1234. [CrossRef]
23. Rovnanik, P.; Rovnanikova, P.; Vysvaril, M. Rheological properties and microstructure of binary waste red brick powder/metakaolin geopolymer. *Constr. Build. Mater.* **2018**, *188*, 924–933. [CrossRef]
24. Bassani, M.; Tefa, L.; Coppola, B.; Palmero, P. Alkali-activation of Aggregate Fines from Construction and Demolition Waste: Valorisation in View of Road Pavement Subbase Applications. *J. Clean. Prod.* **2019**, *234*, 71–84. [CrossRef]
25. Vafaei, M.; Allahverdi, A. Strength development and acid resistance of geopolymer based on waste clay brick powder and phosphorous slag. *Struct. Concr.* **2019**, *20*, 1596–1606. [CrossRef]
26. Yang, D.Y.; Liu, M.; Zhang, Z.B.; Yao, P.P.; Ma, Z.M. Properties and modification of sustainable foam concrete including eco-friendly recycled powder from concrete waste. *Case Stud. Constr. Mater.* **2022**, *16*, e00826. [CrossRef]
27. Xiao, J.Z.; Ma, Z.M.; Sui, T.B.; Akbarnezhad, A.; Duan, Z. Mechanical properties of concrete mixed with recycled powder produced from construction and demolition waste. *J. Clean. Prod.* **2018**, *188*, 720–731. [CrossRef]
28. Xuan, D.X.; Tang, P.; Poon, C.S. MSWIBA-based cellular alkali-activated concrete incorporating waste glass powder. *Cem. Concr. Compos.* **2019**, *95*, 128–136. [CrossRef]
29. Huseien, G.F.; Hamzah, H.K.; Sam, A.R.M.; Khalid, N.H.A.; Shah, K.W.; Deogrescu, D.P.; Mirza, J. Alkali-activated mortars blended with glass bottle waste nano powder: Environmental benefit and sustainability. *J. Clean. Prod.* **2019**, *243*, 118636. [CrossRef]
30. Font, A.; Soriano, L.; Tashima, M.M.; Monzo, J.; Borrachero, M.V.; Paya, J. One-partEEco-cellular concrete for the precast industry: Functional features and life cycle assessment. *J. Clean Prod.* **2020**, *269*, 122203. [CrossRef]
31. Wan, X.; Li, H.; Che, X.; Xu, P.; Li, C.; Yu, Q. A study on the application of recycled concrete powder in an alkali-activated cementitious system. *Processes* **2023**, *11*, 203. [CrossRef]
32. Tefa, L.; Bassani, M.; Coppola, B.; Palmero, P. Strength development and environmental assessment of alkali-activated construction and demolition waste fines as stabilizer for recycled road materials. *Constr. Build. Mater.* **2021**, *289*, 123017. [CrossRef]
33. Ma, Z.; Tang, Q.; Wu, H.; Xu, J.; Liang, C. Mechanical properties and water absorption of cement composites with various fineness and contents of waste brick powder from C&D waste. *Cem. Concr. Compos.* **2020**, *114*, 103758.
34. Aiken, T.A.; Gu, L.; Kwasny, J.; Huseien, G.F.; McPolin, D.; Sha, W. Acid resistance of alkali-activated binders: A review of performance, mechanisms of deterioration and testing procedures. *Constr. Build. Mater.* **2022**, *342*, 128057. [CrossRef]
35. Swanepoel, J.; Strydom, C. Utilisation of fly ash in a geopolymeric material. *Appl. Geochem.* **2002**, *17*, 1143–1148. [CrossRef]
36. Fernández-Jiménez, A.; Palomo, A.; Criado, M. Microstructure development of alkali-activated fly ash cement: A descriptive model. *Cem. Concr. Res.* **2005**, *35*, 1204–1209. [CrossRef]
37. Ghanbari, M.; Hadian, A.; Nourbakhsh, A.; MacKenzie, K. Modeling and optimization of compressive strength and bulk density of metakaolin-based geopolymer using central composite design: A numerical and experimental study. *Ceram. Int.* **2017**, *43*, 324–335. [CrossRef]

Disclaimer/Publisher's Note: The statements, opinions and data contained in all publications are solely those of the individual author(s) and contributor(s) and not of MDPI and/or the editor(s). MDPI and/or the editor(s) disclaim responsibility for any injury to people or property resulting from any ideas, methods, instructions or products referred to in the content.

Article

Modification of Concrete Composition Doped by Sewage Sludge Fly Ash and Its Effect on Compressive Strength

Tomasz Kalak [1,*], Patrycja Szypura [1], Ryszard Cierpiszewski [1] and Malgorzata Ulewicz [2]

[1] Department of Industrial Products and Packaging Quality, Institute of Quality Science, Poznan University of Economics and Business, Niepodleglosci 10, 61-875 Poznan, Poland; ryszard.cierpiszewski@ue.poznan.pl (R.C.)

[2] Faculty of Civil Engineering, Czestochowa University of Technology, Akademicka 3 Street, 42-200 Czestochowa, Poland

* Correspondence: tomasz.kalak@ue.poznan.pl

Abstract: The sustainable development of construction materials is an essential aspect of current worldwide trends. Reusing post-production waste in the building industry has numerous positive effects on the environment. Since concrete is one of the materials that people manufacture and use the most, it will continue to be an integral element of the surrounding reality. In this study, the relationship between the individual components and parameters of concrete and its compressive strength properties was assessed. In the experimental works, concrete mixes with different contents of sand, gravel, Portland cement CEM II/B-S 42.5 N, water, superplasticizer, air-entraining admixture, and fly ash from the thermal conversion of municipal sewage sludge (SSFA) were designed. According to legal requirements in the European Union, SSFA waste from the sewage sludge incineration process in a fluidized bed furnace should not be stored in landfills but processed in various ways. Unfortunately, its generated amounts are too large, so new management technologies should be sought. During the experimental work, the compressive strength of concrete samples of various classes, namely, C8/10, C12/15, C16/20, C20/25, C25/30, C30/37, and C35/45, were measured. The higher-class concrete samples that were used, the greater the compressive strength obtained, ranging from 13.7 to 55.2 MPa. A correlation analysis was carried out between the mechanical strength of waste-modified concretes and the composition of concrete mixes (the amount of sand and gravel, cement, and FA), as well as the water-to-cement ratio and the sand point. No negative effect of the addition of SSFA on the strength of concrete samples was demonstrated, which translates into economic and environmental benefits.

Keywords: concrete; compressive strength; sewage sludge fly ash; destructive testing

Citation: Kalak, T.; Szypura, P.; Cierpiszewski, R.; Ulewicz, M. Modification of Concrete Composition Doped by Sewage Sludge Fly Ash and Its Effect on Compressive Strength. *Materials* **2023**, *16*, 4043. https://doi.org/10.3390/ma16114043

Academic Editor: Carlos Leiva

Received: 10 March 2023
Revised: 22 May 2023
Accepted: 24 May 2023
Published: 29 May 2023

Copyright: © 2023 by the authors. Licensee MDPI, Basel, Switzerland. This article is an open access article distributed under the terms and conditions of the Creative Commons Attribution (CC BY) license (https://creativecommons.org/licenses/by/4.0/).

1. Introduction

Construction-related advancements necessitate the ongoing development of building materials, particularly concrete. The competition between various construction materials is intensifying as globalization advances. In order to find and create products with the best strength and durability, there is therefore a lot of research being conducted in the area of concrete durability. Concrete is used in the construction of buildings because it is strong, durable, and has superior strength characteristics. Sand, gravel, and cement are just a few of the ingredients used to make concrete. In order to impart particular performance attributes, extra chemical admixtures are also used in the concrete mix [1]. The United States (72 million metric tons (MMT)), the Netherlands (54 MMT), Turkey (12 MMT), India (12 MMT), France (11 MMT), Italy (10 MMT), Bulgaria (8.4 MMT), Spain (5.7 MMT), and Poland (5.5 MMT) were the world's top producers of sand and gravel in 2021 [2]. For the past ten years, 4.1 billion metric tons of cement have been produced worldwide [3]. China (2100 MMT), India (370 MMT), Vietnam (120 MMT), the United States (95 MMT), and Turkey (85 MMT) were the principal countries producing the most cement in the world in 2022 [4]. By-products from the cement-making process, namely,

CO_2, NO_x, and SO_2, are emitted into the atmosphere. China, which released 853 Mt of CO_2 into the atmosphere in 2021, is the world leader in CO_2 emissions into the atmosphere from cement production. India (149 MMT), Vietnam (54.1 MMT), Turkey (44.4 MMT), the United States (41.2 MMT), Saudi Arabia (28.7 MMT), Indonesia (28.6 MMT), Japan (23.8 MMT), South Korea (23.7 MMT), and Iran (22.5 MMT) are the countries that come behind it [5]. Different industrial by-products are utilized in concrete as partial substitutes for cement to lower CO_2 emissions. Thanks to this, the landfilling of waste is reduced, which becomes a new raw material used by the construction industry in accordance with the principles of sustainable development and the circular economy [6–10].

Fly ash (FA) and waste glass sludge are two of the industrial wastes that are most commonly utilized to produce concrete. Landfilling the waste causes significant environmental issues, such as the tainting of ground or surface waters, an increase in soil alkalinity, and detrimental impacts on vegetation and other living organisms. Thus, using waste materials has environmental benefits by reducing landfilling and lowering CO_2 emissions into the atmosphere [1,11,12]. Furthermore, waste materials are added to increase mechanical qualities such as compressive and tensile strength [13].

Fly ash (FA), a by-product of combustion operations, is one of the useful waste materials added to concrete mixtures. The fine-grained dust, which is made up of grains of erratic sizes and shapes, is a useful mineral additive. FA increases the workability of concrete mixes, but it also makes concrete more resistant to the aggressive sulfate environment, slows down the hardening process, makes concrete more resistant to high temperatures (up to 600 °C), reduces frost resistance, reduces shrinkage, and makes hardened concrete stronger and more durable. The adroit application of FA's features results in concrete with improved properties that are suited to receivers' requirements. An additional benefit is the reduced price of concrete compared to that made using only cement. A further advantage is that the concrete costs are less than those of concrete produced only with cement. Numerous researchers have discussed in the literature the impact of adding different forms of FA on the characteristics of concrete [14–19]. FA waste from the burning of hard coal [14,15], biomass [17,19,20], and sewage sludge from municipal facilities [18,21–23] was used to modify mortars and concrete. Concrete [22,23] and cement [24] were enhanced with dried sewage sludge. Dried sewage sludge was added to concrete in levels ranging from 50 to 15% as a substitute for cement. It has been demonstrated that the replacement of cement with FA in concrete increased the compressive strength after 28 days of curing, but the attained strength of waste-modified concrete was less than that of control samples [22]. These wastes were also substituted for sand in concrete from 2.5 up to 12.5%. According to reports, replacing sand with waste up to 7.5% does not affect the characteristics of concrete's strength [23]. Fly ash from sewage sludge (SSFA) in the amount of 15 to 30% can be added to cement mortars without significantly changing their mechanical properties [25]. The heat of hydration in cement mortars is affected by adding SSFA in amounts ranging from 2.5 to 20%. Along with an increase in the exchange of cement with waste, the rate of heat release was slowed down, especially during the early stages of mortar binding [26]. In this study, SSFA waste was added to concrete mixtures, which has positive social and economic effects. Fly ash is produced from the incineration of sewage sludge from local municipalities. The combustion technology is a modern and advanced solution for the thermal treatment of sewage sludge. Low emissions of pollutants, high energy and combustion efficiency, the ability to recover energy, compatibility with a variety of fuels, and simplicity of the installation operation are only a few of the benefits. This kind of waste production is steadily rising everywhere, including Poland. According to statistics, this country produced 1048.7 thousand tons of sewage sludge in 2019; only 159.7 thousand of those tons were thermally converted, and roughly 113.3 thousand tons were kept in landfills [27,28]. Since it is difficult to effectively use such a large amount of sewage sludge in land reclamation or other applications, the share of thermal treatment methods will rise. As a result, there will be an increase in the amount of SSFA created, which requires the development of effective management techniques.

Numerous financial and environmental advantages come from adding SSFA to cement and concrete. These include removing sewage sludge to free up space in landfills that already exist, cutting costs and raising the quality of many building materials, lowering waste disposal costs, including those associated with landfilling, lowering the use of a significant amount of primary raw materials, implementing a sustainable development economy system by turning sewage sludge waste into new useful products, conserving energy, and lowering the emissions of NO_x, CO_2, and other pollutants. Public education or a greater understanding of the advantages of waste recycling may be an additional benefit. The usage of the SSFA additive has benefits as well as drawbacks, such as its decreased pozzolanic activity, which results in a reduction in strength qualities and an increase in water consumption, although these drawbacks can be offset by grinding, appropriate modification, and treatment. Another limitation is the lower reactivity of these materials compared to cement flour [28–35].

There are reports of laboratory attempts to utilize SSFA waste for use in regular concrete and cement mortars in the literature. However, the findings of this research are unclear and do not suggest that they might be implemented in the manufacturing of regular concrete products. Therefore, a study was undertaken to determine compressive strength, one of the most crucial parameters for modified concrete. SSFA may be used as an addition in the creation of concrete mixes in line with the applicable Polish legislation (Journal of Laws of 2016, item 108) [36] and the Directive of the European Parliament and of the Council (EU/2010/75) [37]. The SSFA produced using circulating fluidized bed combustion (CFBC) technology should be considered a possible material for use in the manufacturing of building materials, given the significance of zero-emissions, sustainability, and the circular economy [38–42].

The aim of the research was to determine the influence of SSFA waste generated during the combustion process of municipal sewage sludge in a fluidized bed furnace on the compressive strength of different classes of concrete. The correlation of the mechanical strength of waste-modified concretes with respect to the composition of concrete mixes (amount of sand and gravel, cement, SSFA), water-to-cement ratio, and sand point was analyzed.

The use of fly ash as a waste product from the incineration of local municipal sewage sludge utilizing modern circulating fluidized bed combustion (CFBC) technology is the scientific novelty of the research. Using this technology to thermally treat sewage sludge has a number of advantages, including a sizable decrease in waste weight, the recovery of thermal energy during combustion, high process efficiency, low pollutant emissions, the reduction of odors, system stability, and low susceptibility to changes in sludge composition. SSFA may be added to concrete, bricks, asphalt, ceramics, or other materials in addition to serving as a filter medium. It has a high content of mineral substances, and its quality and composition depend on the chemical composition of the sludge, the conditions of the combustion process, flue gas cleaning, and the cooling rate. According to the new regulations in the European Union, landfilling of waste should be avoided, and methods of safe and ecological disposal are to be sought. In Poland, the problem is the excessive amount of sewage sludge produced; hence, more and more thermal processing installations are being built. These activities are in line with the trends of sustainable development and the circular economy. Hence, this research work is a response to regional problems in Poland related to waste management and brings suggestions that may be a partial solution to these problems. In Central Europe, the construction industry is a booming sector that can partially absorb excess SSFA and convert it into new building materials.

The present studies in this manuscript are limited to the compressive strength of concrete and the analysis of correlations with individual components. The limitation also applies to a period of up to 28 days. Because compressive strength is a feature that serves one of the most crucial roles in concrete structures, tensile and flexural strength tests were left out. Because the experiments were concentrated on the strength characteristics of the concrete samples, other tests, such as durability evaluations, were not conducted.

2. Experimental Procedure
2.1. SEM-EDS Analysis

The elemental composition of SSFA samples was examined with a scanning electron microscope (SEM) Hitachi S-3700N (accuracy: <0.4 nm for all fields of view and pixel sizes;

manufacturer: Hitachi High-Technologies Corporation, Tokyo, Japan) with an attached Noran SIX energy dispersive X-ray spectrometer (EDS) microanalyzer (ultra-dry silicon drift type with resolution (FWHM) 129 eV, accelerating voltage: 20.0 kV; manufacturer: Thermo Electron Corporation, Schönwalde-Glien, Germany).

2.2. Transmission Electron Microscopy (TEM)

High-resolution transmission electron microscopy (HRTEM) and scanning transmission electron microscopy (STEM) were carried out using a JEOLARM 200F (accuracy: the highest point resolution is around 0.05 nm). SSFA samples for analysis using transmission electron microscopy were prepared by deposition of catalysts onto holey carbon films supported by a copper TEM grid.

2.3. XRD Analysis

X-ray diffraction measurements were made using the D8 Advance diffractometer (apparatus accuracy is approximately 100% for phase identification and 86% for three-step-phase-fraction quantification; manufacturer: Bruker AXS Advanced X-ray Solutions Gmbh, Karlsruhe, Germany). In configuration, the diffractometer is equipped with a Johansson monochromator ($\lambda_{Cu\,K\alpha 1}$ = 1.5406 Å) and a LynxEye silicon strip detector. The minimum measurement angle is 0.6° 2Θ deg. The XRD powder diffraction method requires the delivered sample to be carefully powderized. A standard measuring dish has a container for powder with a diameter of 25 mm and a depth of 1.5 mm. Before measurement, the sample powder needs to be lightly pressed.

2.4. Preparation of Concrete Mixes for Tests

Samples for testing the compressive strength were taken in accordance with the PN-EN 12390-3:2019-07 standard [43]. The fresh concrete mixes (tested recipes C8/10, C12/15, C16/20, C20/25, C25/30, C30/37, and C35/45), produced at the concrete mixing plant, were taken into a container whose walls were previously lubricated with agents, preventing the concrete mixture from sticking to the container. The concrete mixture was transferred from the container into a mold with dimensions of 150 mm × 150 mm × 150 mm. Two layers of the mixture, each no taller than 20 cm, were vibrated throughout the molding process. The amount of compaction of the concrete mix in the mold and venting determine how long the concrete mix is vibrated for. It was taken care not to shake the mixture for an extended period of time, as this could cause ingredient segregation or the laitance effect of the cement to show on the surface (bleeding). The sample's surface was vibrated and then lapped to level it.

In the next stage, the samples were then completely immersed in water at 20 °C for a minimum of 24 h for maturation in molds. Then, before placing them in the testing machine, the samples were dried by wiping excess water from the surface. They were cured in a construction laboratory for 28 days. On the 28th day, the samples were subjected to a compressive strength test using a strength press. All procedures were performed in accordance with the European standards SIST EN 12390-1:2021, SIST EN 12390-2:2019, SIST EN 12390-3:2019, and SIST EN 12390-4:2019 [44–46].

The samples' weights and side length measurements were used to calculate their densities. The testing device's bearing surfaces as well as the surfaces of the concrete samples were thoroughly cleaned and dried. The sample was set up in the press so that the compressive forces were perpendicular to the direction of sample formation.

2.5. Compressive Strength Test

Compressive strength tests were conducted using an Ele International Automatic Compression Machine (accuracy: it was calibrated to within 0.5% of reading, from 1% to 100% of machine capacity; manufacturer: ELE International, Carter Lanes, Klin Farm, Milton Keynes, United Kingdom) that meets the requirements of the PN-EN 12390-4:2020-03 standard [47]. The sample was loaded, and the rate at which the clamping force increased was between 0.2 and 1.0 MPa/s. The apparatus recorded the value of the force with which it acted on the sample's surface after achieving the sample's maximum strength value. The compressive strength was calculated using the following formula:

$$CS = \frac{F}{A} \quad (1)$$

where CS is the compressive strength (MPa), F is the maximum force at the point of failure (N), and A is the cross-sectional surface area (m^2).

The correctness of the test performance was determined using the schemes of correct and incorrect destruction of samples defined in the European Standard SIST EN 12390-2:2009 [44].

3. Results and Discussion

3.1. Characterization of the Concrete Materials

The investigation was conducted using seven different types of recipes for concrete mixtures of various strengths. Table 1 lists the characteristics and compositions of the concrete mixtures utilized in the study. The percentage of grain masses in the aggregate with a diameter of up to 2 mm is known as the sand point. The w/c ratio is used to determine the correct amount of water and cement in the concrete mixture. It is important to ensure appropriate properties of the mixture (consistency and workability) and functional properties of the concrete (strength, water absorption, and tightness).

Table 1. Composition and properties of concrete mixes used in the tests.

Concrete Composition and Properties	Recipe Number of Mixes						
	510,471	515,471	520,470	525,410	530,412	537,412	545,460
	Concrete Class						
	C8/10	C12/15	C16/20	C20/25	C25/30	C30/37	C35/45
Sand 0/2 mm (kg)	990	810	730	695	690	620	620
Gravel 2/8 mm (kg)	360	455	521	497	525	555	562
Gravel 8/16 mm (kg)	468	560	645	610	640	670	642
CEM II/B-S 42.5 N (kg)	205	230	265	275	320	340	380
Water (kg)	180	174	170	169	166	160	163
Air content in the mixture (%)	2	1.9	1.9	1.9	1.9	2	4.5
Fly ash (kg)	90	80	75	70	65	60	55
Superplasticizer Sikament 400/30	—	—	2.12	3.03	3.52	3.4	4.56
Air-entraining admixture Sika® Luftporenbildner LPS A-94	—	—	—	—	—	—	1.14
Sand point (%)	0.54	0.44	0.39	0.38	0.37	0.33	0.32
Water to cement ratio w/c	0.88	0.76	0.64	0.61	0.52	0.47	0.43

The recipe number is the designation of the concrete composition, which determines the class of concrete and is characterized by a certain strength. Each digit in the recipe number has a specific meaning:

- The first digit indicates the type of cement; in this case, each recipe starts with 5, which means that Portland slag cement CEM II/B-S 42.5 N was used. Requirements of the cement are as follows: SO$_3$ content $\leq 3.5\%$, Cl content $\leq 0.1\%$, initial setting time ≥ 60 min, change in volume ≤ 10 mm, compressive strength after 2 days ≥ 10 MPa, compressive strength after 28 days between 42.5 and 62.5 MPa, Na$_2$O$_{eq}$ content $\leq 0.8\%$.
- The next two digits indicate the class of concrete, e.g., 10 means class C8/10.
- The digit 4 repeated in each of the concrete recipes indicates that aggregate with grain sizes up to 16 mm was used for research.
- The last two digits are successive numbers assigned as a result of creating a recipe.
- The following ingredients were used to make samples of concrete mixes:
    - Sand, 0/2 mm (from the Prusim sand mine, Poland);
    - Gravel, 2/8 mm and 8/16 mm (from the KSM Rakowice—Górażdże Kruszywa mine, Poland);

- CEM II/B-S 42.5 N cement (from the Górażdże Cement, Poland), symbol designation: CEM II—multi-component Portland cement, 42.5—compressive strength class after 28 days determined in accordance with the Polish standard PN-EN 196-1, S—granulated blast furnace slag, N—normal early strength cement;
- Water from municipal water supplies;
- Fly ash waste produced as a by-product of the combustion of municipal sewage sludge;
- Sikament 400/30 (Sika Poland Sp. zo.o., Warsaw, Poland) plasticizer;
- LPS A94 (Sika Poland Sp. zo.o.) air-entraining admixture.

The physico-chemical properties of 0/2 sand and 2/8 and 8/16 gravel are shown in Tables 2 and 3.

Table 2. The physico-chemical properties of 0/2 mm sand.

Compounds, Parameters	Content and Results
SiO_2	>99.1%
Fe_2O_3	320 ppm
Al_2O_3	2600 ppm
CaO	240 ppm
MgO	70 ppm
Clay	0.2%
$CaCO_3$	0.4%
Moisture	<0.1%
pH	7.0
Density	2.54 g/cm^3

Table 3. The properties of 0/2 sand and 2/8 and 8/16 gravel.

Property	Sand, 0/2 mm	Gravel, 2/8 mm	Gravel, 8/16 mm
The content of mineral dust	0.65% category f_3	0.46% category f_3	0.42% category f_3
The content of organic substances	Absence	Absence	Absence
Bulk density	1.71 kg/dm^3	1.67 kg/dm^3	1.78 kg/dm^3
Flatness index	—	6.2%, category FI_{10}	5.6%, category FI_{10}
Absorptivity	—	0.58%	0.55%

The chemical composition of CEM II/B-S 42.5 N was determined using the X-ray Diffraction (XRD) method, and the results are presented in Table 4. The physical properties of the material are as follows: specific surface area 4589 cm^2/g, specific gravity 2.91 kg/dm^3, initial setting time 240 min, 2-day compressive strength 22.3 MPa, 28-day compressive strength 56.4 MPa.

Table 4. Chemical composition of CEM II/B-S 42.5 N cement.

Compound	CaO	SiO_2	Al_2O_3	Fe_2O_3	MgO	Na_2O	K_2O	Na_2O_{eq}
Content (%)	53.41	28.38	6.76	2.54	4.26	0.45	0.62	0.74

Sikament 400/30 superplasticizer is a homogeneous dark brown liquid consisting of sulfonated and non-sulfonated polycondensates of sodium and magnesium salts and carbohydrates (naphthalene, lignosulfonate, and gluconate). Sikament 400/30 has the following characteristics: density of 1.1 kg/m3, pH of 5.0 ± 1.0, chloride ion content ≤0.1%, and sodium oxide equivalent ≤4.0%. The superplasticizer admixture's activity in concrete mixtures is based on the electrostatic repulsion principle. This enables the concrete mix and concrete to have the following properties: intensive wetting of the cement grains by forming a double ionic layer; a significant reduction in friction force; the ability to reduce the amount of mixing water or to liquefy the mix with a constant amount of mixing water; an improvement in workability; the facilitation of the achievement of high water tightness, frost resistance, and reduced water absorption; longer consistency retention; and high durability.

The Sika® Luftporenbildner LPS A-94 air-entraining admixture (Sika Poland Sp. zo.o., Warsaw, Poland) is a homogeneous dark brown liquid consisting of modified tensides. The properties are the following: density of 1 kg/dm^3, pH of 7.0 ± 1.0, chloride ion content ≤0.1%, and sodium oxide equivalent ≤0.6%. The process by which the additive in concrete mixes works is based on the production of extremely small air bubbles in the cement grout. Correct pore size and arrangement, increased workability due to air entrainment, and high levels of frost resistance in water and salt solutions are all made possible by this process.

Fly ash, used in the studies, was obtained as a result of the incineration of municipal sewage sludge (SSFA) using circulating fluidized bed combustion (CFBC) technology. This technology uses a recirculating loop, i.e., air is blown through a bed of sand at a speed sufficient to suspend particles on the air stream (approx. 1–3 m/s^2) at a temperature of approx. 800–900 °C. Up to 95% of pollutants can be blocked from entering the atmosphere thanks to this solution. Grain fractions are reported in Table 5 following the granulation analysis. SSFA is a mixture of irregularly shaped and sized particles that frequently form different agglomerates. An examination of the bulk density revealed that it was 0.81 ± 0.02 g/cm^3. On a vibrating table, the fly ash was further compressed to a density of 1.52 ± 0.03 g/cm^3. The outcomes of these investigations are crucial for the potential addition of fly ash to building materials. SEM-EDS analysis revealed that it was primarily comprised of SiO_2, Al_2O_3, P_2O_5, CaO, CO_2, Fe_2O_3, K_2O, and other substances (Table 6). The fact that the multi-component SSFA mixture was not a homogeneous product needs to be underlined here. Depending on where in the silo tank the samples were gathered, there may have been a small difference in the quantitative composition of the samples taken. The mixture was nevertheless expected to exhibit the same crystalline phases according to X-ray diffraction analysis. Similar results of SSFA composition analysis can be found in the literature [48–50].

Table 5. Granulometric composition of SSFA.

FA grain fractions (mm)	0–0.212	0.212–0.5	0.5–0.71	0.71–1.0	1.0–1.7	>1.7
Granulometric composition (%)	90.17 ± 1.2	8.64 ± 1.2	0.51 ± 0.07	0.68 ± 0.06	0	0

Table 6. Elemental composition of SSFA (SEM-EDS analysis).

Element	Ca	O	P	Al	Si	Fe	C	Mg	K	Zn	Ti	Na
Weight content (%)	7.3	46.8	6.8	12.9	13.2	3.0	1.59	2.34	4.35	0.46	0.35	0.85
Atomic content (%)	3.86	62.0	4.63	10.1	9.98	1.13	2.81	2.03	2.36	0.15	0.15	0.79
Compound Content (%)	CaO 10.21	-	P_2O_5 15.52	Al_2O_3 24.42	SiO_2 28.32	Fe_2O_3 4.27	CO_2 5.84	MgO 3.87	K_2O 5.24	ZnO 0.57	TiO_2 0.85	Na_2O 1.15

The literature reports that the amount of CaO in the mixture affects the setting time. Fly ash is appropriate for concrete mixes when the CaO level is between 5% and 15%, according to Diaz et al. [51]. It was reported that setting time decreases as CaO content rises. The SSFA used in this study showed a CaO content of 10.2%. Measurements were made pointwise on a specific SSFA sample during the analysis using the SEM-EDS method. However, because the substance is a heterogeneous mixture, the oxide content may differ slightly depending on where a measurement site is located on the sample. Calculations utilizing Formula (2) can be used to predict the final setting time of concrete mixtures [52].

$$ST = 4838.7 \cdot e^{-0.262 \cdot CaO} \tag{2}$$

where ST is setting time (min), e is exponential number, and CaO is CaO content in fly ash (%, m/m).

According to the formula, the predicted setting time of a concrete mix with a CaO content of 10.2% in SSFA is 333.4 min.

Figure 1 displays the TEM image of the ash sample. The particles have a flocculent nature, uneven forms, and various diameters. It is possible to see shades of varying intensities; lighter zones correspond to material layers that are thinner, and darker zones correspond to layers that are thicker. Other authors in the literature have also found similar findings for ash TEM images [53–55]. The temperature of combustion, the air circulation process, and the cooling rate all impact the morphology of SSFA particles. Because mineral substances were not partially melted in a fluidized bed boiler at temperatures between 800 and 900 °C, irregular shapes, rough structures, or occasionally sharp edges may be the outcome. On the other hand, the enormous surface area created by the SSFA particles' rough and loose texture may have a negative impact on the amount of water required when employed as an additive in concrete compositions [56–59].

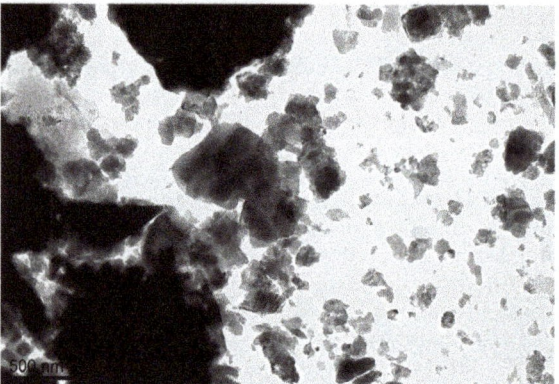

Figure 1. TEM image of SSFA (scale bar: 500 nm).

3.2. Analysis of Strength Tests

Compressive strength was measured on six cubic samples: C8/10, C12/15, C16/20, C20/25, C25/30, C30/37, and C35/45. The compressive strength of concrete is a parameter determining its durability, and the appropriate selection of its class allows it to be used in a building structure exposed to specific external influences. All test results, including sample weight, density, destructive force at maximum sample load, and compressive strength (Figure 2), are presented in Tables S1–S7 (Supplementary Materials). Table 7 provides a summary of all test results. A destructive force that should not be less than the guaranteed value is also given for each class of concrete. Because of the relationship, as the destructive force is reduced, the concrete's strength is likewise reduced. There is a particular minimum strength value that each kind of concrete should meet. It is important that the concrete achieves the guaranteed strength without significantly exceeding it, as this could indicate an inadequate composition of the concrete mix [1,60].

Figure 2. Stages of destruction of the concrete sample during the compressive strength test.

Table 7. Average strength test results of concrete mixes.

Class Concrete	Sample Weight (kg)	Density (kg/dm^3)	Destructive Force (kN)	Compressive Strength (Mpa)
C8/10	7.385 ± 0.015	2.188 ± 0.005	307.83 ± 9.58	13.67 ± 0.41
C12/15	7.588 ± 0.023	2.248 ± 0.007	501.67 ± 85.53	22.3 ± 3.82
C16/20	7.771 ± 0.041	2.303 ± 0.012	661 ± 33.17	29.38 ± 1.46
C20/25	7.673 ± 0.053	2.274 ± 0.016	713.67 ± 15.99	31.72 ± 0.70
C25/30	7.655 ± 0.055	2.268 ± 0.016	821.5 ± 30.62	36.52 ± 1.39
C30/37	7.723 ± 0.051	2.288 ± 0.015	1068.3 ± 89.30	47.47 ± 3.95
C35/45	7.783 ± 0.047	2.306 ± 0.014	1240.83 ± 26.69	55.15 ± 1.18

According to the results in Table 7, variable proportions of concrete ingredients, including sand, gravel, cement, water, SSFA, and chemical admixtures, have a considerable impact on the strength of concrete mixes. It is clear that the majority of the volume of concrete is made up of the aggregate used to create it, which has a big impact on the properties of both freshly mixed concrete and hardened concrete. Sand serves as a fine aggregate, and when utilized for weaker concretes of the C8/10 and C12/15 classes vs. concretes with enhanced strength of the C20/25–C35/45 classes, a substantial quantitative difference was seen. Due to the fact that the quantitative selection of specific fractions in 2/8 and 8/16 gravel depended on the criteria that the concrete was to fulfill, such as the intended use of environmental conditions or workability, there was no dependence on the class of concrete in this situation. In the construction industry, the aggregate-to-cement ratio is often chosen based on the experience of the production plant or from graphs and tables created using data from laboratory tests. The primary fluidizing agents utilized in the creation of medium- and high-strength concretes are known as superplasticizers. These raw materials used in the research enabled up to a 40% reduction in the amount of mixing water while retaining consistency. Superplasticizers improve workability, increase the strength and durability of concrete, and improve the aesthetics of the surface.

The relationship between the class of concrete samples and their compressive strength is shown in Figure 3. These findings support the hypothesis that the compressive strength of concrete improves as the concrete class increases, namely, from 13.7 MPa (C8/10) to 55.2 MPa (C35/45). The amount of added CEM II/B-S 42.5 N cement is inextricably linked with the compressive strength; it depends on the concrete class and its demand for 1 m^3 of concrete; however, it should not exceed 400 kg. The cement content proves the durability of concrete to the extent that it affects the water-to-cement (w/c) ratio. The relationship between the amount of cement and SSFA used and the compressive strength of concrete is presented in Figure 4.

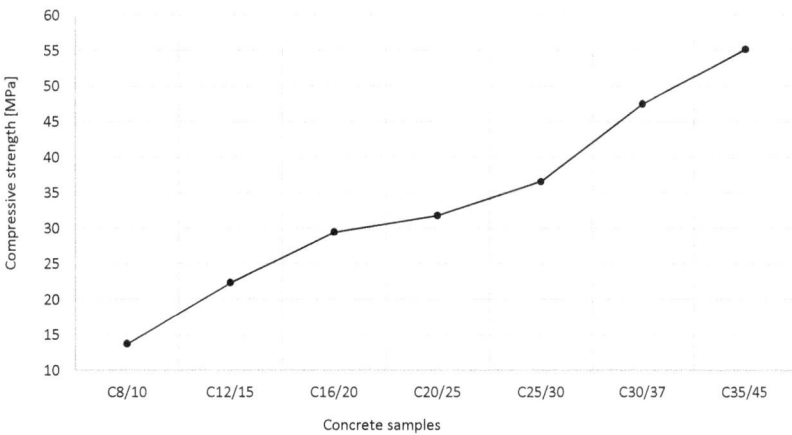

Figure 3. The impact of concrete class sample on compressive strength.

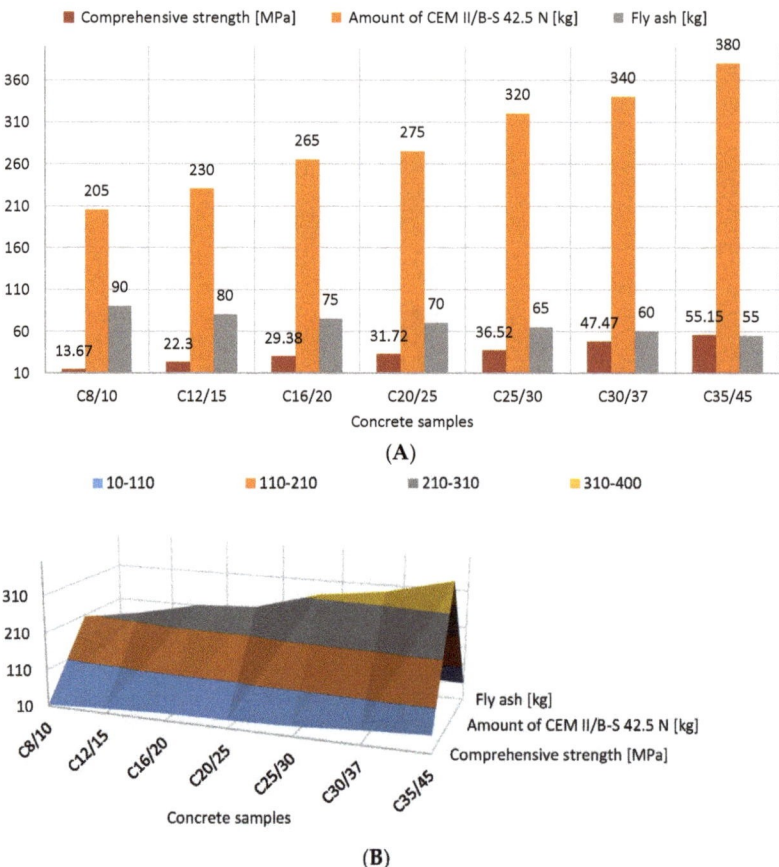

Figure 4. The relationship between the amount of CEM II/B-S 42.5 N cement and SSFA and the compressive strength of concrete mixes (version 2D: (**A**), 3D: (**B**)).

It demonstrates that the correct amount of cement should be used to produce the right class of concrete in order to achieve a strength greater than the one that is guaranteed. According to these experiments, increasing the cement addition led to an increase in compressive strength. The addition of SSFA did not weaken the strength properties of concrete, so it can be successfully used. Other researchers who conducted research on the addition of FA published findings that were comparable [16–19].

Figure 5 illustrates how the amount of sand in concrete mixtures affects compressive strength. Sand was shown to improve compressive strength in concrete mixes from 13.7 MPa (990 kg of sand) to 55.2 MPa (620 kg of sand). The link between the concrete's compressive strength and the sand point is depicted in Figure 6. The correct sand point for the aggregate should be chosen in order to ensure proper placement and compaction of concrete mixtures while accounting for transport. It can be said that as the concrete strength class increases, so does the water content in the aggregate, which results in a drop in the percentage of the fine aggregate fraction (0–2 mm). The decrease in the sand point, whose defined value influences the strength of the concrete, makes this advantageous. For the addition of 2/8 and 8/16 gravel, however, an inverse relationship was observed. The compressive strength reported increased as more of these ores were introduced (Figure 7).

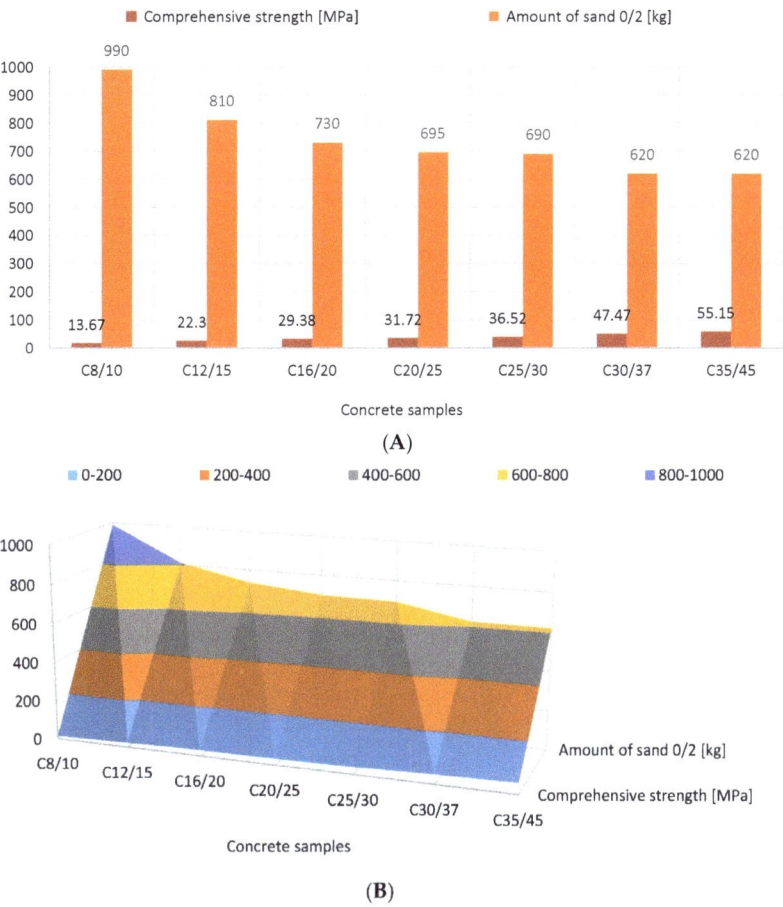

Figure 5. Relationship between amount of 0/2 mm sand and compressive strength of concrete (version 2D: (**A**), 3D: (**B**)).

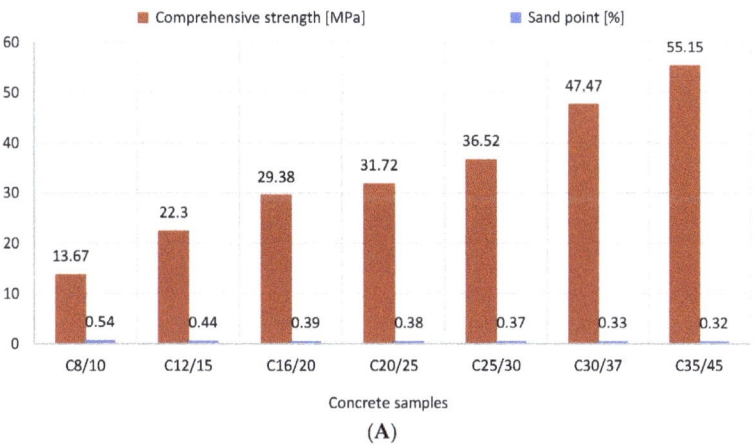

Figure 6. *Cont.*

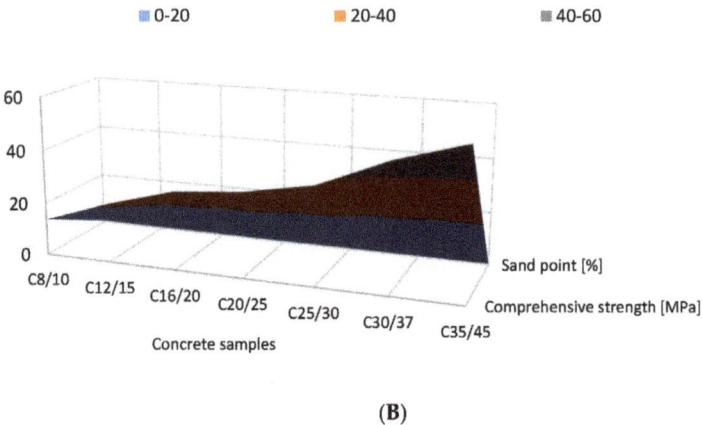

(B)

Figure 6. Relationship between sand point and compressive strength of concrete (version 2D: (**A**), 3D: (**B**)).

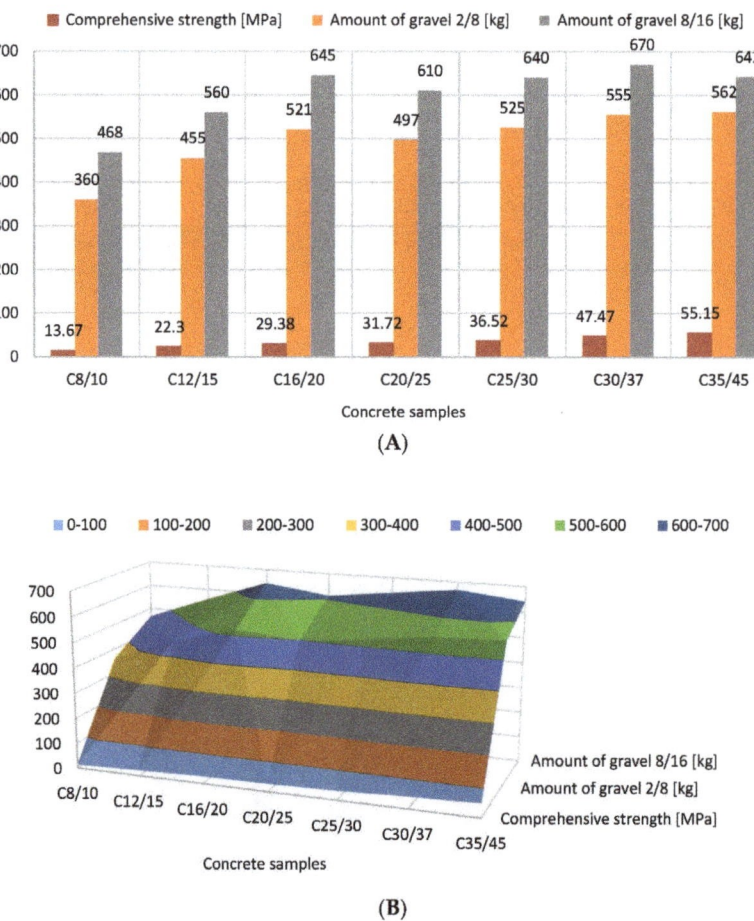

(B)

Figure 7. Relationship between amount of 2/8 and 8/16 mm gravel and compressive strength of concretes (version 2D: (**A**), 3D: (**B**)).

One of the most crucial factors affecting the strength of properly compacted concrete is the w/c ratio. When water is added to concrete mixtures to promote fluidity, the w/c value rises, and the strength of the concrete is compromised. As a result, it is advised to aim for low w/c ratios in order to produce concrete that is both strong and of high quality. The correlation between the water-to-cement ratio (w/c) and the compressive strength of concrete is depicted in Figure 8. It is clearly visible that mixtures with low water-to-cement ratios showed an increase in strength compared to mixtures with high water-to-cement ratios. In actual construction practice, mixes with a low water-to-cement ratios strengthen far more quickly than mixes with high water-to-cement ratios. The w/c ratio rises as the amount of water in the concrete mix grows; in other words, the more water that is in the concrete, the greater the w/c ratio, which has a negative effect on the strength of the concrete.

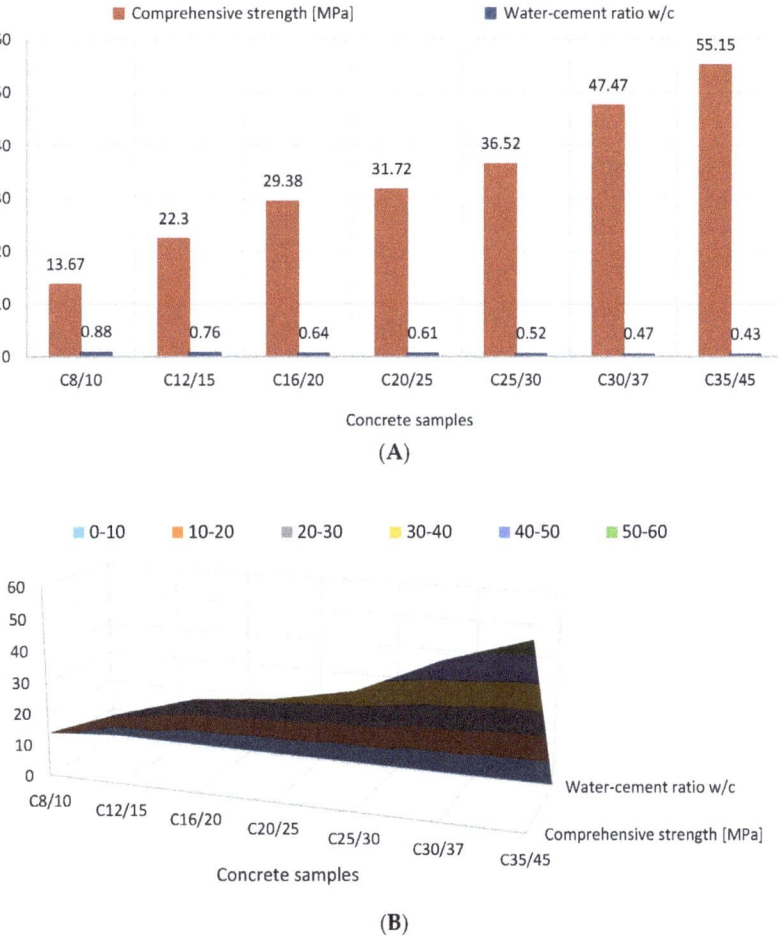

Figure 8. The relationship between the water-to-cement (w/c) ratio and the concrete compressive strength (version 2D: (**A**), 3D: (**B**)).

4. Conclusions

These investigations allowed us to show the relationship between the mechanical strength of waste-modified concretes and various parameters, including the composition of concrete mixes (amount of sand and gravel, cement, SSFA), water-to-cement (w/c) ratio, and sand point. Current Polish and European regulations allow the use of fly ash obtained from the combustion of sewage sludge as an additive for the preparation of concrete mixes [36,37]. Therefore, the addition of SSFA was used as a waste material resulting from circulating fluidized bed combustion (CFBC) technology. According to a review of the experimental studies, the following results were found:

1. The compressive strength of concrete was shown to increase as the class of concrete increased while keeping the appropriate composition of components (from 13.7 MPa (C8/10) to 55.2 MPa (C35/45)).
2. Concrete mixes with low w/c ratios and high cement contents showed increases in compressive strength compared to mixes with high w/c ratios and low cement contents.
3. The addition of SSFA as a by-product of municipal sewage sludge incineration did not reduce the compressive strength, so it can be successively used. There are no legal regulations regarding the requirements for the physical and chemical properties of SSFA from the combustion of municipal sewage sludge, and there are also no legal specifications regarding the possibility of their use in the production of concrete. Oxides of calcium, phosphorus, aluminum, silicon, and iron had the highest concentrations in SSFA samples. Utilizing the waste materials lowers the price of making concrete and prevents it from going to landfills.
4. According to these investigations, compressive strength increased when sand content decreased (from 990 kg to 620 kg) and sand point value decreased (from 0.54% to 0.32%). On the other hand, the compressive strength increased as the amount of 2/8 gravel (from 360 kg to 562) and 8/16 gravel (from 468 kg to 670 kg) increased.

The results of the experimentation revealed how specific components and parameters affected the compressive strength of concrete samples. The components of the concrete mixes and their mutual ratios should be selected in such a way that the execution of the concrete is as economical as possible while maintaining the minimal standards of quality, in particular, strength, durability, and consistency. The experimental outcomes supported the hypothesis that the thermal transformation of municipal sewage sludge into fly ash using CFBC technology improved the strength characteristics of concrete. The offered findings serve as the foundation for further study in this field. Further research on the influence of composition variability on the technical parameters of concretes containing fly ash from sewage sludge seems necessary, as does a correlation analysis between these variables and the strength properties of concrete samples.

Supplementary Materials: The following supporting information can be downloaded at: https://www.mdpi.com/article/10.3390/ma16114043/s1, Table S1: Strength test results of C8/10 class concrete; Table S2: Strength test results of C12/15 class concrete; Table S3: Strength test results of C16/20 class concrete; Table S4: Strength test results of C20/25 class concrete; Table S5: Strength test results of C25/30 class concrete; Table S6: Strength test results of C30/37 class concrete; Table S7: Strength test results of C35/45 class concrete.

Author Contributions: Conceptualization, R.C.; methodology, T.K. and R.C.; formal analysis, T.K., P.S. and R.C.; investigation, P.S. and T.K.; resources, T.K. and R.C.; data curation, T.K. and R.C.; writing—original draft preparation, T.K.; writing—review and editing, T.K., R.C. and M.U.; visualization, T.K.; supervision, T.K. and R.C.; project administration, R.C. All authors have read and agreed to the published version of the manuscript.

Funding: This research did not receive a specific grant from any a funding agency in the public, commercial, or not-for-profit sectors.

Institutional Review Board Statement: Not applicable.

Informed Consent Statement: Not applicable.

Data Availability Statement: Data are contained within the article.

Acknowledgments: The authors would like to express special thanks to the Nanobiomedical Centre, Adam Mickiewicz University in Poznań (Poland), for taking TEM images in nanoscale of the research samples.

Conflicts of Interest: The authors declare no conflict of interest.

References

1. Umar, T.; Yousaf, M.; Akbar, M.; Abbas, N.; Hussain, Z.; Ansari, W.S. An Experimental Study on Non-Destructive Evaluation of the Mechanical Characteristics of a Sustainable Concrete Incorporating Industrial Waste. *Materials* **2022**, *15*, 7346. [CrossRef] [PubMed]
2. Garside, M. Worldwide Industrial Sand and Gravel Production by Country 2021. Available online: https://www.statista.com/ (accessed on 24 February 2023).
3. Garside, M. Cement Production Worldwide from 1995 to 2022. Available online: https://www.statista.com/ (accessed on 24 February 2023).
4. Garside, M. Major Countries in Worldwide Cement Production in 2022. Available online: https://www.statista.com/ (accessed on 24 February 2023).
5. Tiseo, I. Carbon Dioxide Emissions from the Manufacture of Cement Worldwide in from 1990 to 2021, by Select Country. Available online: https://www.statista.com/ (accessed on 24 February 2023).
6. Anderson, T.R.; Hawkins, E.; Jones, P.D. CO_2, the greenhouse effect and global warming: From the pioneering work of Arrhenius and Callendar to today's Earth System Models. *Endeavour* **2016**, *40*, 178–187. [CrossRef] [PubMed]
7. Karalar, M.; Özkılıç, Y.O.; Deifalla, A.F.; Aksoylu, C.; Arslan, M.H.; Ahmad, M.; Sabri, M.M.S. Improvement in Bending Performance of Reinforced Concrete Beams Produced with Waste Lathe Scraps. *Sustainability* **2022**, *14*, 12660. [CrossRef]
8. Qaidi, S.; Najm, H.M.; Abed, S.M.; Özkılıç, Y.O.; Al Dughaishi, H.; Alosta, M.; Sabri, M.M.S.; Alkhatib, F.; Milad, A. Concrete Containing Waste Glass as an Environmentally Friendly Aggregate: A Review on Fresh and Mechanical Characteristics. *Materials* **2022**, *15*, 6222. [CrossRef]
9. Çelik, A.I.; Özkılıç, Y.O.; Zeybek, Ö.; Özdöner, N.; Tayeh, B.A. Performance Assessment of Fiber-Reinforced Concrete Produced with Waste Lathe Fibers. *Sustainability* **2022**, *14*, 11817. [CrossRef]
10. Amin, M.; Zeyad, A.M.; Tayeh, B.A.; Agwa, I.S. Engineering properties of self-cured normal and high strength concrete produced using polyethylene glycol and porous ceramic waste as coarse aggregate. *Constr. Build. Mater.* **2021**, *299*, 124243. [CrossRef]
11. Hussain, Z.; Pu, Z.; Hussain, A.; Ahmed, S.; Shah, A.U.; Ali, A.; Ali, A. Effect of fiber dosage on water permeability using a newly designed apparatus and crack monitoring of steel fiber–reinforced concrete under direct tensile loading. *Struct. Health Monit.* **2021**, *21*, 2083–2096. [CrossRef]
12. Ahmed, S.; Hussain, A.; Hussain, Z.; Pu, Z.; Ostrowski, K.A.; Walczak, R. Effect of Carbon Black and Hybrid Steel-Polypropylene Fiber on the Mechanical and Self-Sensing Characteristics of Concrete Considering Different Coarse Aggregates' Sizes. *Materials* **2021**, *14*, 7455. [CrossRef]
13. Ismail, M.H.; Megat Johari, M.A.; Ariffin, K.S.; Jaya, R.P.; Wan Ibrahim, M.H.; Yugashini, Y. Performance of High Strength Concrete Containing Palm Oil Fuel Ash and Metakaolin as Cement Replacement Material. *Adv. Civ. Eng.* **2022**, *2022*, 6454789. [CrossRef]
14. Szcześniak, A.; Zychowicz, J.; Stolarski, A. Influence of Fly Ash Additive on the Properties of Concrete with Slag Cement. *Materials* **2020**, *13*, 3265. [CrossRef]
15. Bakoshi, T.; Kohno, K.; Kawasaki, S.; Yamaji, N. Strength and durability of concrete using bottom ash as replacement for fine aggregate. *ACI Spec. Publ.* **1998**, *SP-179*, 159–172.
16. Marthong, C.; Agrawal, T.P. Effect of Fly Ash Additive on Concrete Properties. *Int. J. Eng. Res. Appl.* **2012**, *2*, 1986–1991.
17. Saha, A.K. Effect of class F fly ash on the durability properties of concreto. *Sustain. Environ. Res.* **2018**, *28*, 25–31. [CrossRef]
18. Rutkowska, G.; Ogrodnik, P.; Żółtowski, M.; Powęzka, A.; Kucharski, M.; Krejsa, M. Fly Ash from the Thermal Transformation of Sewage Sludge as an Additive to Concrete Resistant to Environmental Influences in Communication Tunnels. *Appl. Sci.* **2022**, *12*, 1802. [CrossRef]
19. Jura, J.; Ulewicz, M. Assessment of the Possibility of Using Fly Ash from Biomass Combustion for Concrete. *Materials* **2021**, *14*, 6708. [CrossRef]
20. Popławski, J.; Lelusz, M. Assessment of Sieving as a Mean to Increase Utilization Rate of Biomass Fly Ash in Cement-Based Composites. *Appl. Sci.* **2023**, *13*, 1659. [CrossRef]
21. Yusuf, R.O.; Noor, Z.Z.; Din, M.F.M.; Abba, A.H. Use of sewage sludge ash (SSA) in the production of cement and concrete—A review. *Int. J. Glob. Environ. Issues* **2012**, *12*, 214–228. [CrossRef]
22. Mojapelo, K.S.; Kupolati, W.K.; Ndambuki, J.M.; Sadiku, E.R.; Ibrahim, I.D. Utilization of wastewater sludge for lightweight concrete and the use of wastewater as curing medium. *Case Stud. Construct. Mater.* **2021**, *15*, e00667. [CrossRef]
23. Monzó, J.; Payá, J.; Borrachero, M.V.; Peris-Mora, E. Mechanical behavior of mortars containing sewage sludge ash (SSA) and Portland cements with different tricalcium aluminate content. *Cem. Concr. Res.* **1999**, *29*, 87–94. [CrossRef]
24. Rusănescu, C.O.; Voicu, G.; Paraschiv, G.; Begea, M.; Purdea, L.; Petre, I.C.; Stoian, E.V. Recovery of Sewage Sludge in the Cement Industry. *Energies* **2022**, *15*, 2664. [CrossRef]
25. Amminudin, A.L.; Ramadhansyah, P.J.; Doh, S.I.; Mangi, S.A.; Haziman, W.I.M. Effect of Dried Sewage Sludge on Compressive Strength of Concrete. *IOP Conf. Ser. Mater. Sci. Eng.* **2020**, *712*, 012042. [CrossRef]
26. Haustein, E.; Kuryłowicz-Cudowska, A.; Łuczkiewicz, A.; Fudala-Ksiazek, S.; Cieślik, B.M. Influence of Cement Replacement with Sewage Sludge Ash (SSA) on the Heat of Hydration of Cement Mortar. *Materials* **2022**, *15*, 1547. [CrossRef] [PubMed]
27. Central Statistical Office. Municipal Sewage Sludge, Warsaw 2020. Available online: https://stat.gov.pl/ (accessed on 24 March 2023).

28. Pan, S.-C.; Lin, C.-C.; Tseng, D.H. Reusing sewage sludge ash as adsorbent for copper removal from wastewater. *Resour. Conserv. Recycl.* **2003**, *39*, 79–90. [CrossRef]
29. Pan, J.R.; Huang, C.; Kuo, J.-J.; Lin, S.-H. Recycling MSWI bottom and fly ash as raw materials for Portland cement. *Waste Manag.* **2008**, *28*, 1113–1118. [CrossRef] [PubMed]
30. Chen, L.; Lin, D.F. Applications of sewage sludge ash and nano-SiO2 to manufacture tile as construction material. *Constr. Build. Mater.* **2009**, *23*, 3312–3320. [CrossRef]
31. Ginés, O.; Chimenos, J.M.; Vizcarro, A.; Formosa, J.; Rosell, J.R. Combined use of MSWI bottom ash and fly ash as aggregate in concrete formulation: Environmental and mechanical considerations. *J. Hazard. Mater.* **2009**, *169*, 643–650. [CrossRef]
32. Juric, B.; Hanzic, L.; Ilic, R.; Samec, N. Utilization of municipal solid waste bottom ash and recycled aggregate in concrete. *Waste Manag.* **2006**, *26*, 1436–1442. [CrossRef]
33. Monzó, J.; Payá, J.; Borrachero, M.V.; Girbés, I. Reuse of sewage sludge ashes (SSA) in cement mixtures: The effect of SSA on the workability of cement mortars. *Waste Manag.* **2003**, *23*, 373–381. [CrossRef]
34. Malliou, O.; Katsioti, M.; Georgiadis, A.; Katsiri, A. Properties of stabilized/solidified admixtures of cement and sewage sludge. *Cem. Concr. Compos.* **2007**, *29*, 55–61. [CrossRef]
35. Ogrodnik, P.; Rutkowska, G.; Szulej, J.; Żółtowski, M.; Powęzka, A.; Badyda, A. Cement Mortars with Addition of Fly Ash from Thermal Transformation of Sewage Sludge and Zeolite. *Energies* **2022**, *15*, 1399. [CrossRef]
36. Regulation of the Minister of Development of 21 January 2016 on the Requirements for the Thermal Waste Treatment Process and the Methods of Handling Waste Resulting from This Process. *J. Laws* **2016**, item108. Available online: http://isap.sejm.gov.pl/ (accessed on 15 February 2023).
37. EU/2010/75. Directive 2010/75/UE of the European Parliament and the Council of 24 November 2010 on Industrial Emissions (Integrated Pollution Prevention and Control). Available online: https://eur-lex.europa.eu/ (accessed on 15 February 2023).
38. Kalak, T.; Cierpiszewski, R. Comparative studies on the adsorption of Pb(II) ions by fly ash and slag obtained from CFBC technology. *Pol. J. Chem. Technol.* **2019**, *21*, 72–81. [CrossRef]
39. Kalak, T.; Marciszewicz, K.; Piepiórka-Stepuk, J. Highly Effective Adsorption Process of Ni(II) Ions with the Use of Sewage Sludge Fly Ash Generated by Circulating Fluidized Bed Combustion (CFBC) Technology. *Materials* **2021**, *14*, 3106. [CrossRef] [PubMed]
40. Kalak, T.; Tachibana, Y. Removal of lithium and uranium from seawater using fly ash and slag generated in the CFBC technology. *RSC Adv.* **2021**, *11*, 21964–21978. [CrossRef] [PubMed]
41. Kalak, T. Efficient use of circulating fluidized bed combustion fly ash and slag generated as a result of sewage sludge incineration to remove cadmium ions. *Des. Water Treat.* **2022**, *264*, 72–90. [CrossRef]
42. Tachibana, Y.; Kalak, T.; Nogami, M.; Tanaka, M. Combined use of tannic acid-type organic composite adsorbents and ozone for simultaneous removal of various kinds of radionuclides in river water. *Water Res.* **2020**, *182*, 116032. [CrossRef]
43. *PN-EN 12390-3:2019-07*; Testing Hardened Concrete—Part 3: Compressive Strength of Test Specimens. Polish Committee for Standardization: Warsaw, Poland, 2019.
44. *SIST EN 12390-2:2009*; The European Standard. Testing Hardened Concrete. Part 2: Making and Curing Specimens for Strength Tests. European Commission: Brussels, Belgium, 2019.
45. *SIST EN 12390-1:2021*; Testing Hardened Concrete—Part 1: Shape, Dimensions and other Requirements for Specimens and Moulds. European Commission: Brussels, Belgium, 2021.
46. *SIST EN 12390-4:2019*; Testing Hardened Concrete—Part 4: Compressive Strength—Specification for Testing Machines. European Commission: Brussels, Belgium, 2019.
47. *PN-EN 12390-4:2020-03*; Testing Hardened Concrete—Part 4: Compressive Strength—Specification for Testing Machines. Polish Committee for Standardization: Warsaw, Poland, 2022.
48. Mahieux, P.-Y.; Aubert, J.-E.; Cyr, M.; Coutand, M.; Husson, B. Quantitative mineralogical composition of complex mineral wastes—Contribution of the Rietveld method. *Waste Manag.* **2009**, *30*, 378–388. [CrossRef]
49. Coutand, M.; Cyr, M.; Clastres, P. Use of sewage sludge ash as mineral admixture in mortars. *Proc. Inst. Civ. Eng.—Constr. Mater.* **2006**, *159*, 153–162. [CrossRef]
50. Zhou, Y.-F.; Li, J.-S.; Lu, J.-X.; Cheeseman, C.; Poon, C.S. Sewage sludge ash: A comparative evaluation with fly ash for potential use as lime-pozzolan binders. *Constr. Build. Mater.* **2020**, *242*, 118160. [CrossRef]
51. Diaz, E.I.; Allouche, E.N.; Eklund, S. Factors Affecting the Suitability of Fly Ash as Source Material for Geopolymers. *Fuel* **2010**, *89*, 992–996. [CrossRef]
52. Wijaya, A.L.; Ekaputri, J.J.; Triwulan. Factors influencing strength and setting time of fly ash based-geopolymer paste. *MATEC Web Conf.* **2017**, *138*, 01010. [CrossRef]
53. Ma, L.; Wei, Q.; Chen, Y.; Song, Q.; Sun, C.; Wang, Z.; Wu, G. Removal of cadmium from aqueous solutions using industrial coal fly ash-nZVI. *Soc. Open Sci.* **2018**, *5*, 171051. [CrossRef] [PubMed]
54. Li, W.; Zhou, S.; Wang, X.; Xu, Z.; Yuan, C.; Yu, Y.; Zhang, Q.; Wang, W. Integrated evaluation of aerosols from regional brown hazes over northern China in winter: Concentrations, sources, transformation, and mixing states. *J. Geophys. Res.* **2011**, *116*, 129642250. [CrossRef]
55. Assi, A.; Bilo, F.; Zanoletti, A.; Ponti, J.; Valsesia, A.; La Spina, R.; Depero, L.E.; Bontempi, E. Review of the Reuse Possibilities Concerning Ash Residues from Thermal Process in a Medium-Sized Urban System in Northern Italy. *Sustainability* **2020**, *12*, 4193. [CrossRef]

56. Wu, M.-F.; Huang, W.-H. Evaluation of Fly Ash from Co-Combustion of Paper Mill Wastes and Coal as Supplementary Cementitious Materials. *Materials* **2022**, *15*, 8931. [CrossRef]
57. Saraber, A.J. Co-combustion and its impact on fly ash quality; full-scale experiments. *Fuel Process. Technol.* **2014**, *128*, 68–82. [CrossRef]
58. Faleschini, F.; Zanini, M.A.; Brunelli, K.; Pellegrino, C. Valorization of co-combustion fly ash in concrete production. *Mater. Des.* **2015**, *85*, 687–694. [CrossRef]
59. Mlonka-Mędrala, A.; Dziok, T.; Magdziarz, A.; Nowak, W. Composition and properties of fly ash collected from a multifuel fluidized bed boiler co-firing refuse derived fuel (RDF) and hard coal. *Energy* **2021**, *234*, 121229. [CrossRef]
60. Abdullah, B.I.; Abdulkadir, M.R. Correlation between destructive and non-destructive tests results for concrete compressive strength. *J. Zankoi Sulaimani* **2016**, *18*, 119–132. [CrossRef]

Disclaimer/Publisher's Note: The statements, opinions and data contained in all publications are solely those of the individual author(s) and contributor(s) and not of MDPI and/or the editor(s). MDPI and/or the editor(s) disclaim responsibility for any injury to people or property resulting from any ideas, methods, instructions or products referred to in the content.

Article

Influence of Post-Consumer Waste Thermoplastic Elastomers Obtained from Used Car Floor Mats on Concrete Properties

Alina Pietrzak and Malgorzata Ulewicz *

Faculty of Civil Engineering, Czestochowa University of Technology, Dabrowskiego 69 Street, PL 42-201 Czestochowa, Poland
* Correspondence: malgorzata.ulewicz@pcz.pl; Tel.: +48-343250935

Abstract: In this paper, the influence of post-consumer thermoplastic elastomer (TPE) additive derived from used car floor mats on the physical and mechanical properties of concrete is presented. Waste elastomer (fractions 0–2 and 2–8 mm) was used as a substitute for sand or fine aggregate in the amount of 2.5, 5.0, 7.5, and 10.0% by weight of cement. For all series, the physical and mechanical properties of concrete (for example, compressive strength, flexural tensile strength, water absorption, density, and frost resistance), as well as its microstructure, were tested. It has been shown that post-consumer elastomer waste from used car floor mats in the amount of 2.5% of cement weight can replace sand and gravel aggregate in concrete without reducing their mechanical strength and without changing their microstructure. The compressive strength (after 28 days) of concretes in which the waste was introduced as a substitute for sand and aggregate was 57.0 and 57.2 MPa, respectively (the strength of the control sample was 57.0 MPa). The use of post-consumer waste in concrete allows for a reduction in the consumption of natural aggregate (the addition of 2.5% of waste material saves natural aggregate approximately 20 kg/m^3), which reduces the cost of concrete production and also has a positive impact on the environment (i.e., it saves cost and space in landfills, where currently used car floor mat are deposited).

Keywords: concrete; compressive strength; frost resistance; microstructure; post-consumer waste; thermoplastic elastomer

Citation: Pietrzak, A.; Ulewicz, M. Influence of Post-Consumer Waste Thermoplastic Elastomers Obtained from Used Car Floor Mats on Concrete Properties. *Materials* **2023**, *16*, 2231. https://doi.org/10.3390/ma16062231

Academic Editor: Carlos Leiva

Received: 6 February 2023
Revised: 2 March 2023
Accepted: 8 March 2023
Published: 10 March 2023

Copyright: © 2023 by the authors. Licensee MDPI, Basel, Switzerland. This article is an open access article distributed under the terms and conditions of the Creative Commons Attribution (CC BY) license (https://creativecommons.org/licenses/by/4.0/).

1. Introduction

Currently, there are a large number of polymeric waste materials on the market that have no practical use. This group includes used car floor mats marked with waste code: 16 01 19 (i.e., waste not included in other groups—used or unusable vehicles, waste from dismantling, inspection, and maintenance of vehicles—plastics) [1]. Considering that in 2020, in Poland alone, 479,900 end-of-life cars were dismantled, with each of them equipped with at least four car mats, and more than 1.92 million pieces of this waste ended up in landfills [2]. On the scale of the European Union or the world, this problem is particularly important, as there is currently no effective way to manage this type of waste. In Poland, in 2023, for depositing used floor car mats at a landfill, an environmental fee must be paid (the method of recycling this waste is unknown) in the amount of PLN 300.17 per 1 ton of waste (price list following the fee of the Minister of Climate and Environment—MP 2022, item 1009). The diverse chemical composition of the main polymer and the diverse composition of the modifying additives used to make car linings at different times make the segregation of this type of waste ineffective and chemical recycling impossible. Car floor mats are usually made of needle velour (polypropylene or polyamide) finished with rubber crumbs (e.g., styrene–butadiene, butadiene–acrylonitrile, polybutadiene, or polyolefin). Moreover, this material contains modifying additives, such as plasticizers, reinforcing fillings, flame retardant agents, UV stabilizers, mineral fillings, antioxidants, and coloring agents. The composition of the floor coverings varies depending on the year of production, the manufacturer, and the technology used. Therefore, currently, such chemically diverse waste is deposited in landfills and an effective management technology is sought for them.

Bearing in mind how important the problem for the environment is the management of used car floor coverings, in this work we attempted to determine the possibility of using this type of waste to produce concrete. The basis for this assumption were numerous review papers [3–8] on the use of polymer waste for concrete production and satisfactory results on the use of post-production waste of thermoplastic elastomers from the production of car floor mats, as presented in our previous paper [9]. It was shown that the addition of waste thermoplastic elastomer from the production process of car floor mats in the amount of 2.5% of cement mass to concrete as an aggregate substitute did not reduce the mechanical properties of concrete. There are many reports in the literature concerning the possibility of using other, various polymer wastes as an additive to concrete. According to the literature [10–18], polyterephthalate (PET) is the most extensively studied in terms of its use in concrete technology. Waste polyterephthalate is used in concrete mainly in two forms: as aggregate (coarse and fine) replacing natural aggregate and as polymer fiber. It was shown that the method of preparation and the shape of PET recyclate affect the parameters of concrete. Recyclate with a smooth, spherical surface has less impact on the workability of concrete than recyclate with a heterogeneous shape. The addition of PET recyclate increases [10,17] or decreases [12,14–16] the compressive strength of concrete. This strength depends to a large extent on the amount of added waste [13]. In addition, the flexural strength of concretes with the addition of waste PET may be higher [10,17] or lower [11,12,15,16] than the control concrete samples. The addition of PET recyclate to concrete reduces the modulus of elasticity, regardless of the tested consistency and the water–cement ratio. Most studies have shown an increase in water absorption with increasing PET waste inside concrete [18].

Moreover, post-production waste and post-consumer polystyrene waste were used for the production of concrete on a laboratory scale. This material was used both in the form of ball-shaped regranulate [19–23] and pellets [24–31] as an aggregate substitute in amounts from 10% to even 100%. The resulting composite was characterized by low density and better thermal parameters compared to ordinary concrete. The density of the modified concrete decreased with the increase in the amount of added polystyrene. The compressive strength of concrete depends, to a large extent, on the amount of waste added. The addition of polystyrene recyclate increases [20–23] or decreases [24–31] the compressive strength of concrete compared to control concrete samples. In addition, recyclate from waste polyethylene and polypropylene was used to produce concrete [32–39]. This material was used as a substitute for aggregate or as concrete reinforcement. Recycling mixtures of this raw material are characterized by varied quality and mechanical properties. Since the properties of fibers made of pure synthetic polymer differ significantly from fibers obtained from recycled material, fibers made of pure polypropylene or polyethylene were more often used in the study. Concretes containing 1% polypropylene fibers show higher compressive and splitting or bending strength than a series of controlled concretes. In addition, the addition of polypropylene recyclate increases the compressive strength [33,35,39] and flexural strength [35,39] of concrete concerning the control samples.

There are also several literature reports on the use of rubber in the production of concrete [40–46]. Concretes with the addition of rubber recyclate are usually characterized by lower values of mechanical parameters than concretes produced without the addition of waste [41,43,45–47], although there are also reports [40] that the addition of this material does not affect the mechanical properties of concrete. The decrease in the compressive strength after adding waste rubber, as shown in the works, is mainly related to the size of the introduced waste material. As shown in [42], the addition of recyclate (rubber from tire treads) in the amount of 5 and 10% with a grain size of 0.59 mm reduces the compressive strength of concrete by 61.5 and 88.5%, respectively. However, after adding waste with a grain size of 0.29 mm, the compressive strength was 70.9 and 97.4%, respectively. The decrease in strength may be related to the defect in binding the rubber waste to the cement matrix and the increased porosity of the matrix. However, References [47–49] show that the use of rubber recyclate in combination with silica fume had a positive effect on the mechanical properties of the concretes obtained.

Although polymeric materials, such as PET, PS, PP, and rubber, were most often used in testing the mechanical properties of concrete, the literature contains several papers on the use of other polymers for this purpose, including PVC [50–53] or polyurethane [54]. The compressive strength of concrete modified with PVC decreases, regardless of whether it was used in the form of granulate or powder [50–53]. In addition, the addition of polyurethane reduces the compressive strength of concrete to the control samples. The greatest decrease in strength (86%) was recorded for the series of concrete in which the aggregate was replaced with waste in the amount of 33.7% [54].

In conclusion, it should be stated that, in recent years, there has been a clear increase in the interest of scientists in the use of polymer waste for the production of concrete. This trend is in line with the growing demand for sustainable construction. This is evidenced by review papers [4,5,7] in which the literature discussing the effects of waste plastic aggregates on the strength characteristics of cementitious composites was collected and summarized. The behavior of such composites in aggressive environments, such as freezing/thawing, impact, abrasion, chemical attack, and exposure to elevated temperatures, is discussed. The direction of this research is the result of the systematically increasing amount of polymer waste and the increase in ecological awareness and the need to protect natural resources (replacing aggregate with waste materials). The obtained positive results of research on the use of the post-production waste of thermoplastic elastomers from the production of car mats, presented by us in work [9], allowed us to put forward the thesis that it will also be possible to use elastomer waste from used mats for this purpose, although they slightly differ in chemical composition from waste from the production process. Waste from the production process can be returned to production (waste recirculation), while there is no recycling method for used floor car mats. The use of this waste as a substitute for aggregate without deterioration of the mechanical parameters of concrete seems to be a good direction in the search for methods for the effective reduction of concrete production costs and the use of useless waste.

2. Materials and Methods

2.1. Materials

The properties of concrete were modified by the use of post-consumer waste (used car floor mats). Used automobile floor mats were obtained from the Maderpol automobile dismantling plant in Silesian Province, located in Mokra (Poland). After preshredding, the post-consumer waste (Figure 1) was crushed to 0–2 and 2–8 mm fractions in a granulator (SG-2417 SHINI). Both fractions were used in the study. The other materials used for the production of concrete were the same as in our previous work [9], because it is intended to help compare the obtained results. In this research, Portland cement CEM I 42.5R (Cemex) meeting the requirements of PN-EN 197-1, sand of 0–2 mm fraction, gravel aggregate of 2–8 and 8–16 mm fractions, concrete admixtures CHRYSO Plast 331 and CHRYSO Air LB (Company CHRYSO Poland), and water from an intake in Czestochowa (which met the requirements of PN-EN 1008:2004) were used.

2.2. Methods

Nine series of concretes were made. An experimental method was used to design the composition of the control concrete (SK) mix. A c/w ratio of 2.22 and a consistency class of S2 were assumed. Portland cement CEM I 42.5R, a gravel–sand aggregate mixture with a grain size of 0–16 mm and a sand point of PP = 33.1%, tap water, and plasticizing admixture CHRYSO Plast 331 at 0.35% by weight of cement and aerating admixture CHRYSO Air LB at 0.2% by weight of cement were used for the control concrete. The control concrete (SK) was modified with post-consumer waste thermoplastic elastomer formed from used car floor mats. The material was added in the amount of 2.5, 5.0, 7.5, and 10% by weight of cement, introducing a volumetric adjustment in the weight of sand (series S1–S4) or the weight of gravel aggregate mixture of 2–8 mm (series S5–S8). The concrete series were designated from S1 to S8. The solid components of the concretes were cement (372.5 kg/m^3), water (167.5 dm^3/m^3, plasticizer (1.3 dm^3/m^3), and air-entraining admixture (0.745 dm^3/m^3), and the content of the variable components is presented in Table 1.

Figure 1. Post-consumer waste from used car liners: preshredded material (**a**); shredded to 2–8 mm fraction (**b**); shredded to 0–2 mm fraction (**c**) [55].

Table 1. Composition of the mixture of control concrete and waste-modified concretes as a substitute for sand and gravel aggregate of grain size 2–8 mm.

Composition	Units	Series								
		SK	S1	S2	S3	S4	S5	S6	S7	S8
Sand	kg/m^3	463.70	443.6	423.6	403.5	384.4	463.7	463.7	463.7	463.7
Gravel 8–16 mm	kg/m^3	776.20	776.2	776.2	776.2	776.2	776.2	776.2	776.2	776.2
Gravel 2–8 mm	kg/m^3	635.10	635.1	635.1	635.1	635.1	615.0	594.9	574.9	554.8
Post-consumer waste	kg/m^3	-	9.31	18.63	27.94	37.25	9.310	18.63	27.94	37.25

All tests of both concrete mixes (consistency) and hardened concrete (compressive strength, bending strength, tensile strength when splitting, strength loss, and weight loss after frost resistance testing) were carried out per PN-EN standards, described in detail in our previous work [9]. The SEM/EDS analysis of the microstructure of the synthesized concrete composites containing post-consumer waste thermoplastic elastomer was also performed. A scanning electron microscope (SEM) equipped with an X-ray energy-dispersive spectroscopy (EDS)-based chemical composition analysis system was used for the study.

3. Results and Discussion

3.1. Characteristics of Post-Consumer Waste

Crushed post-consumer waste of used car mats was used to test the properties of the concretes. Using a WDXRF X-ray spectrometer (Model S8 Tiger from Bruker Company, Billerica, MA, USA), the elemental composition of the thermoplastic elastomer waste used in the study was determined, which is shown in Table 2. The results of the thermogravimetry (TG), differential thermogravimetry (DTG), and differential scanning calorimetry (DSC) of the used waste are shown in Figure 2. The test was carried out in the thermal analysis device Jupiter STA 449 F5 (Netzsch) in a temperature range from 30 °C to 600 °C, with a temperature increase rate of 10 °C/min in an air atmosphere, and with a gas flow rate of 100 cm^3/min. The curves were analyzed to identify the composition and degradation temperatures of the components in the waste material. The TG curve showed a one-step degradation of the waste elastomer. Two peaks were observed in the DSC curves, a broad peak was seen around 90 °C to 360 °C with a maximum of 232.4 °C and a narrower peak with a maximum of 501.4 °C. The curve clearly shows the initial phase of volatilization of oils, plasticizers, and fillers with a low molecular weight and low boiling point (below

230 °C), as well as a greater weight loss in the temperature range of 230–500 °C, caused by the decomposition of the polymeric material (mostly natural and synthetic styrene–butadiene–rubber). Similar thermogravimetric curves were obtained for commercial rubber products [56,57].

Table 2. The elemental composition of post-consumer waste, %.

Elemental Composition (% (m/m))											
Ca	Si	Al	Zn	S	Ti	Ba	Fe	K	Sr	C	Other
5.37	1.36	0.93	0.60	0.37	0.09	0.10	0.05	0.06	0.01	78.10	12.92

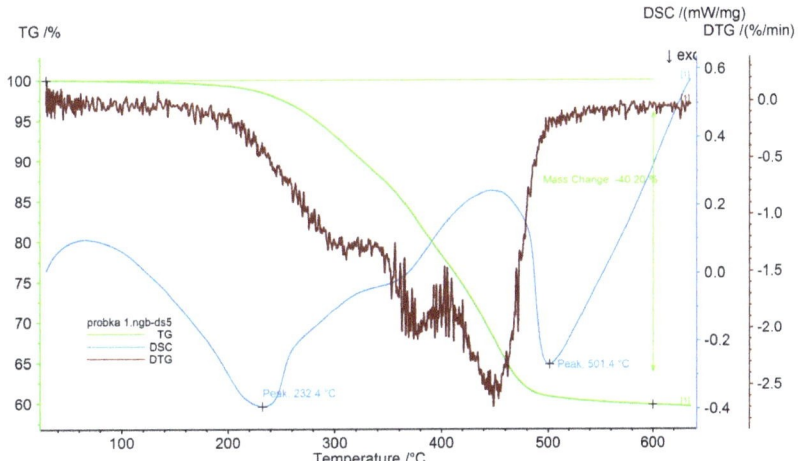

Figure 2. TGA-DTA thermogram of post-consumer thermoplastic elastomer from used car floor mats.

3.2. The Consistency of Concrete Mixes

In the first stage of the research, samples of control concrete (SK) were made, assuming a consistency of S2 (cone slump between 50 and 90 mm) and samples of concretes modified with the post-consumer waste of thermoplastic elastomers. After the concrete mixture was prepared, samples were taken to determine the consistency class and air content of the concrete mixture. For the control mix (SK), the cone slump was 70 mm, which classifies it as S2. The air content of this mix was 3.5%. For concrete mixtures modified with the addition of thermoplastic elastomer waste (S1–S8), a cone slump of 50–75 mm was obtained, which also classifies them in the S2 consistency class (Table 3). The air content of the concrete mixtures modified with the tailings was in the range of 3.90–4.6%.

Table 3. Consistency class and air content of the tested concrete mixtures.

Series	Consistence (mm)/Class	Air Content (%)
SK	70/S2	3.50
S1	70/S2	4.30
S2	50/S2	4.20
S3	50/S2	4.15
S4	50/S2	3.90
S5	65/S2	3.90
S6	60/S2	4.20
S7	70/S2	4.60
S8	50/S2	4.00

3.3. The Mechanical Properties of Concretes

The next stage of the study included the determination of the mechanical strength of concretes modified with waste thermoplastic elastomers. For each series of concrete, 18 cubic specimens of 150 mm sides were made and tested for the compressive strength of the concrete after 7, 28, and 56 days of maturation under laboratory conditions. The standard deviation and 95% confidence interval were determined for the results obtained in each series. The results of compressive strength tests for control and modified concretes are presented in Table 4. After 7 days of concrete curing, a decrease in the compressive strength was observed for all series of waste-modified concretes. Only in the case of concretes containing 2.5% of the waste additive used as a substitute for sand was the hardness comparable to that of the control concrete samples (strength decreased by only 0.64%). The highest decrease in the compressive strength was recorded for concrete samples containing 10% waste, regardless of whether it was used as a sand substitute or as an aggregate substitute, which amounted to 14.16%. A similar tendency of a decrease in the compressive strength was observed after 28 days of maturation of the samples. In the case of concretes containing 2.5% of the waste additive used as a substitute for sand, the compressive strength was comparable to that of the control concrete samples. On the other hand, the decrease in the compressive strength for concrete samples containing 10% of waste used as a sand substitute was 15.09%, and 17.72% for those used as an aggregate substitute. Compressive strength tests after 56 days of maturation of the samples were also carried out for all concrete series. The control concrete (SK) obtained an average compressive strength higher by 8.5% than the 28-day average compressive strength, which was 61.9 ± 1.19 MPa. The concrete of the S1 series, where waste was applied at 2.5%, achieved an average strength compressive strength at the level of the control concrete (62.0 ± 1.9 MPa). However, with the increase in the amount of waste added to the concrete, a decrease in the value of this parameter was observed. The addition of waste to concrete in the amount of 10%, used as a substitute for sand, caused a decrease in the compressive strength by 17.93% and used as a substitute for aggregate with a grain size of 2–8 mm by 12.28%.

Table 4. Compressive strength of the tested concretes and strength class.

Series	Compression Strength (MPa)			Resistance Class
	After 7 Days	After 28 Days	After 56 Days	
SK	46.6	57.0	61.9	C40/50
S1	46.3	57.0	62.0	C40/50
S2	41.1	50.1	56.2	C35/45
S3	40.7	48.4	51.0	C30/37
S4	40.0	47.4	50.8	C30/37
S5	46.2	57.2	62.1	C40/50
S6	43.9	50.7	56.8	C35/45
S7	42.0	48.5	54.9	C30/37
S8	40.0	46.9	54.3	C30/37

Comparing the strength values obtained for concretes (series S5–S8), in which post-consumer waste of thermoplastic elastomers was used as a substitute for aggregate with a grain size of 2–8 mm, we can conclude that they were comparable to the results obtained previously in [9], using post-production waste of these elastomers (waste from the production of car mats). In the case of post-production waste, a decrease in the strength was also observed with an increase in the amount of waste added, ranging from 2.5 to 10%. The compressive strength tested after 7, 28, and 56 days of curing of the modified concrete samples with 2.5% of post-production waste was comparable to the strength of the reference samples (an increase below 1.4%). On the other hand, adding more waste (5.0, 7.5, and 10%) resulted in a decrease in the value of this parameter.

The decrease in the compressive strength of the concrete observed with the increase in the amount of added waste can be explained by the very low bonding force between the surface of waste from thermoplastic elastomers and the cement slurry. Moreover, the hydrophobic nature of polymer waste can inhibit the hydration reaction of cement by limiting the movement (access) of water in the concrete mix. A negative effect on the concrete's compressive strength parameter was also observed after adding other polymeric materials, such as PET [11,12,14–16], EPE [18], EPS [19–22], and rubber [41,43,45–47]. However, some authors report that the addition of PET [10,13], EPS [23–37], or PP [32,34,38] wastes increases the compressive strength of concretes modified with this waste.

The addition of waste thermoplastic elastomer to concrete also affects the flexural strength and tensile strength of the concrete (Table 5). The average flexural strength of the S1–S4 series concretes modified with waste thermoplastic elastomer used instead of sand was lower from 5.85 to 6.96% in relation to the control concrete. On the other hand, for the series of concretes S5 and S6 modified with thermoplastic elastomers from post-consumer waste, used as a substitute for 2–8 mm gravel aggregate, in the amount of 2.5 and 5.0% of the cement mass, an increase in the bending strength was observed concerning the series of control concrete, respectively, by 7.24 and 3.64%. While for the series of concretes S7 and S8, in which waste was used in the amount of 7.5 and 10.0%, there was a decrease in this parameter by 4.7 to 5.8%, respectively. The increase in the flexural strength of concrete after adding waste polymers was also noted by other authors. An increase in the bending strength of concretes modified with PET [11,12,14,15] and PP [34,38] was noted. In [13], it was shown that the flexural strength changes markedly with the increase in the amount of added PET waste. The addition of 1% PET increases this parameter by 18.6%, while the addition of 3% decreases it by 19%. On the other hand, the addition of waste LDPE does not affect the flexural strength of concrete [31].

Table 5. Flexural tensile strength and splitting tensile strength of concrete.

Series	Flexural Strength (MPa)	Splitting Tensile Strength (MPa)
SK	3.59	3.78
S1	3.38	4.15
S2	3.36	3.8
S3	3.36	3.25
S4	3.34	3.06
S5	3.85	4.19
S6	3.71	3.88
S7	3.42	3.61
S8	3.38	3.58

In the concretes of the S1 and S2 concrete series, where post-consumer waste was dosed as a substitute for sand, there was an increase in the average splitting tensile strength. The increase in the average splitting compressive strength was 9.7 and 0.5%. The other series (S3 and S4) showed a decrease in the tested parameter compared to the SK series, by 14.0 and 19.0%, respectively. For the concretes of the S5 and S6 series, where waste was used as a substitute for the 28 mm grain size aggregate in the amount of 2.5 and 5.0%, the average splitting tensile strength was 4.19 ± 0.6 MPa and 3.88 ± 0.1 MPa, respectively. This means that the increase in the splitting strength parameter compared to the control concrete series, for which the strength was 3.78 ± 0.9 MPa, was 10.9 and 2.6%, respectively. For the other series (S7 and S8), there was a decrease in the tested parameter in relation to the SK series, by 4.5% and 5.3%, respectively. As in the case of the compressive strength, both a decrease and an increase in the tensile strength of concrete were observed in the case of the modification with waste polymers [58–60]. It has usually been pointed out that these changes depend on the amount of waste used. The declining trend in the tensile strength splitting was usually not as pronounced as for compressive strength.

3.4. The Other Tested Properties of Concretes

The next stage of testing was to determine the absorbability, density, pressurized water penetration, frost resistance, and abrasion resistance of the concrete (Table 6). The absorbability test was performed after 28 days of maturation of the samples. For the control concrete (SK), the absorbability was 5.4%. The concretes modified with tailings at 2.5, 5.0, 7.5, and 10% showed similar levels of absorbability (from 4.9 to 5.6%). The concretes modified with post-consumer thermoplastic elastomer wastes have achieved absorption below 9%, which means that these concretes can be used in rooms where they will be protected against direct weather conditions (e.g., floors in halls). Literature reports show that water adsorption in concretes modified with PET increases with the amount of added waste and the size of regranulate [5,17]. However, according to [61], the replacement of sand with aggregate in the amount of 3% by both PET and polycarbonate (PC) does not affect the water absorption or apparent porosity of concrete composites modified with these wastes.

Table 6. Tested parameters for individual concrete series.

Series	Water Absorbability (%)	Density (kg/m^3)	Water Penetration (mm)	Abrasion Strength (cm^2/50 cm^2)
SK	5.4	2271	65	7.4
S1	5.5	2245	55	6.5
S2	5.6	2233	63	6.7
S3	5.5	2220	67	6.9
S4	4.9	2219	70	7.3
S5	5.6	2227	60	6.9
S6	5.5	2222	60	6.9
S7	5.6	2212	65	7.3
S8	5.2	2210	68	7.5

The control concrete (SK), due to the fact of its density, which amounted to 2271 kg/m^3, was classified as ordinary concrete following PN-EN 206 + A1:2016-12. In addition, all concrete modified with thermoplastic elastomer post-production waste were classified as ordinary concrete, as their density was in the range of 2000–2600 kg/m^3. The control concrete achieved an average water penetration depth of 65 mm. For the series of concretes modified with waste thermoplastic elastomers, the average depth of the water penetration ranged from 55 mm to 70 mm. The lowest value of the tested parameter (55 mm) was obtained for the S1 series, where 2.5% of post-consumer waste was used, and the highest value (70 mm) was obtained for the S4 series concretes containing 10% of this waste. For the control concrete (SK), the abrasiveness was 7.4 cm^3/50 cm^2. For concretes modified with thermoplastic elastomer waste, the abrasion resistance was obtained, ranging from 6.5 to 7.5 cm^3/50 cm^2.

A frost resistance test was conducted for all concretes. For each batch of concrete, 12 cubic specimens of 100 mm sides were made, 6 specimens were subjected to the frost resistance test, and 6 specimens were left in the water as comparison specimens. For the control concrete, the average decrease in the compressive strength after 150 freeze–thaw cycles was 4.4%, while the average weight loss was 0.34% (Figure 3).

The average decrease in the compressive strength for the series of S1–S3 concretes modified with post-consumer waste thermoplastic elastomers used, as a substitute for sand, was lower than the average decrease in the compressive strength of the control concrete. The decrease in the compressive strength for these concretes was 1.1, 1.3, and 2.1%, respectively. In contrast, the average weight loss for these modified series ranged from 0.10% to 0.30% and was lower than the average weight loss for the control concrete (0.34%). For all series of concretes (S5–S8) modified with post-consumer waste car floor mats, used as a substitute for gravel aggregate with a grain size of 2–8 mm, a lower decrease in the compressive strength after frost resistance testing was obtained than for the control concrete. In addition, as shown previously [9], using 2.5 and 5% post-production waste as an aggregate with a grain size of 2–8 mm, a slight decrease in the strength of concrete after frost resistance tests was observed in relation to the control samples (2.4 and 4.3%, respectively), while the addition of 7.5% and 10% waste caused a greater decrease in the compressive strength, by 8.0% and 11.6%, respectively. A lower decrease in the compressive strength after the frost resistance tests in relation to the control samples was also shown by concretes in which the aggregate was partially replaced with the addition of EPS [23], PP [38], and rubber [43].

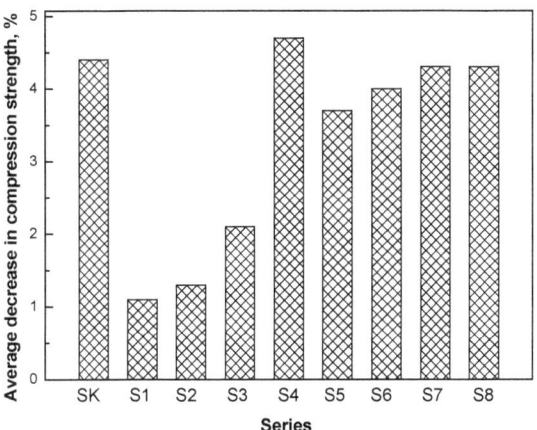

Figure 3. Average decrease in the compressive strength of the concretes after frost resistance testing.

3.5. The Concrete Surface Morphology

In the next stage of the study, the surface layer of the synthesized composites was analyzed, determining mainly their microstructure and elemental composition. Figure 4 shows a microscopic photo of the control concrete (SK) at 80× magnification (Figure 4a), along with maps of the distribution of the dominant elements in this area (Figure 4b). Noticeable in the control concrete (SK) is the lighter structure of the concrete matrix and the darker area representing the silicon-based aggregate. The microstructure at the aggregate–cement matrix interface is compact. As can be seen from the EDS analysis of the concrete surface visible in the photo, in addition to the presence of calcium (33.06%; blue color), a significant content of silicon (20.36%; green color) and iron (4.49%; pink color) can be observed. Aluminum, sulfur, potassium, magnesium, and carbon were present in amounts less than 1.0%. The structures of the concretes modified with post-consumer waste of thermoplastic elastomers from used car floor mats (series S1–S8) did not differ from the structure of the control concrete, as per Figure 5 showing an example structure for the S1 series of concrete. Microscopic images show the compact structure of the cement matrix with both the use of natural aggregate and the use of polymer waste granules. From the EDS analysis, in all the modified concretes, a decrease was observed compared to the control series, in the calcium and silicon content (approximately 4–7%) and an increase in the carbon content (approximately 5% for concrete containing 10% waste), which may be related to the increased amount of post-consumer waste. In the series of concretes where waste thermoplastic elastomers were used, the presence of elements, such as titanium, sodium, and zinc, was also observed, which were not present in the control series. The obtained results are comparable to those previously obtained for concretes with the addition of post-consumer waste of thermoplastic elastomers presented in [9]. Thus, it can be concluded that both post-production and post-consumer wastes of thermoplastic elastomers added in the range of 2.5 to 10% do not significantly affect the change in the microstructure of the tested concretes. In addition, the addition of fly ash from biomass combustion in the amount of 10–30% [62] and waste glass powder in the amount of 25% [63] to concretes does not affect their microstructure. On the other hand, the microstructure of concretes produced with 100% waste glass powder as a substitute for natural sand was characterized by many pores, which resulted from the angular shape of the grain and the smooth surface of the waste sand particles (this resulted in poor bond formation, especially in the interfacial zone). In turn, the microstructure of concretes with waste basalt dust (10–30%) showed a lower porosity in the interfacial transition zone [64].

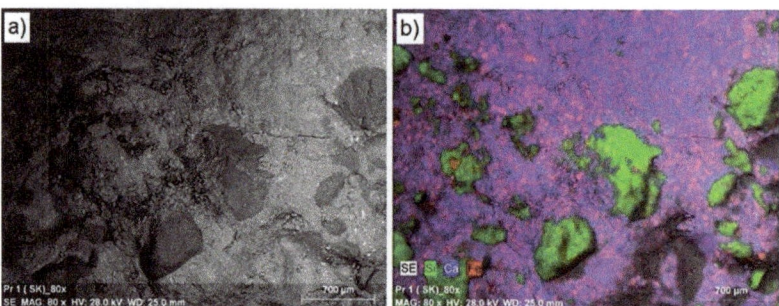

Figure 4. Control concrete microstructure: (**a**) 80 times magnification; (**b**) a map of the location of dominant elements [9].

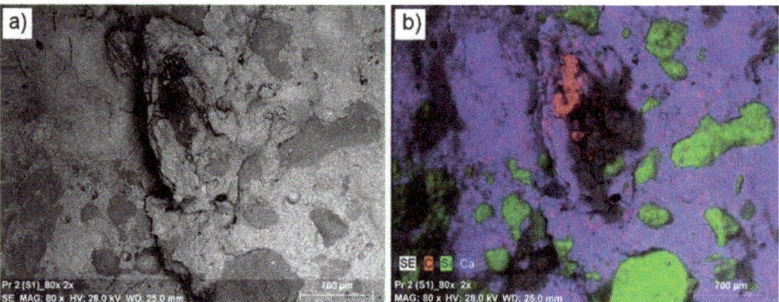

Figure 5. Concrete microstructure S1 series: (**a**) magnification 80 times; (**b**) a map of the location of dominant elements.

4. Conclusions

The study conducted in this paper on the properties and structure of concrete composites containing post-consumer waste thermoplastic elastomer from used car floor mats and the analysis of the results obtained allows us to conclude that the waste used in the amount of 2.5% can be added as a substitute for sand or as a substitute for gravel aggregate of 2–8 mm grain size for the manufacture of concrete and concrete composites. The addition of waste in this amount does not reduce the parameters of concrete (compressive strength, flexural strength, tensile strength by splitting, frost resistance, and abrasion resistance) with respect to the materials produced without the addition of waste thermoplastic elastomer. The use of waste 0–2 mm fraction in the amount of 2.5% by weight of cement as a substitute for sand allows for reducing the consumption of these aggregates by 20 kg/m^3, or approximately 5.0%. The microstructure of concretes made based on CEM I 42.5R cement containing the addition of post-consumer waste thermoplastic elastomers in the amount of 2.5% to 10% is characterized by a very similar microstructure to the control concrete made without the addition of waste polymers.

All the waste-modified concretes achieved a lower density than the control concrete, which ranged from 2000 to 2600 kg/m^3, allowing for the concretes to be classified as ordinary concretes. Both the control concrete and the concretes modified with post-consumer waste thermoplastic elastomer obtained a water absorption rate of approximately 5.0%. Concrete modified with thermoplastic waste obtained from used car mats can be used for elements protected against direct exposure to weather conditions (e.g., industrial floors, tile floors). All concrete modified with post-consumer waste of thermoplastic elastomer met the standard requirements. After performing 150 freeze–thaw cycles, the average decrease in the compressive strength did not exceed 20%, and the average weight loss did not exceed 5%. Thus, it can therefore be concluded that the addition (in the amount of 2.5%) to thermoplastic concrete of an elastomer derived from post-consumer waste, which is currently deposited in landfills, as a substitute for gravel aggregate of 2–8 mm is economically justified and also positively affects the protection of natural resources.

Author Contributions: Conceptualization, M.U. and A.P.; methodology, A.P.; formal analysis, M.U.; investigation, A.P.; resources, M.U. and A.P.; data curation, A.P.; writing—original draft preparation, A.P.; writing—review and editing, M.U.; visualization, A.P.; supervision, M.U. All authors have read and agreed to the published version of the manuscript.

Funding: This research received no external funding.

Institutional Review Board Statement: Not applicable.

Informed Consent Statement: Not applicable.

Data Availability Statement: The data presented in this study are available upon request from the corresponding author.

Acknowledgments: The authors would like to thank Maderpol automobile dismantling plant (Poland) for providing waste materials for the research.

Conflicts of Interest: The authors declare no conflict of interest.

References

1. Directive 2014/955/EU. Journal of the European Union, Brussels. Available online: https://eur-lex.europa.eu/legal-content/EN/TXT/PDF/?uri=CELEX:32014D0955&from=PL (accessed on 1 March 2023).
2. Environment 2022, Statistics Poland, Warsaw. 2022. Available online: https://stat.gov.pl (accessed on 1 March 2023).
3. Mercante, I.; Alejandrino, C.; Ojeda, J.P.; Chini, J.; Maroto, C.; Fajardo, N. Mortar and concrete composites with recycled plastic: A review. *Sci. Technol. Mater.* **2018**, *30*, 69–79. [CrossRef]
4. Ming, Y.; Chen, P.; Li, L.; Gan, G.; Pan, G. A comprehensive review on the utilization of recycled waste fibers in cement-based composites. *Materials* **2021**, *14*, 3643. [CrossRef] [PubMed]
5. Almeshal, I.; Tayeh, B.A.; Alyousef, R.; Alabduljabbar, H.; Mohamed, A.M.; Alaskar, A. Use of recycled plastic as fine aggregate in cementitious composites: A review. *Constr. Build. Mater.* **2020**, *253*, 119146. [CrossRef]
6. Gu, L.; Ozbakkaloglu, T. Use of recycled plastics in concrete: A critical review. *Waste Manag.* **2016**, *51*, 19–42. [CrossRef] [PubMed]
7. Rashad, A.M. A comprehensive overview about recycling rubber as fine aggregate replacement in traditional cementitious materials. *Int. J. Sustain. Built Environ.* **2016**, *5*, 46–82. [CrossRef]
8. Sharma, R.; Bansal, P.P. Use of different forms of waste plastic in concrete—A review. *J. Clean. Prod.* **2016**, *112*, 473–482. [CrossRef]
9. Ulewicz, M.; Pietrzak, A. Properties and Structure of Concretes Doped with Production Waste of Thermoplastic Elastomers from the Production of Car Floor Mats. *Materials* **2021**, *14*, 872. [CrossRef]
10. Ochi, T.; Okubo, S.; Fukui, K. Development of recycled PET fiber and its application as concrete-reinforcing fiber. *Cem. Concr. Compos.* **2007**, *29*, 448–455. [CrossRef]
11. Choi, Y.W.; Moon, D.J.; Chung, J.S.; Cho, S.K. Effects of waste PET bottlers aggregate on the properties of concrete. *Cem. Concr. Res.* **2005**, *35*, 776–781. [CrossRef]
12. Choi, Y.W.; Moon, D.J.; Kim, Y.J.; Lachemi, M. Characteristics of mortar and concrete containing fine aggregate manufactured from recycled waste polyethylene terephthalate bottles. *Constr. Build. Mater.* **2009**, *23*, 2829–2835. [CrossRef]
13. Nibudey, R.; Nagarnaik, P.; Parbat, D.; Pande, A. Strength and fracture properties of post consumed waste plastic fiber reinforced concrete. *Int. J. Civ. Struct. Environ. Infrastruct. Eng. Res. Dev.* **2013**, *3*, 9–16.
14. Rahmani, E.; Dehestani, M.; Beygi, M.H.A.; Allahyari, H.; Nikbin, I.M. On the mechanical properties of concrete containing waste PET particles. *Constr. Build. Mater.* **2013**, *47*, 1302–1308. [CrossRef]
15. Ferreira, L.; De Brito, J.; Saikia, N. Influence of curing conditions on the mechanical performance of concrete containing recycled plastic aggregate. *Constr. Build. Mater.* **2012**, *36*, 196–204. [CrossRef]
16. Silva, R.; De Brito, J.; Saikia, N. Influence of curing conditions on the durability-related performance of concrete made with selected plastic waste aggregates. *Cem. Concr. Compos.* **2013**, *35*, 23–31. [CrossRef]
17. Saikia, N.; de Brito, J. Waste polyethylene terephthalate as an aggregate in concrete. *Mater. Res.* **2013**, *16*, 341–350. [CrossRef]
18. Bhagat, G.V.; Savoikar, P.P. Durability related properties of cement composites containing thermoplastic aggregates—A review. *J. Build. Eng.* **2022**, *53*, 104565. [CrossRef]

19. Sabaa, B.; Ravindrarajah, R.S. Engineering properties of lightweight concrete containing crushed expanded polystyrene waste. In Proceedings of the Symposium MM: Advances in Materials for Cementitious Composites, Boston, MA, USA, 1–3 December 1997.
20. Tang, W.; Lo, Y.; Nadeem, A. Mechanical and drying shrinkage properties of structural-graded polystyrene aggregate concrete. *Cement Concr. Compos.* **2008**, *30*, 403–409. [CrossRef]
21. Kan, A.; Demirboğa, R. A novel material for lightweight concrete production. *Cement Concr. Compos.* **2009**, *31*, 489–495. [CrossRef]
22. Herki, A.; Khatib, J.; Negim, E. Lightweight concrete made from waste polystyrene and fly ash. *World Appl. Sci. J.* **2013**, *21*, 1356–1360.
23. Bengin, M.; Herki, A. Absorption Characteristics of Lightweight Concrete Containing Densified Polystyrene. *Civ. Eng. J.* **2017**, *3*, 595–609.
24. Ravindrarajah, R.S. Bearing strength of concrete containing polystyrene aggregate. *Durability Build. Mater. Compon.* **1999**, *8*, 505–514.
25. Babu, K.G.; Babu, D.S. Behaviour of lightweight expanded polystyrene concrete containing silica fume. *Cem. Concr. Res.* **2003**, *33*, 755–762. [CrossRef]
26. Chen, B.; Liu, J. Properties of lightweight expanded polystyrene concrete reinforced with steel fiber. *Cem. Concr. Res.* **2004**, *34*, 1259–1263. [CrossRef]
27. Babu, D.S.; Babu, K.G.; Wee, T. Properties of lightweight expanded polystyrene aggregate concretes containing fly ash. *Cem. Concr. Res.* **2005**, *35*, 1218–1223. [CrossRef]
28. Haghi, A.; Arabani, M.; Ahmadi, H. Applications of expanded polystyrene (EPS) beads and polyamide-66 in civil engineering, Part One: Lightweight polymeric concrete. *Compos. Interfaces* **2006**, *13*, 441–450. [CrossRef]
29. Babu, D.S.; Babu, K.G.; Tiong-Huan, W. Effect of polystyrene aggregate size on strength and moisture migration characteristics of lightweight concrete. *Cem. Concr. Compos.* **2006**, *28*, 520–527. [CrossRef]
30. Madandoust, R.; Ranjbar, M.M.; Mousavi, S.Y. An investigation on the fresh properties of self-compacted lightweight concrete containing expanded polystyrene. *Constr. Build. Mater.* **2011**, *25*, 3721–3731. [CrossRef]
31. Xu, Y.; Jiang, L.; Xu, J.; Li, Y. Mechanical properties of expanded polystyrene lightweight aggregate concrete and brick. *Constr. Build. Mater.* **2012**, *27*, 32–38. [CrossRef]
32. Chaudhary, M.; Srivastava, V.; Agarwal, V. Effect of waste low density polyethylene on mechanical properties of concrete. *J. Acad. Ind. Res.* **2014**, *3*, 123.
33. Martínez-Barrera, G.; Vigueras-Santiago, E.; Hernández-López, S.; Brostow, W.; Menchaca-Campos, C. Me-chanical improvement of concrete by irradiated polypropylene fibers. *Polym. Eng. Sci.* **2005**, *45*, 1426–1431. [CrossRef]
34. Abdelmoti, H.; Mustafa, A.M. Use of Polypropylene Waste Plastic Pellets as Partial Replacement for Fine Aggregate in Concrete. *Univ. Khartoum Eng. J.* **2019**, *9*, 37–43.
35. Wang, Y.; Wu, H.; Li, V.C. Concrete reinforcement with recycled fibers. *J. Mater. Civ. Eng.* **2000**, *12*, 314–319. [CrossRef]
36. Kojnoková, T.; Markovičová, L.; Nový, F. The changes of LD-PE films after exposure in different media. *Prod. Eng. Arch.* **2020**, *26*, 185–189. [CrossRef]
37. Kakooei, S.; Akil, H.M.; Jamshidi, M.; Rouhi, J. The effects of polypropylene fibers on the properties of reinforced concrete structures. *Constr. Build. Mater.* **2012**, *27*, 73–77. [CrossRef]
38. Sivakumar, A.; Santhanam, M. A quantitative study on the plastic shrinkage cracking in high strength hybrid fibre reinforced concrete. *Cem. Concr. Compos.* **2007**, *29*, 575–581. [CrossRef]
39. Pietrzak, A.; Ulewicz, M. The impact of the length of polypropylene fibers on selected properties of concrete. *Acta Sci. Pol. Arch. Bud.* **2019**, *18*, 21–25. [CrossRef]
40. Sofi, A. Effect of waste tyre rubber on mechanical and durability properties of concrete—A review. *Ain Shams Eng. J.* **2018**, *9*, 2691–2700. [CrossRef]
41. Balaha, M.M.; Badawy, A.A.M.; Hashish, M. Effect of using ground tire rubber as fine aggregate on the behaviour of concrete mixes. *Indian J. Eng. Mater. Sci.* **2007**, *14*, 427–435.
42. Albano, C.; Camacho, N.; Reyes, J.; Feliu, J.L.; Hernández, M. Influence of scrap rubber to Portland I concrete composites: Destructive and non-destructive testing. *Compos. Struct.* **2005**, *71*, 439–446. [CrossRef]
43. Bravo, M.; de Brito, J. Concrete made with used tyre aggregate: Durability-related performance. *J. Clean. Prod.* **2012**, *25*, 42–50. [CrossRef]
44. Taha, M.M.R.; El-Dieb, A.S.; El-Wahab, M.A.A.; Abdel-Hameed, M.E. Mechanical, fracture, and microstructural investigations of rubber concrete. *J. Mater. Civ. Eng.* **2008**, *20*, 640–649. [CrossRef]
45. Lavagna, L.; Nisticò, R.; Sarasso, M.; Pavese, M. An Analytical Mini-review on the compression strength of rubberized concrete as a function of the amount of recycled tires crumb rubber. *Materials* **2020**, *13*, 1234. [CrossRef] [PubMed]
46. Guelmine, L. The freeze-thaw durability of concrete containing the rubber aggregate of tirewaste. *Res. Eng. Struct. Mater.* **2022**, *8*, 253–264.
47. Gesoğlu, M.; Güneyisi, E. Strength development and chloride penetration in rubberized concretes with and without silica fume. *Mater. Struct.* **2007**, *40*, 953–964. [CrossRef]
48. Onuaguluchi, O.; Panesar, D.K. Hardened properties of concrete mixtures containing pre-coated crumb rubber and silica fume. *J. Clean. Prod.* **2014**, *82*, 125–131. [CrossRef]

49. Pelisser, F.; Zavarise, N.; Longo, T.A.; Bernardin, A.M. Concrete made with recycled tire rubber: Effect of alkaline activation and silica fume addition. *J. Clean. Prod.* **2011**, *19*, 757–763. [CrossRef]
50. Haghighatnejad, N.; Mousavi, S.Y.; Khaleghi, S.J.; Tabarsa, A.T.; Yousefi, S. Properties of recycled PVC aggregate concrete under different curing conditions. *Constr. Build. Mater.* **2016**, *126*, 943–950. [CrossRef]
51. Najjar, A.K.; Bashaand, E.A.; Milad, M.B.K. Rigid polyvinyl chloride waste for partial replacement of natural coarse aggregate in concrete mixture. *Int. J. Chem. Environ. Eng.* **2013**, *4*, 399–403.
52. Kou, S.C.; Lee, G.; Poon, C.S.; Lai, W.L. Properties of lightweight aggregate concrete prepared with PVC granules derived from scraped PVC pipes. *Waste Manag.* **2009**, *29*, 621–628. [CrossRef]
53. Senhadji, Y.; Escadeillas, G.; Benosman, A.S.; Mouli, M.; Khelafi, H.; Kaci, S.O. Effect of incorporating PVC waste as aggregate on the physical, mechanical, and chloride ion penetration behavior of concrete. *J. Adhes. Sci. Technol.* **2015**, *29*, 625–640. [CrossRef]
54. Mounanga, P.; Gbongbon, W.; Poullain, P.; Turcry, P. Proportioning and characterization of light weight concrete mixtures made with rigid polyurethane foam wastes. *Cem. Concr. Compos.* **2008**, *30*, 806–814. [CrossRef]
55. Pietrzak, A.; Ulewicz, M. The influence of post-consumer car mats waste on selected concrete parameters. *E3S Web Conf.* **2018**, *49*, 00082. [CrossRef]
56. Garcia, P.S.; de Sousa, F.D.B.; de Lima, J.A.; Cruz, S.A.; Scuracchio, C.H. Devulcanization of ground tire rubber: Physical and chemical changes after different microwave exposure times. *Polym. Lett.* **2015**, *9*, 1015–1026. [CrossRef]
57. Mora-Rodríguez, G.K.; Hernández Carrillo, C.G.; Pineda-Triana, Y. Compositional, thermal and morphological characterization of recycled and modified elastomers for inclusion in commercial cement mixtures. *J. Phys. Conf. Ser.* **2019**, *1386*, 012064. [CrossRef]
58. Nabajyoti Saikia, N.; de Brito, J. Use of plastic waste as aggregate in cement mortar and concreto preparation: A review. *Constr. Build. Mater.* **2012**, *34*, 385–401. [CrossRef]
59. Albano, C.; Camacho, N.; Hernandez, M.; Matheus, A.; Gutierrez, A. Influence of content and particle size of pet waste bottles on concrete behaviour at different w/c ratios. *Waste Manag.* **2009**, *29*, 2707–2716. [CrossRef]
60. Saikia, N.; de Brito, J. Mechanical properties and abrasion behaviour of concreto containing shredded pet waste bottle aggregate as a partial substitution of natural aggregate. *Constr. Build. Mater.* **2014**, *52*, 236–244. [CrossRef]
61. Hannawi, K.; Kamali-Bernard, S.; Prince, W. Physical and mechanical properties of mortars containing PET and PC waste aggregates. *Waste Manag.* **2010**, *30*, 2312–2320. [CrossRef]
62. Jura, J.; Ulewicz, M. Assessment of the Possibility of Using Fly Ash from Biomass Combustion for Concrete. *Materials* **2021**, *14*, 6708. [CrossRef]
63. Olofinnade, O.M.; Ede, A.N.; Ndambuki, J.M.; Ngene, B.U.; Akinwumi, I.I.; Ofuyatan, O. Strength and microstructure of eco-concreteproduced using waste glass as partial andcomplete replacement for sand. *Cogent Eng.* **2018**, *5*, 1483860. [CrossRef]
64. Dobiszewska, M.; Beycioğlu, A. Physical Properties and Microstructure of Concrete with Waste Basalt Powder Addition. *Materials* **2020**, *13*, 3503. [CrossRef] [PubMed]

Disclaimer/Publisher's Note: The statements, opinions and data contained in all publications are solely those of the individual author(s) and contributor(s) and not of MDPI and/or the editor(s). MDPI and/or the editor(s) disclaim responsibility for any injury to people or property resulting from any ideas, methods, instructions or products referred to in the content.

Article

Insight into the Behavior of Mortars Containing Glass Powder: An Artificial Neural Network Analysis Approach to Classify the Hydration Modes

Fouad Boukhelf [1,*], Daniel Lira Lopes Targino [1,2], Mohammed Hichem Benzaama [1], Lucas Feitosa de Albuquerque Lima Babadopulos [2] and Yassine El Mendili [1,3]

[1] Builders Lab, Builders Ecole d'Ingénieurs, ComUE NU, 1 rue Pierre et Marie Curie, 146110 Epron, France
[2] Graduate Program in Civil Engineering—Structures and Civil Construction (PEC), Department of Structural Engineering and Civil Construction (DEECC), Technology Center (CT), Federal University of Ceará (UFC), Bloco 733, Campus do Pici s/n, Fortaleza 60440-900, CE, Brazil
[3] Institut de Recherche en Constructibilité IRC, Ecole Spéciale des Travaux Publics, 28 Avenue du Président Wilson, 94234 Cachan, France
* Correspondence: fouad.boukhelf@builders-ingenieurs.fr; Tel.: +33-231-463-202

Citation: Boukhelf, F.; Targino, D.L.L.; Benzaama, M.H.; Lima Babadopulos, L.F.d.A.; El Mendili, Y. Insight into the Behavior of Mortars Containing Glass Powder: An Artificial Neural Network Analysis Approach to Classify the Hydration Modes. *Materials* **2023**, *16*, 943. https://doi.org/10.3390/ma16030943

Academic Editor: Malgorzata Ulewicz

Received: 20 December 2022
Revised: 14 January 2023
Accepted: 16 January 2023
Published: 19 January 2023

Copyright: © 2023 by the authors. Licensee MDPI, Basel, Switzerland. This article is an open access article distributed under the terms and conditions of the Creative Commons Attribution (CC BY) license (https://creativecommons.org/licenses/by/4.0/).

Abstract: In this paper, an artificial neural network (ANN) model is proposed to predict the hydration process of a new alternative binder. This model overcomes the lack of input parameters of physical models, providing a realistic explanation with few inputs and fast calculations. Indeed, four mortars are studied based on ordinary Portland cement (CEM I), cement with limited environmental impact (CEM III), and glass powder (GP) as the cement substitution. These mortars are named CEM I + GP and CEM III + GP. The properties of the mortars are characterized, and their life cycle assessment (LCA) is established. Indeed, a decrease in porosity is observed at 90 days by 4.6%, 2.5%, 12.4%, and 7.9% compared to those of 3 days for CEMI, CEMIII, CEMI + GP, and CEMIII + GP, respectively. In addition, the use of GP allows for reducing the mechanical strength in the short term. At 90 days, CEMI + GP and CEMIII + GP present a decrease of about 28% and 57% in compressive strength compared to CEMI and CEMIII, respectively. Nevertheless, strength does not cease increasing with the curing time, due to the continuous pozzolanic reactions between $Ca(OH)_2$ and silica contained in GP and slag present in CEMIII as demonstrated by the thermo-gravimetrical (TG) analysis. To summarize, CEMIII mortar provides similar performance compared to mortar with CEM I + GP in the long term. This can later be used in the construction sector and particularly in prefabricated structural elements. Moreover, the ANN model used to predict the heat of hydration provides a similar result compared to the experiment, with a resulting R^2 of 0.997, 0.968, 0.968, and 0.921 for CEMI, CEMIII, CEMI + GP, and CEMIII + GP, respectively, and allows for identifying the different hydration modes of the investigated mortars. The proposed ANN model will allow cement manufacturers to quickly identify the different hydration modes of new binders by using only the heat of hydration test as an input parameter.

Keywords: artificial neural network; data processing; cement; glass powder valorization; life cycle assessment

1. Introduction

With a production of about 10 Gm^3/year, concrete is the most important building material and is, therefore, responsible for some of the world's environmental problems due to the production of more than 4 billion tons/year of cement, which requires production temperatures higher than 1450 °C [1,2]. Indeed, concrete is responsible for 4 to 8% of the world's CO_2 emissions at all stages of production with 50% of the emissions in the construction sector alone [3,4]. Among materials, only coal oil and gas are a more important source of greenhouse gases [5].

One of the reflections that have emerged in recent decades is the substitution of part of the cement by other alternative binders, the supplementary cementitious materials (SCMs), with similar characteristics to cement, especially the pozzolanic and the hydration reactivities. The reduction in cement manufacturing's carbon footprint remains a challenge and will have a strong appeal in the coming years.

Once glass becomes waste, it is landfilled, which is not sustainable, because it does not decompose in the environment. The recovery of glass waste eliminates the unnecessary consumption of limited landfill areas and can reduce energy consumption, among others. As a result, it is necessary to seek solutions that are more environmentally friendly. One of the proposed solutions is to replace part of the Portland cement, the most polluting component of concrete, with wastes and by-products, while trying to keep similar mechanical characteristics. Several industrial by-products have been successfully used as SCMs, including silica fume (SF), ground granulated blast furnace slag (GGBFS), and fly ash (FA) [6–8]. These materials are used to create blended cements that can improve concrete's early durability and/or long-term strength, workability, and cost.

Glass powder (GP), which is studied in this paper, could also be a relevant solution for the creation of less-CO_2-emitting concrete due to its high pozzolanic properties [9–12]. Each year, millions of tons of glass waste are generated worldwide [13,14]. In addition to reducing the amount of cement in concrete, glass waste will be recycled. This could be an important step toward sustainable development [15,16]. Furthermore, glass powder provides additional performance in terms of workability [17], microstructure properties [18,19], mechanical strength performance in the long term [20–22], durability [23], and hygrothermal performance [16,24,25].

In addition, in the context of sustainable applications of industrial by-products in environmentally friendly cementitious binders, Life Cycle Analysis (LCA) is a suitable tool for assessing the environmental impacts of cement production and its associated supply chains [26]. Assessing a literature review of LCA for the concrete and cementitious materials, it is seen that the major pollutant component is ordinary Portland cement (OPC), mainly due to the clinker manufacturing process. Indeed, there are three main steps for the cement manufacturing (cf., Figure 1): (i) the pre-processing, where the raw materials are crushed and milled into a fine powder, to enter the preheater; (ii) the calcination, where temperatures greater than 1400 °C are achieved, forming the clinker; (iii) finally, the addition of gypsum to control the setting time followed by final grinding to transform the clinker into a fine-grained mixture (90% of the particles diameter are on the order of 10 microns).

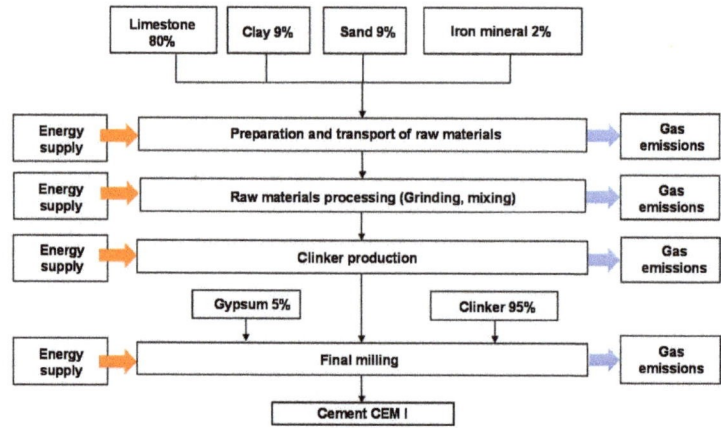

Figure 1. Basic process, "cradle-to-gate" scope and system limits [27].

The main steps of the LCA assessment are: (i) determining the scope and limitations of OPC production and its partial replacement by more environmentally friendly SCMs, evaluating greenhouse gas emissions, and energy and raw material consumption; (ii) selecting output and input data using literature reviews to obtain average values for each necessary parameter [27–33], (iii) using the input data in the evaluation of the environmental impact of the mixtures proposed in this study by preceding the interpolation, and (iv) finally, presenting the results in terms of the decrease or increase in greenhouse gas emissions and energy and raw material consumption. The regulatory parameters of the two cements used, among others, are presented in Table 1, which includes the two types of cement used in this study (CEM I and CEM III). Indeed, the mineral content of CEM I is mainly pure gypsum ($CaSO_4 \cdot 2H_2O$) with a maximum content of 5%, which controls setting time. Regarding CEM III, it contains GGBFS with a content between 36 and 95%. The current LCA assessment will be divided into three types of raw materials (i.e., clinker, GGBFS, and GP).

Table 1. Cement regulations following European Standard (EN 197-1) [27].

Class	Mineral Content		Mineral Components/Inorganic Process Additions	Market Share
	(Allowable Range)	(Average)		
Ordinary Portland cement (CEM I)	≤5%	2.5%	inorganic process addition	30%
Portland composite cement (CEM II)	6–35%	20.5%	ground granulated blast furnace slag (GGBFS), silica fume, pozzolan, fly ash, burnt shale	57%
Blast furnace cement (CEM III)	36–95%	65.5%	GGBFS	5%
Pozzolanic cement (CEM IV)	11–55%	33%	silica fume, pozzolan, fly ash	6%
Composite cement (CEM V)	36–80%	58%	GGBFS, pozzolan, fly ash	3%

To investigate the hydration kinetics of these alternative binders, several models have been proposed in the literature over recent decades. The hydration heat models are dependent on several variables, some of which are uncontrollable. Finding a credible model that accurately describes the hydration heat is a challenging task. Based on the literature, the simulation methods can be classified into three categories: (i) linear regression models [34], (ii) physical models [35], and (iii) intelligent computing models. The linear regression models are constructed by using the least-squares methods to fit experimental data under the assumption that there are linear (or nonlinear) and independent relationships between the heat of hydration and its influencing factors, i.e., cement compositions, hydration age, and cement fineness. In the work of Qin [34], the non-assumptive projection pursuit regression method has been applied to predict the hydration heat of Portland cement.

Concerning the physical models, several models have been proposed in the literature to predict cement hydration heat. Tahersima et al. [35] used a finite element model to predict the hydration heat in a concrete slab-on-grade floor with limestone blended cement. The results showed that the prediction accuracy of finite element results is about 15% more for the maximum temperature rise and 30% more for the peak time. However, the main difficulty of this model lies in the determination of the thermal properties of the material to feed the model. Furthermore, the calculation time takes longer due to discretization. Kondo et al. [36] proposed a single-particle model that considers the circular-shaped layered growth of a uniform thickness of hydrates on a spherical single-reaction cement particle to characterize the hydration kinetics of alite (C_3S). Pommersheim et al. [37,38] suggested an integrated single-particle reaction-diffusion model for the hydration of C_3S. The hydration was modeled by using a classical approach to deal with reactions and diffusion at phase boundaries.

Recently, He et al. [39] developed a numerical method to describe the hydration heat characteristics of Portland cement under atmospheric steam curing conditions. This model has been employed using an iterative algorithm and confirmed with the experimental results. More recently, Nguyen et al. [40] investigated the effects of cement particle distribution on the hydration process of cement paste using a three-dimensional cement hydration computer simulation. The hydration process of cement paste was simulated using the model based on the CEMHYD3D code and the random distribution method was simpler and required less time in generating a pre-hydrated model than the correlation function method. In the work of Chu et al. [41], the CEMHYD3D code was also used to numerically model the hydration of cement paste by following the evolution of all phases and predicting the properties of hydrated cements. Furthermore, Zhao et al. [42] studied the early-age hydration of blended cement and modeled its hydration kinetics by measuring the hydration heat. Within this paper, the classification of hydration modes was discussed based on the Krstulovic–Dabic model [43,44], which was built on the data from the hydration rate. The limitations of numerical methods listed above indicate that there is a knowledge gap that needs to be filled, particularly for machine learning methods.

Indeed, intelligent computing models are usually based on artificial neural networks (ANN) and can also be used to assess the complicated process [45–48]. Nevertheless, a lack of knowledge on ANN applications to predict the hydration process was detected. Based on a literature review, limited works on Machine Learning (ML) and ANN applications applied to thermal data analysis and processing have been found in the literature. In addition, only two studies highlight the use of ANN models to investigate the hydration heat of Portland cement [49,50]. Subasi et al. [50] used a new approach based on an adaptive neuro-fuzzy inference system (ANFIS). Cook et al. [49] used a Random Forest (RF) model to achieve high-accuracy predictions of time-dependent thermal hydration kinetics of OPC-based analysis. Indeed, the largest drawback of RF is that it can become too slow and ineffective for real-time forecasts if there are too many combined decision trees. In general, these algorithms are quick to learn but take a long time to produce predictions after they have been trained. Moreover, ANFIS has a significant processing cost due to its complicated structure and gradient learning. This is a major drawback for applications with a high amount of data [51].

Due to the wide range of factors that influence the performance of the binder hydration process, the dynamic simulation of the hydration is a difficult task. Several variables influence heat hydration and they can occasionally be beyond control. It can be difficult to find a model that accurately describes the behavior of materials as they hydrate. Indeed, complex and non-linear systems influenced by their environment define behavior dynamics. However, the identification of the thermal parameters of the material to feed the model is where the main challenge of the physical models is presented in the literature. In addition, the discretization lengthens calculation times. For this reason, in this paper, an ANN model was used to predict the hydration process. The general objective is to fill the information gap and to present a dynamic model that can overcome the mentioned obstacles, compared to physical models, such as providing a realistic explanation of the heat of hydration, with minimum inputs and faster calculations. In order to contribute on that matter, the aims of this work were: (i) to assess the environmental impact of binders used and highlight the potential of glass powder (GP) valorization in the reduction in cement's carbon footprint; (ii) to investigate the effect of GP on the hydration heat, and on the microstructural and mechanical properties of mortars, in addition to the thermogravimetric analysis results; (iii) to propose a numerical approach to identify and classify the transition modes of the cement hydration process. For this last purpose, a multilayer perceptron regressor (MLP) was proposed to generate a smooth curve model adjusted to the data reading signal and to map each experimental dataset. This method provides an interpolated model with less noise and/or other interferences. Once validated, the numerical analysis of the generated model was used to identify the transition modes. This identification was based on local maxima and minima of the generated shape format [52–55]. This method allows the development of a reliable model that can identify and set the hydration process [56].

2. Materials and Methods

Following the approach presented in Figure 2, the methodology employed in this paper is based on experimental and numerical steps. Each phase is made up of multiple sub-steps, as detailed in the following subsections. The first step concerns the materials characterizations at the laboratory scale and their carbon footprint assessment. The second step focuses on the data processing aspects and the general proposed numerical model dedicated to assessing the hydration process of the mortars used.

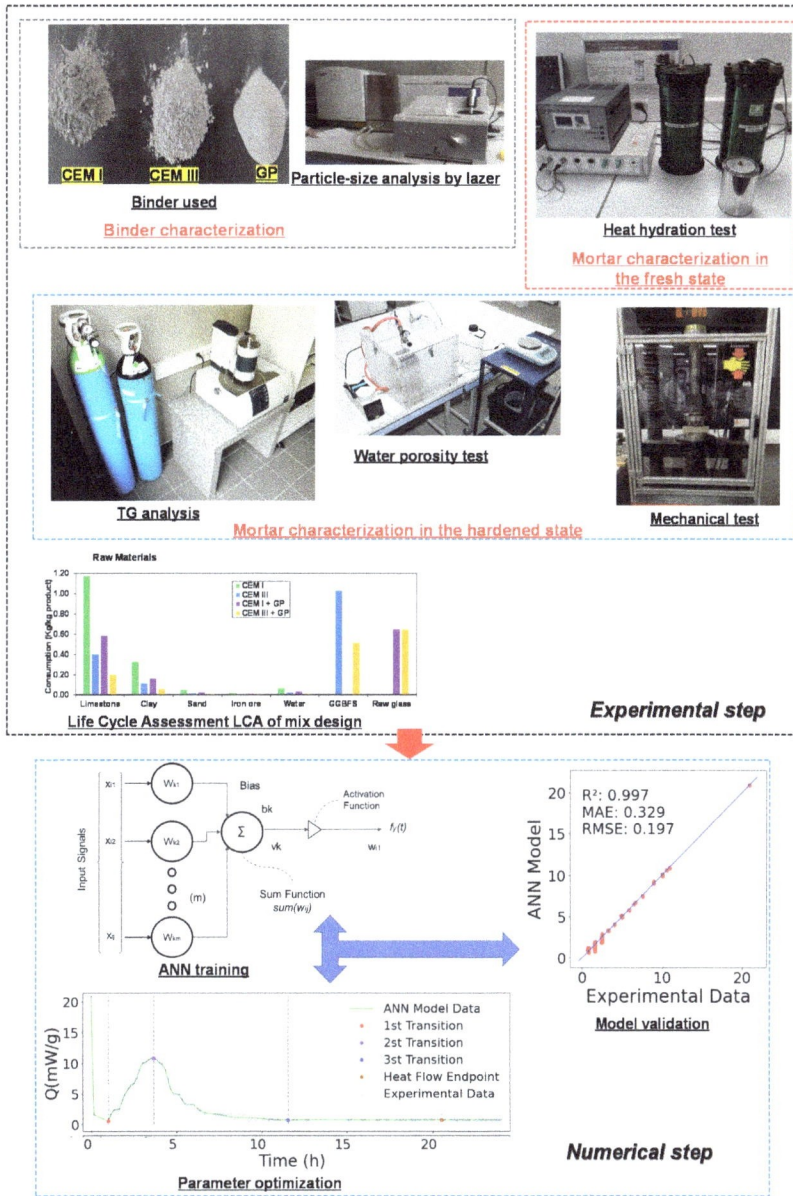

Figure 2. Conceptual study plan.

2.1. Experimental Program

Experiments were carried out to investigate the substitution effect of cement, which is responsible for significant greenhouse gas (GHG) emissions, by other non-biodegradable GHG alternatives, such as glass powder (GP), which represents a significant amount of waste. The use of GP as cement substitution would affect the consistency, the heat of hydration, and the mechanical properties. In the first step, a characterization of the binders' constituents was performed through particle size analysis and density analysis. Afterward, mortars with Portland cement (CEM I 52.5 N), blast furnace cement (CEM III 32.5 N), and glass powder (GP) were mixed. In this study, 50% of the cement mass was replaced by GP. Indeed, Zeybek et al. [57] conducted a series of compressive strength, splitting tensile strength, and flexural strength tests to investigate the effect of waste glass powder on the mechanical properties of concrete. In their study, glass powder was used to replace up to 50% of cement. The study conducted by Kalakada et al. [58] showed also a decrease in compressive strength at high cement substitution. The compression strength of the concrete specimen was reduced by 65% when 50% of the cement was replaced with glass powder. In this study, it was chosen to replace the cement by 50% of GP in mass in view of other experiences in the literature with high substitution rates, but it is important to notice that focus is given on the proper validation of the relevance of the ANN model with materials having a low hydration heat (which is expected at those high substitution rates—up to 50%). The microstructural characterization of the tested mortars was conducted using an isothermal calorimeter, water porosity, and a TG analysis. In addition, the workability and mechanical strength of the mortars were also assessed and compared to control mortars with CEM I and CEM III. To explain the improvement of the mechanical strength and the decrease in the porosity in the long term, the thermogravimetric analysis was performed after 90 days.

2.1.1. Raw Materials

Two types of cement were used in this study, CEM I 52.5 N and CEM III 32.5 N in compliance with the European standard [59]. Glass powder was used as mass substituents of cement. The glass powder was recovered from glass waste of demolition, provided by the waste disposal plant of Caen's metropolitan region (Normandy, France). The density of binders was determined using a helium pycnometer and the particle size distribution was assessed by a laser particle size analyzer in the wet process in compliance with the European standard [60]. Figure 3 shows the particle size distribution of the used binders. The results show that the studied CEM I and CEM III were almost the same size as the literature result [61] and presented a finer particle size when compared to GP. Indeed, the average particle sizes D_{50} of the investigated binders were 16, 13, and 40 µm for CEM I, CEM III, and GP, respectively, as reported in Table 2. Several studies proved that a GP fineness finer than 75 µm would produce a pozzolanic reaction instead of alkali-silica reaction (ASR) [62,63]. Concerning the density results, CEM I presented a high density of about 3150 kg/m^3 compared to those of CEM III and GP, which were 2880 and 2510 kg/m^3, respectively.

Table 2. Particles size and density of binders used.

	CEM I	CEM III	GP
d_{10} (µm)	3	1.5	8
d_{50} (µm)	16	13	40
d_{90} (µm)	40	32	80
d_{max} (µm)	300	70	120
Density (kg/m^3)	3150	2880	2510

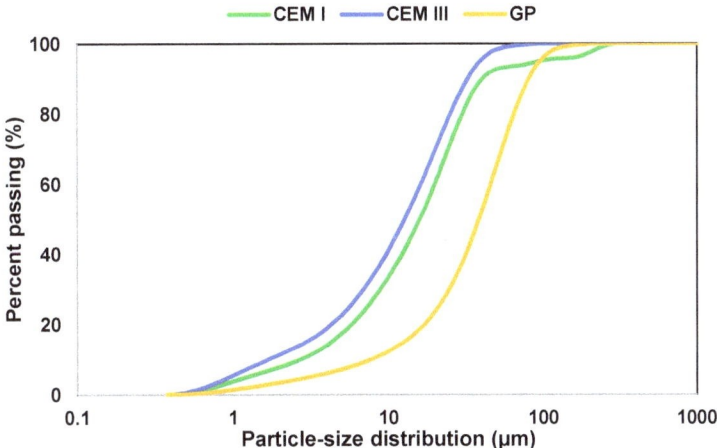

Figure 3. Particle-size distribution of the cements and glass powder used.

2.1.2. Mix Design and Experimental Protocol

The mortar mixture was prepared using the mix design method described by Equations (1)–(4) [64].

$$V_S + V_C + V_{GP} + V_W \approx 1 \text{ m}^3 \tag{1}$$

$$\frac{W}{GP + C} = 0.5 \tag{2}$$

$$\frac{GP}{C + GP} = 0.5 \tag{3}$$

$$\frac{S}{GP + C} = 0.3 \tag{4}$$

where V_S, V_C, V_{GP}, and V_W (m³) are the sand, cement, glass powder, and water volumes, respectively. S, W, GP, and C (kg) are the sand, effective water, glass powder, and cement contents per cubic meter of mortar, respectively. The water-to-binder ratio was taken as equal to 0.5 (Equation (2)). Such values of paste volume and water-to-binder ratio were chosen to obtain a homogeneous mortar mixture by limiting bleeding and segregation and to ensure its good workability [64]. Moreover, the water-to-binder ratio was chosen high enough to mitigate the phenomenon of self-desiccation due to hydration [65,66]. The cement substitution ratio was fixed at 0.5 (Equation (3)) to maintain the glass powder content (GP) as equal to the cement content (C). The volume of sand (V_S) was fixed to obtain a total mortar volume of 1 m³. The mortar mix compositions are given in Table 3. A commercial SikaCem® superplasticizer polycarboxylate was added to all the mixtures in order to obtain a better and comparable consistency. Indeed, due to the low content of the superplasticizer, compared to the mortar constituents, its content was neglected. The superplasticizer/cement ratio varied between 0.36% and 0.54%. The slump measured at the fresh state varied between 12 and 21 mm.

In the fresh state, the workability of mortars was assessed using a flow table test according to the ASTM standard [67]. This test is performed by placing mortar in a mini mortar cone, which has been previously wetted, in two layers on a vibrating table. Then, 15 vibrations are applied with a speed of 1 vibration/second. At the end of this test, the flow is measured. In the hardened state, the porosity, and the compressive and the tensile strengths of all specimens were tested. According to the French standard [68], the prismatic mortars of 4 × 4 × 16 cm³ were cast. The specimens were cured at the standard conditions, i.e., for 24 h at 20 ± 1 °C and relative humidity no less than 60%. Then, the specimens were unmolded and cured in water for different test ages (3, 14, 28, and 90 days) at (22 ± 1) °C. After that, all mortar specimens were taken out of the water for the mechanical properties tests.

Table 3. Mix proportions of mortar (kg/m³).

	CEM I	CEM III	CEM I + GP	CEM III + GP
Sand 0/4	1795	1638	1761	1685
CEM I	500	0	250	0
CEM III	0	500	0	250
GP	0	0	250	250
Water	250	250	250	250
Superplasticizer	1.8	1.5	1.4	1.2
Flow (cm)	12	13	17	21

The test of the water porosity consisted of placing the specimens in saturation benches until the total saturation in water, which was ensured by a vacuum pump [69]. Following 24 h of saturation, the specimens were collected, wiped, and their wet and underwater weights measured. After that, the specimens were put in the oven at 105 °C for 24 h and a second dry weight measurement was taken. The porosity corresponded to [69].

Concerning the mechanical tests, the corresponding loading rates of the compressive and flexural strength were 2.4 ± 0.2 kN/s and 50 ± 10 kN/s, respectively. Finally, the thermogravimetric analysis (TG) was carried out to monitor the solid phase formations on the mortars and the possible pozzolanic reaction (portlandite—CH, calcite—CC, etc.). The device used was the NETZSCH® instrument (STA 449 F5 Jupiter, Selb, Germany). It consists of a sealed chamber to control the atmosphere of the sample by injection of helium gas, an oven to manage the temperature, a weighing module (microbalance), a thermocouple to measure the temperature, and a computer to control and record data. During the test, the temperature should be increased from room temperature to 1200 °C with a constant rate of 10 °C/min. This velocity is widely used in the literature and it is recommended by the French standard [69]. The obtained TG results were used to calculate the portlandite rate (CH) and the chemically bound water rate (w_b) by Equations (5) and (6) [70].

$$CH\% = \frac{m_{410} - m_{490}}{m_{490}} \times \frac{M_{Ca(OH)_2}}{M_{H_2O}} \times 100\% \quad (5)$$

$$w_b\% = \frac{m_{40} - m_{490}}{m_{490}} \times 100\% \quad (6)$$

where m_{40}, m_{410}, and m_{490} are the sample weights at 40 °C, 410 °C, and 490 °C, respectively. $M_{Ca(OH)_2}$ and M_{H_2O} are the molecular masses of portlandite (CH) and water (H_2O), respectively. To compare the results, CH rate and w_b rate were normalized to 100 g of anhydrous material.

2.2. Data Processing Model

In this section, a binder hydration monitoring method is proposed. This method is based on an artificial neural network (ANN) model processing the data from the hydration heat tracking test. Indeed, a multilayer perceptron regression (MLP) was used to proceed to an initial treatment of the data considered as signals in order to transform them into a time series regression. Afterward, a numerical analysis of the generated model was carried out in order to identify the different transition modes. The identification of the transition modes was based on the local maxima and minima values of the generated model and related to the hydration process.

2.2.1. Artificial Neural Network Modeling

In the literature, the most used data-driven model for building and material behavior is the artificial neural network (ANN) [71–73]. Numerous types of ANNs have been developed over the years with varying characteristics. A multilayer perceptron (MLP) is a class of ANN, which is applied for both classification and regression problems, it is a classical type of neural network, very flexible, and can be used generally to learn a map from inputs to outputs, among other applications. It is classified as supervised learning, using a backpropagation technique for training (cf., Figure 4). All studies in the literature highlight good results in long short-term memory (LSTM) applications showing versatility in a multitude of scenarios [74–77].

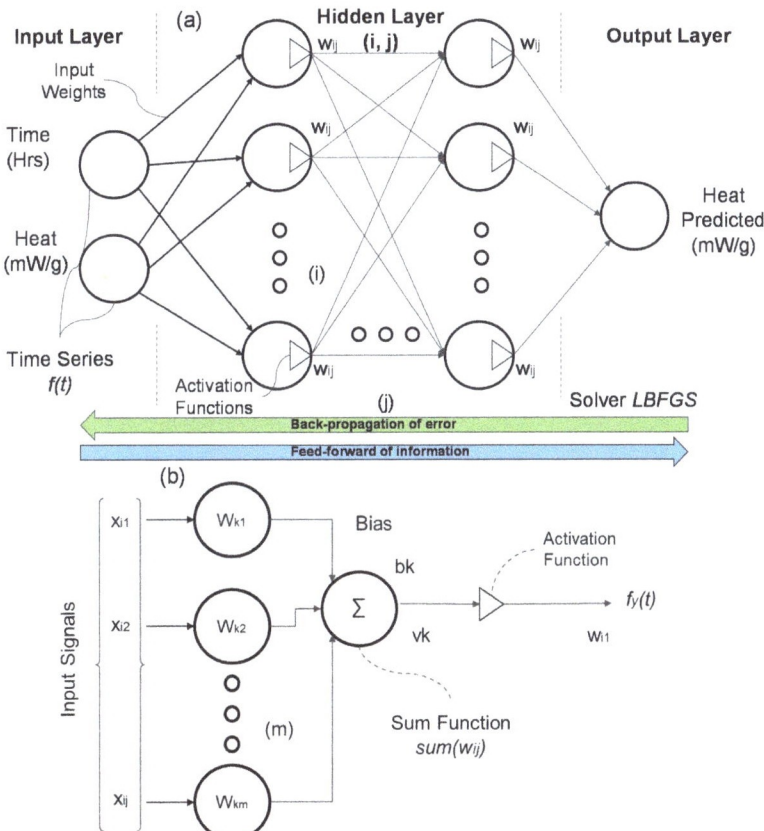

Figure 4. Simplified representation of an artificial neural network and its components: (**a**) Design Concept of a Multi-Layer Perceptron Model (MLP) and (**b**) Neuron simplified structure.

The MLP model consists of three parts or nodes organized in the input layer, hidden layers, and output layer. Except for the input nodes, all the other concepts apply non-linear activation functions in which the size of the hidden layer is usually determined by hyperparametric optimization [75]. The details of the inputs and internal weights are optimized depending on the applied problem and dataset.

As seen in Figure 4b, the neuronal structure is also composed of three elements [78,79]: (i) 'synapses' or connecting links that are associated with particular weights; (ii) the processing unit that handles the input signals weighted (summing junction) by the synaptic weight and adjusts them by adding a bias value (b_k); (iii) an activation function (φ) that limits the signal amplitude at the neuron's output. Figure 4b demonstrates a typical neuronal structure, applied for the hidden and output layer. The input layer differs by the absence of the activation function and the single input synapse, connected to the input data. The neuron k can be described mathematically by Equations (7) and (8).

$$u_k = \sum_{j=1}^{m} x_j \times w_{kj} \tag{7}$$

$$v_k = u_k + b_k \tag{8}$$

where u_k is the linear combiner of the input signal and v_k is the weighted sum of the input values adjusted by the b_k. The activation function is a set of mathematical equations that define the output signal of the neuron k. The activation function may be divided into four distinct groups [80], as in Equations (9)–(12). Equation (9) represents the threshold function, which is equivalent to a Heaviside function. Equation (10) is an identification function (linear function). Equation (11) is a logistic sigmoid function (having a sigmoidal nonlinearity with a as its slope parameter). Equation (12) is a hyperbolic tangent function (another kind of nonlinearity).

$$\varphi(v) = \begin{cases} 1 \ if \ v > 0 \\ 0 \ if \ v = 0 \\ -1 \ if \ v < 0 \end{cases} \tag{9}$$

$$\varphi(V) = 0 \tag{10}$$

$$\varphi(V) = \log f(V) = \frac{1}{1+e^{-\alpha v}} \tag{11}$$

$$\varphi(V) = \tanh(V) = \frac{2}{1+e^{-\alpha v}} \tag{12}$$

The sum of the weighted and bias-corrected inputs passes through the activation function φ to obtain the final output signal (y_k), as shown in Equation (13).

$$y_k = \varphi(\sum_{j=1}^{m} x_j \times w_{kj} + b_k) \tag{13}$$

The implementation of the ANN model is carried out in three steps. The initial stage is to train the model to minimize the absolute error by altering the weight parameters, based on comparisons between experimental and model outputs, which should be as close as possible. The LBFGS (Limited-Memory Broyden–Fletcher–Goldfarb–Shanno) is a popular optimization algorithm for estimating weight parameters in machine learning. It uses a quasi-Newton method [81–83]. As the original method (LBFGS), it uses an estimate of the inverse Hessian matrix, but in that case, LBFGS stores only some chosen vectors, to be able to represent the model with a required approximation.

2.2.2. Evaluation Metrics

In order to evaluate the final model's performance, the correlation coefficient (R^2), the root mean square error ($RMSE$), and the mean absolute error (MAE) are employed [84–86]. The aim is to achieve a model that best fits the labeled training data by evaluating all the combined applied metrics. All these metrics are performance indicators commonly used to measure the efficiency of the machine learning models generated. These indicators' definitions are given in Equations (14)–(16), respectively.

$$R^2 = 1 - \frac{\sum_{i=1}^{n}(y_i - \hat{y}_i)^2}{\sum_{i=1}^{n}(y_i - \overline{y})^2} \tag{14}$$

$$RMSE = \sqrt{\frac{1}{n} \times \sum_{i=1}^{n}(\hat{y}i - yi)^2} \qquad (15)$$

$$MAE = \sum_{i=1}^{n}\left|\frac{yi - \hat{y}}{yi}\right| \qquad (16)$$

2.2.3. Data Processing/Numerical Model

In addition to the characterization of the microstructure and mechanical properties at several ages, a conceptual framework was built to describe and visualize our study methods after reviewing the relevant current literature [42,44–48]. Each phase is composed of several steps, as explained in detail in the following sections. These main phases are summarized in Figure 5 and as below:

- Collecting data: the experimental data were collected based on heat energy measures within the time, during the cement hydration process. That was the base dataset used in the initial data processing, heat energy within time. Four datasets obtained from CEM I, CEM III, CEM I + GP and CEM III + GP were analyzed. Each dataset contained X data points, in a total of Y datapoints being analyzed;
- Model estimation: concept of the initial data processing, testing, and optimizing all the parameters necessary for the execution of the algorithm, such as MLP hidden layers size, activation functions, and optimization algorithm [71,72,82,85];
- Evaluation of conceived model: based on the mentioned metrics parameters (R^2, RMSE, MAE), the best parameters setup was evaluated to proceed with the data classification [85,86];
- Mode's identification: based on the generated model data (hydration heat Q vs. time plot), a numerical peak analysis was proceeded to identify the transition zones and, thus, the transition modes of binder hydration with time [42].

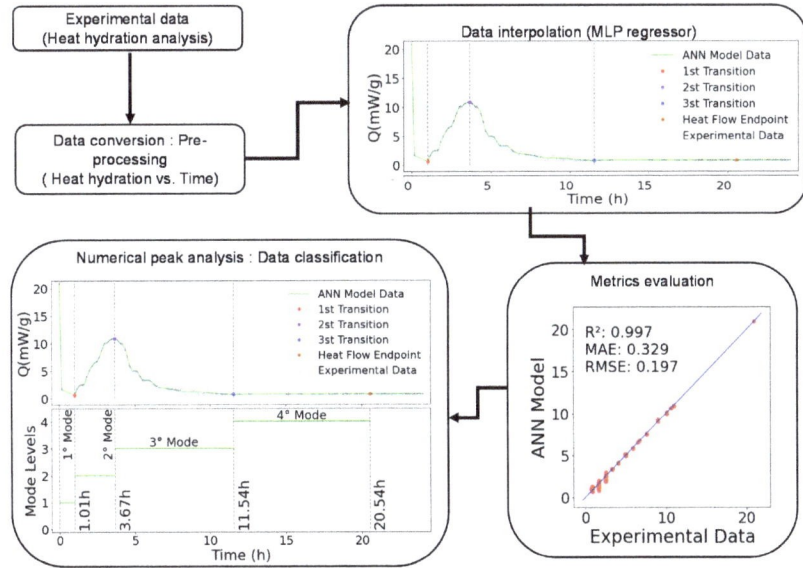

Figure 5. Computational procedure for hydration process classification proposed model.

As a result, the binder hydration modes were obtained, through a staircase levels plot, with respective time transitions [42,45,48]. In addition, within the chemical compositions, the main difference aspects can be highlighted in the behavior, and the total time of each mode for different samples can be compared.

3. Results and Discussion

3.1. Life Cycle Analysis LCA

Tables 4–6 present the LCA inventory obtained after interpolations and considerations reported in the literature. GGBFS is highlighted by the literature with a potential emission reduction of 78.84% in the production of CEM III [64]. Regarding GP production, its manufacturing process shown in Table 4 is not on an industrial scale according to [65]. For masses, units given are in kg of raw material by kg of produced binder (kg/kg product). For raw materials in volume, such as water, units given are in m^3 of raw material by kg of produced binder (m^3/kg product). For energy input, it is either given in terms of kWh of energy by kg of produced binder (kWh/kg product) or in kg of utilized fuel source by kg of produced binder (kg/kg product).

Table 4. Cement regulations following European Standard (EN 197-1) [59].

Raw Materials	Units	CEM I (Clinker)	CEM III (GGBFS)	CEM I + GP	CEM III + GP
Limestone	kg/kg product	1.1737	0.4049	0.5868	0.2025
Clay	kg/kg product	0.3307	0.1141	0.1653	0.0570
Sand	kg/kg product	0.0503	0.0174	0.0252	0.0087
Iron ore	kg/kg product	0.0203	0.0070	0.0102	0.0035
Water	m3/kg product	0.0668	0.0230	0.0334	0.0115
GGBFS	kg/kg product	-	1.0316	-	0.5158
Raw glass	kg/kg product	-	-	0.6480	0.6480

Table 5. Energy consumption inventory input and calculations.

Energy	Units	CEM I (Clinker)	CEM III (GGBFS)	CEM I + GP	CEM III + GP
Electricity (kWh)	kWh/kg product	0.0687	0.0687	0.0403	0.0403
Petroleum coke	kg/kg product	0.0463	0.0463	0.0231	0.0231
Hard coal	kg/kg product	0.0559	0.0559	0.0280	0.0280
Diesel	kg/kg product	3.04×10^{-7}	3.04×10^{-7}	1.52×10^{-7}	1.52×10^{-7}
Fuel oil	kg/kg product	0.0117	0.0117	0.0058	0.0058
Natural Gas (MJ)	MJ/kg product	0.2777	0.2777	0.8768	0.8768
Light distribution	kg/kg product	0.0043	0.0043	0.0022	0.0022

Table 6. General emissions inventory input and calculations.

Emissions	Units	CEM I (Clinker)	CEM III (GGBFS)	CEM I + GP	CEM III + GP
CO_2	kg/kg product	0.0687	0.0687	0.0403	0.0403
NO_x	kg/kg product	0.0463	0.0463	0.0231	0.0231
SO_2	kg/kg product	0.0559	0.0559	0.0280	0.0280
$H_2O(g)$	kg/kg product	3.04×10^{-7}	3.04×10^{-7}	1.52×10^{-7}	1.52×10^{-7}
Particulates	kg/kg product	0.0117	0.0117	0.0058	0.0058
Waste	kg/kg product	0.2777	0.2777	0.8768	0.8768

Figure 6 shows the different raw materials' consumption. The results show a considerable reduction in natural resources replaced by waste. In general, this also implies a reduction in land use, extraction, and other environmental aspects related to the extraction of mineral raw materials using by-products, avoiding dumping and similar impacts.

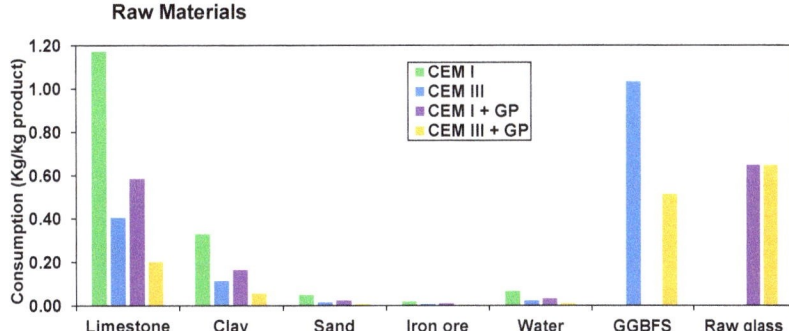

Figure 6. Raw materials' consumption.

Figure 7 presents the results for the relevant energy demands. Apart from the electricity produced by different renewable resources (hydroelectric, wind, and thermoelectric), all other resources are derived from fossil origins. The fuel that has the greatest impact on clinker production is petroleum coke and hard coal, followed by fuel oil. The proposed formulations with a 50% replacement of cement will contribute to a direct 50% reduction in emissions once the clinker manufacturing process is significantly reduced. Regarding the increase in natural gas consumption, LCA of GP reveals that the results presented are not applied on an industrial scale, while scalability in this area could increase performance by reducing consumption.

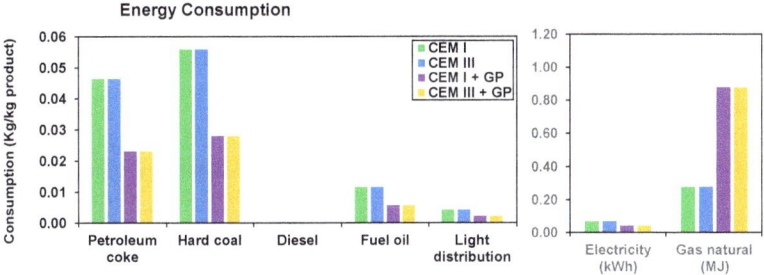

Figure 7. Energy consumption results.

Figure 8 shows the GHG emission results for the formulations studied. The results show a significant reduction in GHG for the formulations containing GP. Regarding mainly CO_2 emissions, CEM III + GP shows a low impact mainly due to the 78.83% reduction of the CEM III formulation and its 50% replacement by GP. Finally, the same reduction profile is observed for the case of heavy metals, i.e., nitrous oxide (NO_x) and sulfur dioxide (SO_2). According to the literature, NO_x is more detrimental to climate change than CO_2, while some authors indicate that it could be 300 times more detrimental to the climate than CO_2.

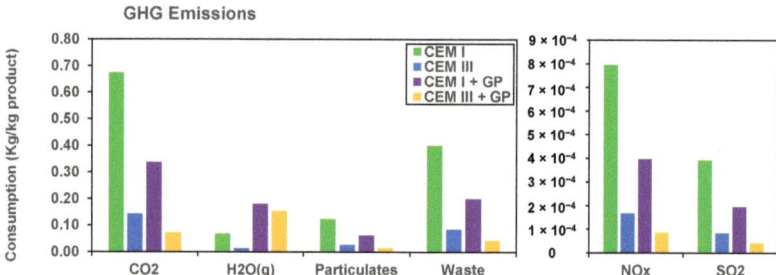

Figure 8. GHG Emissions results.

3.2. Experimental Results

Table 3 presents the flow values measured of the tested mortars. The flow values vary between 12 and 21 mm. Moreover, an improvement in workability is obtained of up to 41.6% and 23.5% for CEM I + GP and CEM III + GP compared to mortars with CEM I and CEM III, respectively. This result is attributed to the higher fineness of GP compared to cement. Therefore, the substitution of cement by GP decreases the water demand for hydration process, i.e., by increasing the amount of free water and workability of the pastes [24].

Figure 9 presents the evolution of the porosity, and the compressive and the tensile strengths (from flexural tests) at several curing ages (3, 14, 28, and 90 days). In general, CEM III presents less water porosity in comparison with CEM I. This phenomenon is explained by the presence of slag in CEM III, which leads to the pores clogging by creating other ranges of non-connected pores [87]. However, after 3 days of curing, the mortars containing GP show high porosity compared to those with only cement. This phenomenon is due to the fineness of GP used, which changes the microstructure and the pore-size distribution [15,24]. A proportional decrease in porosity is observed with increasing age for all mix designs. At 90 days, this decrease is of the order of 4.6%, 2.5%, 12.4%, and 7.9% for mortars based on CEM I, CEM III, CEM I + GP, and CEM III + GP, respectively, compared to the results at 3 days. This result seems to be in good agreement with those found previously by other authors [15,24,49,50].

When it comes to mechanical strength, for all mortars, an increase is observed and this is due to the continuous hydration of the binder with the curing time [22]. In addition, the use of GP with a replacement of 50% of the cement decreases the mechanical strength in the short term. This is due to a dilution effect, which is the immediate consequence of the substitution of a more reactive powder (cement) with a less reactive powder (GP) in the short term [12]. Nevertheless, the mechanical strength does not cease increasing with curing time [15,22,24]. This phenomenon is due to continuous pozzolanic reactions between portlandite ($Ca(OH)_2$) and silica originating from GP, as highlighted by Idir et al. [55], in addition to the slag present in CEM III, as also demonstrated below by the TG analysis. After 90 days of curing time, the CEM I + GP presents a decrease of about 28% and 17.7% in compressive and tensile strength, respectively, compared to CEM I. For CEM III + GP, the decrease is about 57% and 56% in compressive and tensile strength, respectively, compared to CEM III. These results are in agreement with those reported by Adesina et al. [88]. Nevertheless, these values are adequate for the use of CEM I + GP in the construction sector and particularly in prefabricated structural elements.

Porosity (%)				
CEM I	19.4	19.2	19.1	18.5
CEM III	16.1	15.8	15.7	15.7
CEM I + GP	24.9	24.5	24.3	21.8
CEM III + GP	31.7	31.2	29.5	29.2
	3d	14d	28d	90d

Compressive strength (MPa)				
CEM I	12.1	28.3	43.6	46.5
CEM III	8.7	19.2	25.7	35.1
CEM I + GP	10.1	16.5	27.4	33.5
CEM III + GP	4.3	8.2	10.6	15.1
	3d	14d	28d	90d

Tensile strength (MPa)				
CEM I	5.6	5.9	7.2	7.9
CEM III	5.2	5.1	6.8	7.1
CEM I + GP	3.5	3.6	3.9	6.5
CEM III + GP	1.5	1.2	2.8	3.1
	3d	14d	28d	90d

Figure 9. Porosity, and compressive and tensile strengths of mortars used.

To investigate the pozzolanic reaction of GP with portlandite (CH), a thermogravimetric (TG) analysis is conducted at 90 days. Figure 10 shows the three main phases formed in mortars with CEM I and CEM III. The mortar with CEM III has less portlandite and calcite (CC) amounts than CEM I mortar does. This is due to the activated pozzolanic reaction of the blast furnace slag present in the CEM III cement. Nevertheless, in the case of mortars with GP, the two peaks related to portlandite and calcite are reduced, showing the consumption of CH by GP to produce additional CSH hydrates [24]. An increase in the first peak is observed. Indeed, the three phases correspond generally to three processes according to [89,90]. The first band below 200 °C is due to the loss of free water as well as the dehydration of calcium silicate hydrate (CSH), ettringite (A_{Ft}), calcium monosulfoaluminate, hemicarboaluminate, or calcium monocarboaluminate (all these material phases are denoted as A_{Fm}). The second peak around 400–500 °C is contributed to the dehydroxylation of portlandite (CH), while the decarbonization of calcium carbonates (CC) takes place at 700–800 °C. From dTG patterns, the A_{Ft} and A_{Fm} material phases are difficult to dissociate, as the corresponding bands are large, and the decomposition temperatures of these phases are very narrow and overlay each other.

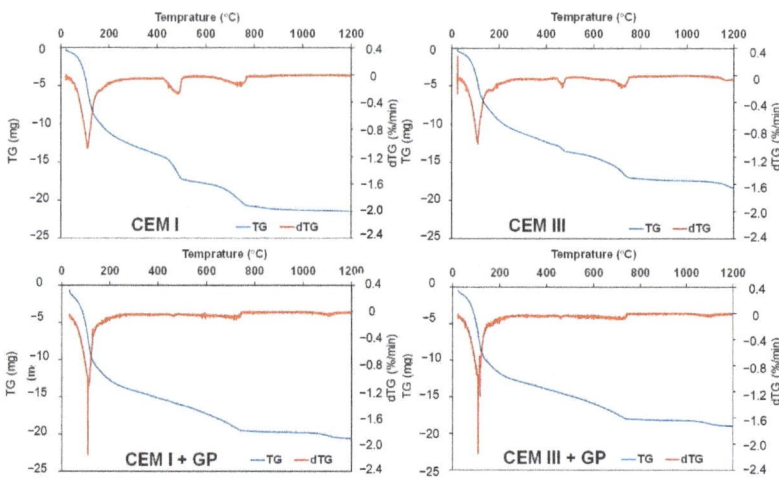

Figure 10. TG and dTG curves of the studied mortars.

Based on TG results, the normalized amounts of the bound water (w_b) and portlandite (CH) are both presented in Figure 11. Indeed, the w_b rate indicates the volume of reaction products (including CSH, A_{Ft}, A_{Fm}, and CH) formed during cement hydration, while normalized CH can be interpreted as an indication of changes in the assemblage of reaction products (CH consumption due to the formation of additional CSH hydrates) as highlighted by Damidot et al. [91]. Figure 11 shows that for the mortars containing low clinker content, the w_b rate is lower than for the standard mortar (CEM I).

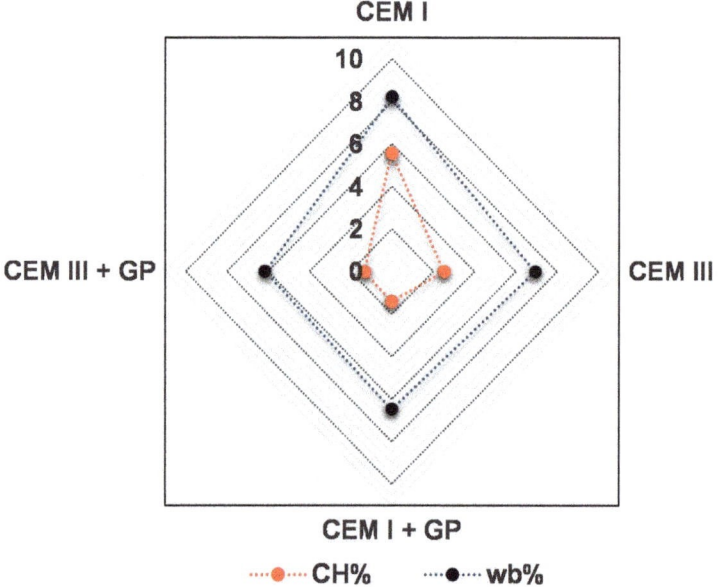

Figure 11. CH% and w_b% of mortars studied.

The increase in w_b in the mortars with GP is related to pozzolanic reactions. Indeed, silica reacts chemically with calcium hydroxide (CH) and produces additional CSH phases.

Due to CH consumption, the normalized CH value for mortar with GP is lower than that for the standard mortars (CEM I and CEM III). These data show the same trend as those reported in literature [92,93], where w_b and CH values for both pure and blended pastes with silica fume were estimated and measured.

3.3. Data Processing/Numerical Results

Figure 12 shows the experimental data and those predicted by the ANN model for the heat of hydration in mW/g. Indeed, the models translating the hydration behavior of CEM I, CEM I + GP, CEM III, and CEM III + GP mortars provide a good agreement with the data found experimentally with a difference of relevance more important for the MAE and RMSE metrics. For the RMSE, the lowest results are obtained with the CEM III + GP results with a 0.92 R^2 score, with the further metrics appointed in the same evaluation for near 0.35 for MAE, and with an improvement for the RMSE, at a value of 0.20, the best result for all datasets.

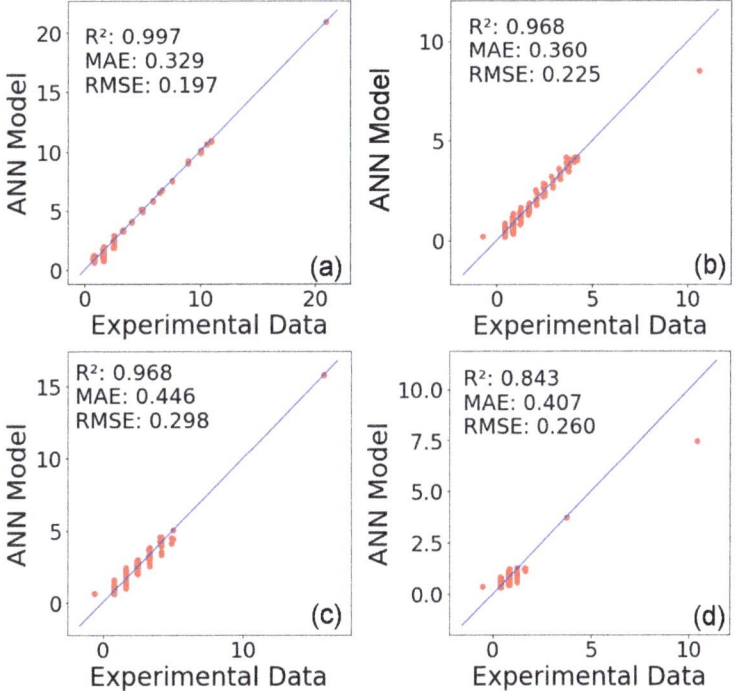

Figure 12. ANN models versus experimental data (for the heat of hydration in mW/g) with error metrics for (**a**) CEM I, (**b**) CEM I + GP, (**c**) CEM III, and (**d**) CEM III + GP.

The hydration modes identification of CEM I, CEM I + GP, CEM III, and CEM III + GP are shown in Figure 13a,b and Figure 14a,b, respectively. In general, the proposed model identifies four main modes [32], including the initial reaction and slow reaction period (Mode 1), acceleration period (Mode 2), deceleration period (Mode 3), and stability period (Mode 4) in the history of heat flow. As shown in Figures 13 and 14, the standard mortar with CEM I started an earlier acceleration period than the other mortars studied with a higher heat release. This mode corresponds to the induction phenomenon interpreted by the geochemistry dissolution theory that the primary mechanism controlling the kinetics up to the end of the induction period is the undersaturation of this surface layer [94]. In addition, tri-calcium aluminate (C_3A) reacts promptly with water from an early age to form calcium hydroaluminate (3CaO–Al$_2$O$_3$–Ca(OH)$_2$–nH$_2$O or hydroxy–A$_{Fm}$), as mentioned by Nguyen et al. [32].

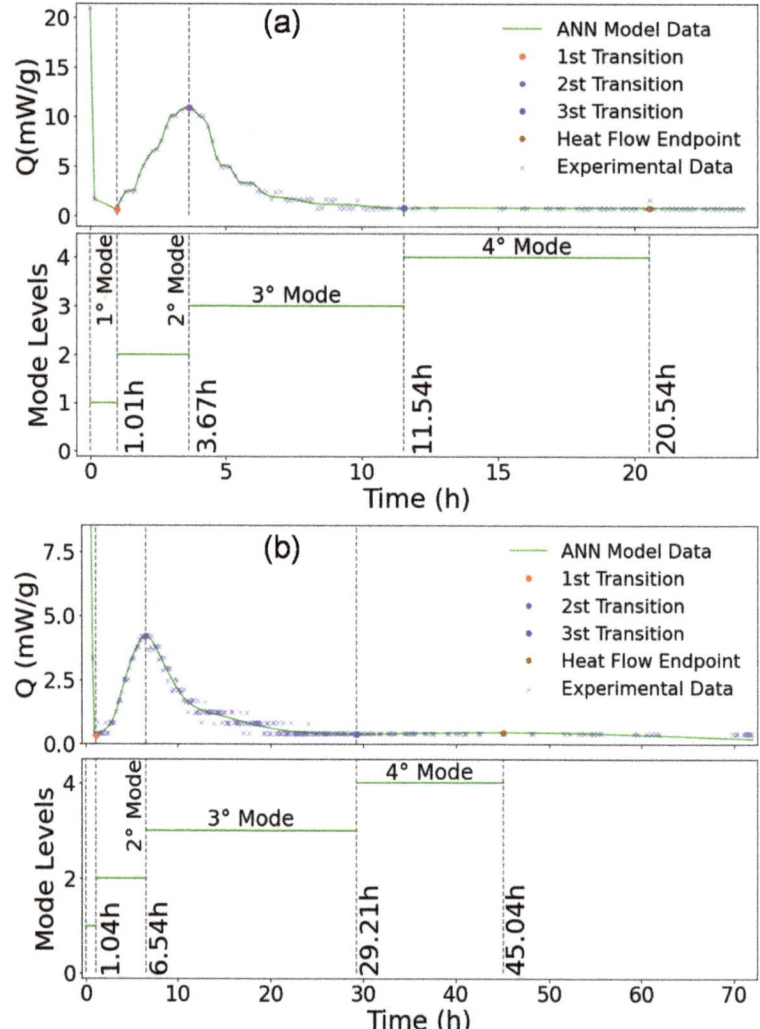

Figure 13. Heat of hydration, ANN model generated, modes, and transitions identification: (**a**) CEM I and (**b**) CEM I + GP.

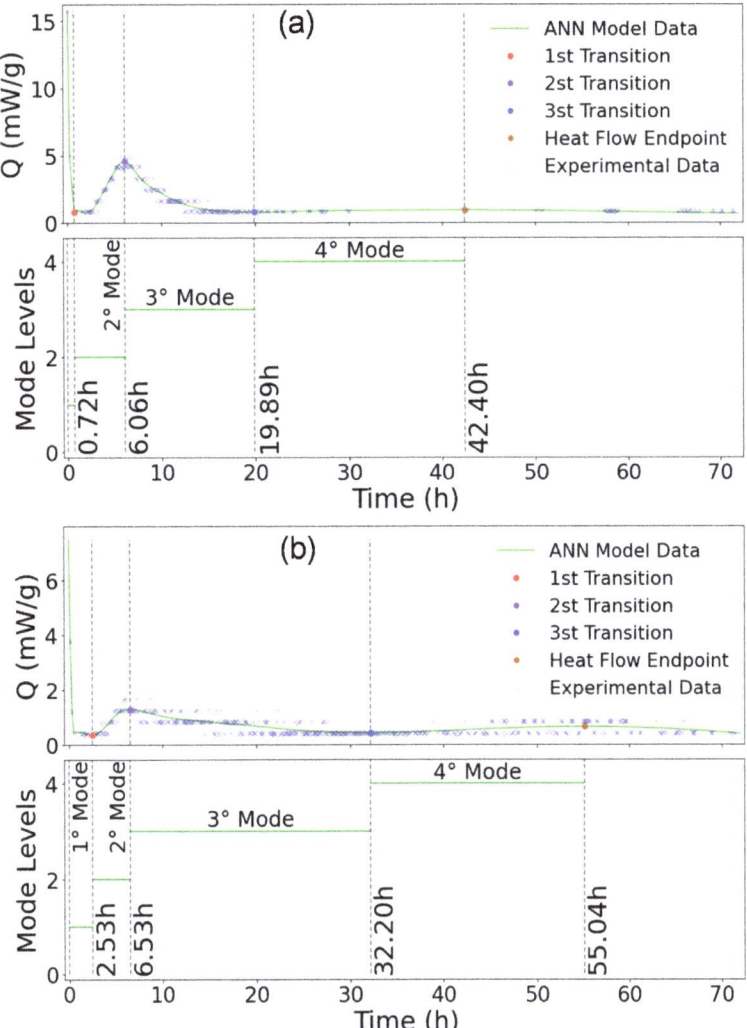

Figure 14. Heat of hydration, ANN model generated, modes, and transitions identification: (**a**) CEM III and (**b**) CEM III + GP.

As shown in Figures 13b and 14a, CEP I + GP and CEM III mortars present almost the same behavior in term of peaks position. Nevertheless, these peaks start at 6.78 h and 7.58 h from the first water contact of the binder for the CEM III and CEM I + GP mortars, respectively. This time delay can be explained by the fact that the containing slag in the CEM III is more reactive than the glass powder.

In the second mode, the peak value decreases and shifts toward a longer hydration time when the CEM III and GP are used. This phenomenon could be attributed to the consumption of C_3S (Alite), which is consistent with the hydration heat rate. It is known that the hydration rate of C_3S is much higher than that of C_2S (Belite) due to the higher solubility and lower activation energy of C_3S compared to C_2S [95]. Indeed, the solubility of C_3S is between 5 and 8 times higher than that of C_2S [95]. Therefore, it could be concluded that Mode 2 is strongly affected by the hydration of C_3S, and this state is in good agreement with the literature [96]. At the end of the induction period, CSH and Portlandite (CH) begin to develop fast. There is still considerable controversy about the trigger for this increase in reaction rate. Indeed, in pastes, it seems that the CH precipitation is the trigger.

Mode 3 corresponds to the renewed dissolution of C_3A with the formation of ettringite (A_{Ft}). Due to this retardation of the C_3A reaction with sulfate, the main reaction of this phase should occur after the main peak of the reaction of alite in a properly sulfated Portland cement. However, in CEM I, the formation of ettringite continues after the exhaustion of sulfate in the solution [97]. In conclusion, the ANN model makes it possible to determine when the initial hydration phase is completed, i.e., after 20.54 h, 42.40 h, 45.04 h, and 55.04 h for mortar based on CEM I, CEM III, CEM I + GP, and CEM III + GP, respectively.

4. Conclusions

An ANN model is used for the first time to identify the different hydration modes of binders using only the heat hydration test as an input parameter. The method is based on a MLP regressor that maps the input signal and generates a fitted model. Afterward, a numerical approach is proceeded to identify the different hydration modes with respective times.

Indeed, ANN model detects four modes during the hydration process. In addition, the model proposed provides similar results in terms of heat of hydration in comparison to the experimental results with R^2 = 0.997, 0.968, 0.968, and 0.921 for CEM I, CEM III, CEM I + GP, and CEM III + GP, respectively. In addition to the mechanical strength achieved, the replacement of 50% of cement by GP is significantly positive for the LCA results. Indeed, the GHG emissions present the most significant results, by reducing 50% and 80% of the emissions of CO_2 and $NO_x + SO_2$ for the CEM I + GP and CEM III + GP mixtures, respectively.

It is revealed in this study that waste glass can be employed as a partial cement substitute in mortar and would be beneficial for GHG emissions of concrete. The present study identifies waste glass as suitable, accessible in large quantities, a local eco-material, and suitable for concrete construction from an economic and environmental standpoint. Nevertheless, when waste glass is employed as partial replacement for cement, the strength values are reduced, which needs to be investigated in the lab for proper mix design and use of the materials. To address this shortcoming, future research will focus on combining waste glass with other recycling fibers. Moreover, the proposed ANN model allows cement manufacturers to quickly identify the different hydration modes of new binders by using only the heat hydration test as an input parameter.

Author Contributions: Conceptualization, investigation, data curation, writing—original draft preparation, methodology, F.B., D.L.L.T., M.H.B. and Y.E.M.; software, D.L.L.T.; validation, formal analysis, writing—review and editing, visualization, supervision, resources, L.F.d.A.L.B. and Y.E.M. All authors have read and agreed to the published version of the manuscript.

Funding: This research was funded by the European Regional Development Fund for having sponsored this study in the frame of a BLUEPRINT to a Circular Economy project (Interreg V A France (Channel) England, Project n°206).

Institutional Review Board Statement: Not applicable.

Informed Consent Statement: Not applicable.

Data Availability Statement: The experimental and computational data presented in the present paper are available from the corresponding author upon request.

Acknowledgments: Authors thank the European Regional Development Fund for financing the research through the BLUEPRINT project.

Conflicts of Interest: The authors declare no conflict of interest.

References

1. Miller, S.A.; Horvath, A.; Monteiro, P.J.M. Readily Implementable Techniques Can Cut Annual CO_2 Emissions from the Production of Concrete by over 20%. *Environ. Res. Lett.* **2016**, *11*, 074029. [CrossRef]
2. Gencel, O.; Karadag, O.; Oren, O.H.; Bilir, T. Steel Slag and Its Applications in Cement and Concrete Technology: A Review. *Constr. Build. Mater.* **2021**, *283*, 122783. [CrossRef]
3. Prakash, R.; Thenmozhi, R.; Raman, S.N.; Subramanian, C. Characterization of Eco-friendly Steel Fiber-reinforced Concrete Containing Waste Coconut Shell as Coarse Aggregates and Fly Ash as Partial Cement Replacement. *Struct. Concr.* **2020**, *21*, 437–447. [CrossRef]
4. Sundaresan, S.; Ramamurthy, V.; Meyappan, N. Improving Mechanical and Durability Properties of Hypo Sludge Concrete with Basalt Fibres and SBR Latex. *Adv. Concr. Constr.* **2021**, *12*, 327–337. [CrossRef]
5. WBCSD World Business Council for Sustainable Development and International Energy Agency Cement Technology Roadmap 2009: Carbon Emissions Reductions up to 2050. Available online: https://www.iea.org/reports/cement-technology-roadmap-carbon-emissions-reductions-up-to-2050 (accessed on 11 November 2022).
6. Shen, D.; Jiao, Y.; Kang, J.; Feng, Z.; Shen, Y. Influence of Ground Granulated Blast Furnace Slag on Early-Age Cracking Potential of Internally Cured High Performance Concrete. *Constr. Build. Mater.* **2020**, *233*, 117083. [CrossRef]
7. Antoni; Chandra, L.; Hardjito, D. The Impact of Using Fly Ash, Silica Fume and Calcium Carbonate on the Workability and Compressive Strength of Mortar. *Procedia Eng.* **2015**, *125*, 773–779. [CrossRef]
8. Bani Ardalan, R.; Joshaghani, A.; Hooton, R.D. Workability Retention and Compressive Strength of Self-Compacting Concrete Incorporating Pumice Powder and Silica Fume. *Constr. Build. Mater.* **2017**, *134*, 116–122. [CrossRef]
9. Chandra Paul, S.; Šavija, B.; Babafemi, A.J. A Comprehensive Review on Mechanical and Durability Properties of Cement-Based Materials Containing Waste Recycled Glass. *J. Clean. Prod.* **2018**, *198*, 891–906. [CrossRef]
10. Serpa, D.; Santos Silva, A.; de Brito, J.; Pontes, J.; Soares, D. ASR of Mortars Containing Glass. *Constr. Build. Mater.* **2013**, *47*, 489–495. [CrossRef]
11. Shayan, A.; Xu, A. Value-Added Utilisation of Waste Glass in Concrete. *Cem. Concr. Res.* **2004**, *34*, 81–89. [CrossRef]
12. Zidol, A.; Tognonvi, M.T.; Tagnit-Hamou, A. Effect of Glass Powder on Concrete Sustainability. *New J. Glas. Ceram.* **2017**, *7*, 34–47. [CrossRef]
13. Shi, C.; Zheng, K. A Review on the Use of Waste Glasses in the Production of Cement and Concrete. *Resour. Conserv. Recycl.* **2007**, *52*, 234–247. [CrossRef]
14. Ling, T.-C.; Poon, C.-S.; Wong, H.-W. Management and Recycling of Waste Glass in Concrete Products: Current Situations in Hong Kong. *Resour. Conserv. Recycl.* **2013**, *70*, 25–31. [CrossRef]
15. Bouchikhi, A.; Benzerzour, M.; Abriak, N.-E.; Maherzi, W.; Mamindy-Pajany, Y. Study of the Impact of Waste Glasses Types on Pozzolanic Activity of Cementitious Matrix. *Constr. Build. Mater.* **2019**, *197*, 626–640. [CrossRef]
16. Du, Y.; Yang, W.; Ge, Y.; Wang, S.; Liu, P. Thermal Conductivity of Cement Paste Containing Waste Glass Powder, Metakaolin and Limestone Filler as Supplementary Cementitious Material. *J. Clean. Prod.* **2021**, *287*, 125018. [CrossRef]
17. Yin, W.; Li, X.; Sun, T.; Chen, Y.; Xu, F.; Yan, G.; Xu, M.; Tian, K. Utilization of Waste Glass Powder as Partial Replacement of Cement for the Cementitious Grouts with Superplasticizer and Viscosity Modifying Agent Binary Mixtures: Rheological and Mechanical Performances. *Constr. Build. Mater.* **2021**, *286*, 122953. [CrossRef]
18. Szudek, W.; Gołek, Ł.; Malata, G.; Pytel, Z. Influence of Waste Glass Powder Addition on the Microstructure and Mechanical Properties of Autoclaved Building Materials. *Materials* **2022**, *15*, 434. [CrossRef]
19. Gołek, Ł.; Szudek, W.; Błądek, M.; Cięciwa, M. The Influence of Ground Waste Glass Cullet Addition on the Compressive Strength and Microstructure of Portland Cement Pastes and Mortars | Wpływ Dodatku Mielonej Stłuczki Szklanej Na Wytrzymałość Oraz Mikrostrukturę Zaczynów i Zapraw z Cementu Portlandzki. *Cem. Wapno Bet.* **2020**, *2020*, 480–494. [CrossRef]
20. Szydłowski, J.; Szudek, W.; Gołek, Ł. Effect of Temperature on the Long-Term Properties of Mortars Containing Waste Glass Powder and Ground Granulated Blast Furnace Slag. *Cem. Wapno Bet.* **2021**, *26*, 264–278. [CrossRef]
21. Gołek, Ł. Glass Powder and High-Calcium Fly Ash Based Binders—Long Term Examinations. *J. Clean. Prod.* **2019**, *220*, 493–506. [CrossRef]
22. Gołek, Ł. New Insights into the Use of Glass Cullet in Cement Composites—Long Term Examinations. *Cem. Concr. Compos.* **2022**, *133*, 104673. [CrossRef]
23. de Castro, S.; de Brito, J. Evaluation of the Durability of Concrete Made with Crushed Glass Aggregates. *J. Clean. Prod.* **2013**, *41*, 7–14. [CrossRef]
24. Boukhelf, F.; Cherif, R.; Trabelsi, A.; Belarbi, R.; Bachir Bouiadjra, M. On the Hygrothermal Behavior of Concrete Containing Glass Powder and Silica Fume. *J. Clean. Prod.* **2021**, *318*, 128647. [CrossRef]
25. Chand, G.; Happy, S.K.; Ram, S. Assessment of the Properties of Sustainable Concrete Produced from Quaternary Blend of Portland Cement, Glass Powder, Metakaolin and Silica Fume. *Clean. Eng. Technol.* **2021**, *4*, 100179. [CrossRef]
26. Boesch, M.E.; Vadenbo, C.; Saner, D.; Huter, C.; Hellweg, S. An LCA Model for Waste Incineration Enhanced with New Technologies for Metal Recovery and Application to the Case of Switzerland. *Waste Manag.* **2014**, *34*, 378–389. [CrossRef]
27. Sánchez, A.R.; Ramos, V.C.; Polo, M.S.; Ramón, M.V.L.; Utrilla, J.R. Life Cycle Assessment of Cement Production with Marble Waste Sludges. *Int. J. Environ. Res. Public Health* **2021**, *18*, 10968. [CrossRef] [PubMed]
28. Valderrama, C.; Granados, R.; Cortina, J.L.; Gasol, C.M.; Guillem, M.; Josa, A. Implementation of Best Available Techniques in Cement Manufacturing: A Life-Cycle Assessment Study. *J. Clean. Prod.* **2012**, *25*, 60–67. [CrossRef]

29. Chen, C.; Habert, G.; Bouzidi, Y.; Jullien, A. Environmental Impact of Cement Production: Detail of the Different Processes and Cement Plant Variability Evaluation. *J. Clean. Prod.* **2010**, *18*, 478–485. [CrossRef]
30. Huntzinger, D.N.; Eatmon, T.D. A Life-Cycle Assessment of Portland Cement Manufacturing: Comparing the Traditional Process with Alternative Technologies. *J. Clean. Prod.* **2009**, *17*, 668–675. [CrossRef]
31. Crossin, E. The Greenhouse Gas Implications of Using Ground Granulated Blast Furnace Slag as a Cement Substitute. *J. Clean. Prod.* **2015**, *95*, 101–108. [CrossRef]
32. Guignone, G.; Calmon, J.; Vieira, G.; Zulcão, R.; Rebello, T. Life Cycle Assessment of Waste Glass Powder Incorporation on Concrete: A Bridge Retrofit Study Case. *Appl. Sci.* **2022**, *12*, 3353. [CrossRef]
33. Zulkarnain, I.; Lai, L.S.; Syakir, M.I.; Rahman, A.A.; Yusuff, S.; Hanafiah, M.M. Life Cycle Assessment of Crushed Glass Abrasive Manufacturing from Recycled Glass. *IOP Conf. Ser. Earth Environ. Sci.* **2021**, *880*, 012054. [CrossRef]
34. Qin, C.; Gong, J.; Xie, G. Modeling Hydration Kinetics of the Portland-Cement-Based Cementitious Systems with Mortar Blends by Non-Assumptive Projection Pursuit Regression. *Thermochim. Acta* **2021**, *705*, 179035. [CrossRef]
35. Tahersima, M.; Tikalsky, P. Finite Element Modeling of Hydration Heat in a Concrete Slab-on-Grade Floor with Limestone Blended Cement. *Constr. Build. Mater.* **2017**, *154*, 44–50. [CrossRef]
36. Kondo, R. Kinetics and Mechanism of the Hydration of Cements. In Proceedings of the Fifth International Symposium on the Chemistry of Cement, Tokyo, Japan, 7–11 October 1968; pp. 203–248.
37. Pommersheim, J.M.; Clifton, J.R. Mathematical Modeling of Tricalcium Silicate Hydration. *Cem. Concr. Res.* **1979**, *9*, 765–770. [CrossRef]
38. Pommersheim, J.M.; Clifton, J.R. Mathematical Modeling of Tricalcium Silicate Hydration. II. Hydration Sub-Models and the Effect of Model Parameters. *Cem. Concr. Res.* **1982**, *12*, 765–772. [CrossRef]
39. He, J.; Long, G.; Ma, K.; Xie, Y.; Ma, C. Hydration Heat Evolution of Portland Cement Paste during Unsteady Steam Curing Process: Modelling and Optimization. *Thermochim. Acta* **2020**, *694*, 178784. [CrossRef]
40. Nguyen, V.T.; Lee, S.Y.; Chung, S.-Y.; Moon, J.-H.; Kim, D.J. Effects of Cement Particle Distribution on the Hydration Process of Cement Paste in Three-Dimensional Computer Simulation. *Constr. Build. Mater.* **2021**, *311*, 125322. [CrossRef]
41. Chu, D.C.; Kleib, J.; Amar, M.; Benzerzour, M.; Abriak, N.-E. Recycling of Dredged Sediment as a Raw Material for the Manufacture of Portland Cement—Numerical Modeling of the Hydration of Synthesized Cement Using the CEMHYD3D Code. *J. Build. Eng.* **2022**, *48*, 103871. [CrossRef]
42. Zhao, Y.; Gao, J.; Chen, G.; Li, S.; Luo, X.; Xu, Z.; Guo, Z.; Du, H. Early-Age Hydration Characteristics and Kinetics Model of Blended Cement Containing Waste Clay Brick and Slag. *J. Build. Eng.* **2022**, *51*, 104360. [CrossRef]
43. Krstulović, R.; Dabić, P. A Conceptual Model of the Cement Hydration Process. *Cem. Concr. Res.* **2000**, *30*, 693–698. [CrossRef]
44. Dabič, P.; Krstulovič, R.; Rušič, D. A New Approach in Mathematical Modelling of Cement Hydration Development. *Cem. Concr. Res.* **2000**, *30*, 1017–1021. [CrossRef]
45. Xu, Z.; Shen, J.; Qu, Y.; Chen, H.; Zhou, X.; Hong, H.; Sun, H.; Lin, H.; Deng, W.; Wu, F. Using Simple and Easy Water Quality Parameters to Predict Trihalomethane Occurrence in Tap Water. *Chemosphere* **2022**, *286*, 131586. [CrossRef]
46. Hong, H.; Zhang, Z.; Guo, A.; Shen, L.; Sun, H.; Liang, Y.; Wu, F.; Lin, H. Radial Basis Function Artificial Neural Network (RBF ANN) as Well as the Hybrid Method of RBF ANN and Grey Relational Analysis Able to Well Predict Trihalomethanes Levels in Tap Water. *J. Hydrol.* **2020**, *591*, 125574. [CrossRef]
47. Lin, H.; Dai, Q.; Zheng, L.; Hong, H.; Deng, W.; Wu, F. Radial Basis Function Artificial Neural Network Able to Accurately Predict Disinfection By-Product Levels in Tap Water: Taking Haloacetic Acids as a Case Study. *Chemosphere* **2020**, *248*, 125999. [CrossRef]
48. Chen, Y.; Yu, G.; Long, Y.; Teng, J.; You, X.; Liao, B.-Q.; Lin, H. Application of Radial Basis Function Artificial Neural Network to Quantify Interfacial Energies Related to Membrane Fouling in a Membrane Bioreactor. *Bioresour. Technol.* **2019**, *293*, 122103. [CrossRef]
49. Cook, R.; Han, T.; Childers, A.; Ryckman, C.; Khayat, K.; Ma, H.; Huang, J.; Kumar, A. Machine Learning for High-Fidelity Prediction of Cement Hydration Kinetics in Blended Systems. *Mater. Des.* **2021**, *208*, 109920. [CrossRef]
50. Subasi, A.; Yilmaz, A.S.; Binici, H. Prediction of Early Heat of Hydration of Plain and Blended Cements Using Neuro-Fuzzy Modelling Techniques. *Expert Syst. Appl.* **2009**, *36*, 4940–4950. [CrossRef]
51. Salleh, M.N.M.; Talpur, N.; Kashif, H. Adaptive Neuro-Fuzzy Inference System: Overview, Strengths, Limitations, and Solutions Adaptive Neuro-Fuzzy Inference System: Overview, Strengths, Limitations, and Solutions. In Proceedings of the International Conference on Data Mining and Big Data, Fukuoka, Japan, 27 July–1 August 2017; pp. 527–535.
52. Luo, D.; Wei, J. Hydration Kinetics and Phase Evolution of Portland Cement Composites Containing Sodium-Montmorillonite Functionalized with a Non-Ionic Surfactant. *Constr. Build. Mater.* **2022**, *333*, 127386. [CrossRef]
53. Riding, K.A.; Poole, J.L.; Folliard, K.J.; Juenger, M.C.G.; Schindler, A.K. Modeling Hydration of Cementitious Systems. *ACI Mater. J.* **2012**, *109*, 225–234. [CrossRef]
54. Meinhard, K.; Lackner, R. Multi-Phase Hydration Model for Prediction of Hydration-Heat Release of Blended Cements. *Cem. Concr. Res.* **2008**, *38*, 794–802. [CrossRef]
55. Kolani, B.; Buffo-Lacarrière, L.; Sellier, A.; Escadeillas, G.; Boutillon, L.; Linger, L. Hydration of Slag-Blended Cements. *Cem. Concr. Compos.* **2012**, *34*, 1009–1018. [CrossRef]
56. Wang, X.-Y. Analysis of Hydration Kinetics and Strength Progress in Cement–Slag Binary Composites. *J. Build. Eng.* **2021**, *35*, 101810. [CrossRef]

57. Zeybek, Ö.; Özkılıç, Y.O.; Karalar, M.; Çelik, A.İ.; Qaidi, S.; Ahmad, J.; Burduhos-Nergis, D.D.; Burduhos-Nergis, D.P. Influence of Replacing Cement with Waste Glass on Mechanical Properties of Concrete. *Materials* **2022**, *15*, 7513. [CrossRef] [PubMed]
58. Kalakada, Z.; Doh, J.H.; Zi, G. Utilisation of Coarse Glass Powder as Pozzolanic Cement—A Mix Design Investigation. *Constr. Build. Mater.* **2020**, *240*, 117916. [CrossRef]
59. *EN 197-1*; Ciment—Partie 1: Composition, Spécifications et Critères de Conformité Des Ciments Courants. European Standard: Pilsen, Czech Republic, 2012.
60. *NF EN 933-1*; Essais Pour Déterminer Les Caractéristiques Géométriques Des Granulats—Partie 1: Détermination de La Granularité—Analyse Granulométrique Par Tamisage. European Standard: Pilsen, Czech Republic, 2012.
61. Kheir, J.; Hilloulin, B.; Loukili, A.; De Belie, N. Chemical Shrinkage of Low Water to Cement (w/c) Ratio CEM I and CEM III Cement Pastes Incorporating Silica Fume and Filler. *Materials* **2021**, *14*, 1164. [CrossRef]
62. Idir, R.; Cyr, M.; Tagnit-Hamou, A. Use of Fine Glass as ASR Inhibitor in Glass Aggregate Mortars. *Constr. Build. Mater.* **2010**, *24*, 1309–1312. [CrossRef]
63. Idir, R.; Cyr, M.; Tagnit-Hamou, A. Pozzolanic Properties of Fine and Coarse Color-Mixed Glass Cullet. *Cem. Concr. Compos.* **2011**, *33*, 19–29. [CrossRef]
64. Bordy, A.; Younsi, A.; Aggoun, S.; Fiorio, B. Cement Substitution by a Recycled Cement Paste Fine: Role of the Residual Anhydrous Clinker. *Constr. Build. Mater.* **2017**, *132*, 1–8. [CrossRef]
65. Jensen, O.M. Thermodynamic Limitation of Self-Desiccation. *Cem. Concr. Res.* **1995**, *25*, 157–164. [CrossRef]
66. Persson, B. Self-Desiccation and Its Importance in Concrete Technology. *Mater. Struct.* **1997**, *30*, 293–305. [CrossRef]
67. *ASTM C230/230M*; Standard Specification for Flow Table for Use in Tests of Hydraulic Cement. ASTM: West Conshohocken, PA, USA, 2014.
68. *NF EN196–1*; Methods of Testing Cement—Part 1: Determination of Strength. European Standard: Pilsen, Czech Republic, 2006.
69. AFPC-AFREM. *Durabilité Des Bétons: Méthodes Recommandées Pour La Mesure Des Grandeurs Associées à La Durabilité*; Laboratoire matériaux et durabilité des constructions, Institut national des sciences Appliquées, Université Paul Sabatier: Toulouse, France, 2009.
70. Kalpokaitė-Dičkuvienė, R.; Pitak, I.; Baltušnikas, A.; Lukošiūtė, S.I.; Denafas, G.; Česnienė, J. Cement Substitution by Sludge-Biomass Gasification Residue: Synergy with Silica Fume. *Constr. Build. Mater.* **2022**, *326*, 126902. [CrossRef]
71. Shah, R.; Pandit, R.K.; Gaur, M.K. Thermal Comfort Analysis through Development of Artificial Neural Network Models: An Experimental Study in Cwa Climate. *Mater. Today Proc.* **2022**, *57*, 2018–2025. [CrossRef]
72. al-Swaidani, A.M.; Khwies, W.T.; Al-Bali, M.; Lala, T. Development of Multiple Linear Regression, Artificial Neural Networks and Fuzzy Logic Models to Predict the Efficiency Factor and Durability Indicator of Nano Natural Pozzolana as Cement Additive. *J. Build. Eng.* **2022**, *52*, 104475. [CrossRef]
73. Di Benedetto, R.M.; Botelho, E.C.; Janotti, A.; Ancelotti Junior, A.C.; Gomes, G.F. Development of an Artificial Neural Network for Predicting Energy Absorption Capability of Thermoplastic Commingled Composites. *Compos. Struct.* **2021**, *257*, 113131. [CrossRef]
74. Gupta, P.; Sinha, N.K. Neural Networks for Identification of Nonlinear Systems: An Overview. In *Soft Computing and Intelligent Systems*; Elsevier: Amsterdam, The Netherlands, 2000; pp. 337–356.
75. Abirami, S.; Chitra, P. Energy-Efficient Edge Based Real-Time Healthcare Support System. *Adv. Comput.* **2020**, *117*, 339–368.
76. Menzies, T.; Kocagüneli, E.; Minku, L.; Peters, F.; Turhan, B. Using Goals in Model-Based Reasoning. In *Sharing Data and Models in Software Engineering*; Elsevier: Amsterdam, The Netherlands, 2015; pp. 321–353.
77. Davies, E.R. Biologically Inspired Recognition Schemes. In *Machine Vision*; Elsevier: Amsterdam, The Netherlands, 2005; pp. 725–755.
78. Yang, Z.R.; Yang, Z. Artificial Neural Networks. In *Comprehensive Biomedical Physics*; Elsevier: Amsterdam, The Netherlands, 2014; pp. 1–17.
79. Hallinan, J.S. Computational Intelligence in the Design of Synthetic Microbial Genetic Systems. *Methods Microbiol.* **2013**, *40*, 1–37.
80. Rajaoarisoa, L. Large-Scale Building Thermal Modeling Based on Artificial Neural Networks: Application to Smart Energy Management. In *Artificial Intelligence Techniques for a Scalable Energy Transition*; Springer International Publishing: Cham, Switzerland, 2020; pp. 15–44.
81. Erway, J.B.; Marcia, R.F. On Solving Large-Scale Limited-Memory Quasi-Newton Equations. *Linear Algebra Appl.* **2017**, *515*, 196–225. [CrossRef]
82. Borhani, M. Multi-Label Log-Loss Function Using L-BFGS for Document Categorization. *Eng. Appl. Artif. Intell.* **2020**, *91*, 103623. [CrossRef]
83. Ali, M.M.M. Modified Limited-Memory Broyden-Fletcher-Goldfarb-Shanno Algorithm for Unconstrained Optimization Problem. *Indones. J. Electr. Eng. Comput. Sci.* **2021**, *24*, 1027. [CrossRef]
84. Nguyen, K.T.; Nguyen, Q.D.; Le, T.A.; Shin, J.; Lee, K. Analyzing the Compressive Strength of Green Fly Ash Based Geopolymer Concrete Using Experiment and Machine Learning Approaches. *Constr. Build. Mater.* **2020**, *247*, 118581. [CrossRef]
85. Naser, M.Z.; Thai, S.; Thai, H.-T. Evaluating Structural Response of Concrete-Filled Steel Tubular Columns through Machine Learning. *J. Build. Eng.* **2021**, *34*, 101888. [CrossRef]
86. Dung, C.V.; Anh, L.D. Autonomous Concrete Crack Detection Using Deep Fully Convolutional Neural Network. *Autom. Constr.* **2019**, *99*, 52–58. [CrossRef]
87. Cherif, R.; Hamami, A.E.A.; Aït-Mokhtar, A. Global Quantitative Monitoring of the Ion Exchange Balance in a Chloride Migration Test on Cementitious Materials with Mineral Additions. *Cem. Concr. Res.* **2020**, *138*, 106240. [CrossRef]
88. Adesina, A.; Das, S. Influence of Glass Powder on the Durability Properties of Engineered Cementitious Composites. *Constr. Build. Mater.* **2020**, *242*, 118199. [CrossRef]

89. Vance, K.; Aguayo, M.; Oey, T.; Sant, G.; Neithalath, N. Hydration and Strength Development in Ternary Portland Cement Blends Containing Limestone and Fly Ash or Metakaolin. *Cem. Concr. Compos.* **2013**, *39*, 93–103. [CrossRef]
90. Stepkowska, E.T.; Blanes, J.M.; Franco, F.; Real, C.; Pérez-Rodrıíguez, J.L. Phase Transformation on Heating of an Aged Cement Paste. *Thermochim. Acta* **2004**, *420*, 79–87. [CrossRef]
91. Damidot, D.; Lothenbach, B.; Herfort, D.; Glasser, F.P. Thermodynamics and Cement Science. *Cem. Concr. Res.* **2011**, *41*, 679–695. [CrossRef]
92. Ni, C.; Wu, Q.; Yu, Z.; Shen, X. Hydration of Portland Cement Paste Mixed with Densified Silica Fume: From the Point of View of Fineness. *Constr. Build. Mater.* **2021**, *272*, 121906. [CrossRef]
93. Lavergne, F.; Ben Fraj, A.; Bayane, I.; Barthélémy, J.F. Estimating the Mechanical Properties of Hydrating Blended Cementitious Materials: An Investigation Based on Micromechanics. *Cem. Concr. Res.* **2018**, *104*, 37–60. [CrossRef]
94. Scrivener, K.L.; Juilland, P.; Monteiro, P.J.M. Advances in Understanding Hydration of Portland Cement. *Cem. Concr. Res.* **2015**, *78*, 38–56. [CrossRef]
95. Bentz, D.P. Three-Dimensional Computer Simulation of Portland Cement Hydration and Microstructure Development. *J. Am. Ceram. Soc.* **1997**, *80*, 3–21. [CrossRef]
96. Kishi, T.; Maekawa, K. Multi-Component Model for Hydration Heat of Portland Cement. *Doboku Gakkai Ronbunshu* **1995**, *1995*, 97–109. [CrossRef] [PubMed]
97. Bullard, J.W.; Jennings, H.M.; Livingston, R.A.; Nonat, A.; Scherer, G.W.; Schweitzer, J.S.; Scrivener, K.L.; Thomas, J.J. Mechanisms of Cement Hydration. *Cem. Concr. Res.* **2011**, *41*, 1208–1223. [CrossRef]

Disclaimer/Publisher's Note: The statements, opinions and data contained in all publications are solely those of the individual author(s) and contributor(s) and not of MDPI and/or the editor(s). MDPI and/or the editor(s) disclaim responsibility for any injury to people or property resulting from any ideas, methods, instructions or products referred to in the content.

Article

Influence of Waste Basalt Powder Addition on the Microstructure and Mechanical Properties of Autoclave Brick

Paulina Kostrzewa-Demczuk [1,*], Anna Stepien [1], Ryszard Dachowski [1] and Rogério Barbosa da Silva [2]

1. Department of Building Engineering Technologies and Organization, Faculty of Civil Engineering and Architecture, Kielce University of Technology, al. Tysiąclecia PP 7, 25-314 Kielce, Poland
2. St. Department Heitor Alencar Furtado, Federal Technological University of Paraná, 5000-Curitiba-Paraná, Curitiba 81280-340, Brazil
* Correspondence: pkostrzewa@tu.kielce.pl

Abstract: In the production of building materials, there has been an increased interest in the use of by-products and industrial waste in recent years. Such modifications make it possible to solve not only technical and economic problems, but also environmental problems. This article describes the use of basalt powder waste in sand-lime products (silicates). The aim of the study was to manage basalt powder waste and to investigate the changes it causes in sand-lime products. The article describes the planning of the experiment, which directly determines the number of samples and their composition, which was necessary to conducting a full analysis and correctly illustrating the relationships occurring in the samples. Basic tests were carried out: compressive strength, density and water absorption, as well as optical tests and scanning microscopy. Based on the research conducted, it was concluded that the use of basalt powder as a component of sand-lime products has positive effects. Studies show that the best results are achieved with a proportion of powder in the raw material mass of about 10%—the compressive strength reaches almost 30 MPa, which is almost twice that of traditional silicate.

Keywords: basalt powder; sand-lime products; silicate bricks; autoclaving; post-production waste

1. Introduction

Construction is a sector of the economy that causes a heavy burden on the environment, alongside other industries. Proper management of waste building materials is not only a serious logistical problem, but also a pressing social problem. This involves both restrictions on waste storage sites and stringent environmental standards related to the quantity and quality of waste generated. One of the ways of managing waste and by-products is their recycling and use for the production of building materials, which is consistent with the principle of sustainable development [1]. In the production of building materials, there has been an increased interest in the use of by-products and industrial waste in recent years. Such modifications make it possible to solve not only technical and economic problems, but also environmental problems. Mining waste [2] and industrial waste have been used in the modification of sand-lime products. The main purpose of these modifications is not only to reduce the cost of manufacturing products, but above all to care for the natural environment through waste management and energy saving at the production stage. Thanks to these modifications, products are created that are substitutes for previously used materials with similar physical and mechanical properties.

Autoclaved sand-lime bricks have been produced for decades. Traditionally, the ingredients of calcium silicate products are mainly sand, lime and water. The quality of the ingredients used, as well as the conditions of their production, translate into the final properties of autoclaved products. The theoretical composition of the raw material mixtures for autoclaved calcium silicate products is 92% quartz sand and 8% quicklime, which results in a molar ratio of C/S = 0.09 (C/S is CaO/SiO_2 ratio). The proportions presented apply to mixtures that consist of quartz sand with 100% SiO_2 content and lime with 100% reactivity. Research conducted in industrial silicate production plants shows that the conditions presented above almost never occur in reality. This is due to the use of sand of natural origin for the production of silicates (usually locally sourced, containing various impurities) and quicklime obtained in industrial plants, where it does not reach 100% reactivity. Accordingly, the actual composition of the autoclaved calcium silicate products is different from the theoretical composition mentioned above. When designing a mixture, it is important to select the mutual proportions of the components, taking into account their reactivity, so that the molar ratio of 0.09 is maintained.

Lime is the binding substance for calcium silicate products. Literature [3] indicate that with the increase in the content of lime in silicates, their compressive strength increases. From an economic point of view, lime is the most expensive component of the mix and in order to optimize the production and price of silicate bricks, the content of 10% of this component in the mix is not exceeded. The suitability of lime for the production of silicates is specified in the PN-EN 459-1 standard [4]. The PN-EN 459-1 standard defines unslaked ground quicklime with the symbol CL (90, 80, 70) as useful (with the content of CaO + MgO in the range of 70–90%, MgO below 5%, CO_2 within 4–12% and SO_3 below 2%). Lime should be slaked at a temperature of at least 60 °C for 10–30 min. Another important aspect related to lime is its granulation, expressed by the size of residues on sieves with a square mesh size of 0.09 mm (proportion not greater than 7%) and 0.2 mm (proportion not greater than 2%) [5].

Sand is the main component of autoclaved bricks, which is related to its relatively low cost and the chemical reactions that take place in the sand-lime mixture under appropriate conditions. The result of these reactions is the production of artificial rock, a very durable material that can be freely formed in the autoclaving process. Quartz sand affects the quality of the final autoclaved product. In addition to the SiO_2 content, the shape, size and mineral content of the sand can also affect the quality of silicate bricks. The use of sands with a silica content of less than 80% is avoided. Quartz sands used for the production of silicate autoclaved products should not contain alkalis (Na_2O and K_2O) above 0.5%, Al_2O_3 content above 5%, Fe_2O_3 above 1.5% and MgO above 3% [5]. Silicate bricks are produced from semi-dry sands, which makes it necessary to ensure the appropriate granulation of the masses used for production. This ensures the appropriate degree of mass compaction and the subsequent strength of silicate products. Quartz sand should be characterized by a continuous grain size curve, without the dominance of any fraction. Proper graining of quartz sands used in the technology of sand-lime products is considered sand containing the following grain fractions: 2.5–0.5 mm in a proportion not exceeding 30%, and 0.5–0.05 mm in a proportion of at least 65%. Generally speaking, the sands used for the production of silicates belong to the group of fine-grained sands.

Natural sand is one of the main components of building wall materials (bricks, mortars and concrete mixes). In many countries, there is currently a significant shortage of natural sand of appropriate quality that can be used for the production of concrete [6]. In 2022, the prices of natural resources in Poland increased significantly, even by 30% [7]. This is mainly due to the crisis related to the COVID-19 pandemic and the war in Ukraine. The prices of energy, raw materials and fuels increased, and thus also the cost of transport and production. A similar situation is observed around the world. Therefore, other materials are being sought that may be a partial replacement for sand. The use of various types of industrial waste in the production of concrete not only reduces the consumption of natural resources, but also allows for the management of waste, which in many cases is burdensome for the environment. The literature is full of examples of the use of waste dust of marble [8,9], quartz [10], basalt [11–14], granite [15] or lime [9,10] as a substitute for sand in concrete. Waste basalt powder has been used in laboratory scale tests as a partial replacement of sand in the proportion of 0–30% of the sand mass in the case of mortars and in the proportion of 0–20% of the sand mass in the case of concretes [1]. The results of the tests conducted indicate that basalt powder can be used in the production of cement mortars and concretes as a substitute for natural sand [1]. In the area of autoclaved bricks, basalt components were added in the form of aggregate [16,17], basalt fiber [17–20] and basalt powder [21,22]. However, these studies focused primarily on autoclaved aerated concrete (AAC) bricks and not on silicates. Scientific cooperation with industrial production plants producing autoclaved bricks—silicates shows that these plants are looking for an additive or filler for bricks that will be ecological, will be characterized by a high content of silica in the composition and a low price (with the potential to reduce the price of ready-made bricks). Basalt dust is a component that meets the above requirements and has a large implementation potential.

In Europe, silicate products have been a valued and widely used building material for many years. More than 50% of European residential buildings are made of silicates. In Poland, for many years, silicates were an extremely underestimated and rarely used building material. Despite over 100 years of tradition and a dominant position in countries such as Germany, and especially the Netherlands, where this material "reigns" on the building materials market, sand-lime products have been rediscovered in our country only in recent years. Today, these products are becoming more and more popular among investors and designers.

The technical literature on the modification of sand-lime materials (silicates) states that the change in the composition and structure of the mixture is usually aimed at increasing the compressive strength of the finished product. The properties of sand-lime products depend on several important technological factors, mainly the quantity and quality of lime used in the silicate mix, the fineness of the mix components, the chemical composition of individual components, the amount of pressing pressure of the mix during the formation of semi-finished products, as well as the temperature, pressure and time of autoclaving of formed elements. By introducing additives (fillers) to the sand-lime mixture, new ions are often introduced, the quantity and quality of which have a significant impact on the formation of both the crystalline and amorphous C-S-H phase, as well as new phases not present in traditional products [23].

One of the effective ways of modifying sand-lime products in terms of improving their strength parameters is the use of ground mineral additives with a high specific surface area, pozzolanic and/or hydraulic properties, in the raw material mixture. An example of such a solution is the use of ground chalcedonite [24,25], ground limestone [26], dolomite powder [27], metakaolinite [28], barium [29], bentonite [30], thermally activated Carboniferous shale [31] and silica fume [32]. In recent years, the impact of waste glass in silicates has been extensively studied [33–37].

Known from the literature and patent descriptions are modifications of the composition of the sand-lime mixture, in which the mass substitute is mineral aggregates, i.e., barite [38–40], basalt [38], graphite [41] and igneous rocks of natural origin—preferably diabase [42].

Preliminary studies show an improvement in the physico-mechanical properties of sand-lime products with the addition of basalt powder [21]. This work is devoted to the study of the impact of waste material in the form of basalt powder on the parameters of silicate brick and the possibility of its management in the production of sand-lime products. The mixture of sand and lime was treated as one factor, because in the industrial conditions in which the tests were conducted, such a mixture was provided by the production plant. The mixture of sand and lime had constant proportions of both components. Basalt powder was a fine filler, the use of which in a very large amount would result in an increase in the water demand of the mixture, changes in the consistency of the mixture, problems during molding and, as indicated by preliminary tests [21], a general deterioration of the product properties.

2. Materials and Methods

2.1. Components of Raw Material Mass

As already mentioned in the introduction, the main components of autoclaved bricks (silicates) are sand, lime and water. In this work, in addition to the basic ingredients of the silicate raw material mass, basalt powder was used. This section describes the parameters and origin of the raw materials used in the research.

The lime used in the technological process comes from Trzuskawica S.A. Plant located in Sitkówka, Poland (the chemical composition of lime is given in Table 1) and it was a lime denoted as CL 90-Q, R5, P1, the main component of which is CaO.

Table 1. Chemical composition of basalt powder, sand and quicklime.

Oxides Composition	Basalt Powder [%]	Sand [%]	Quicklime [%]
SiO_2	44.32	99.31	<QL
CaO	9.99	<QL	94.72
Al_2O_3	12.87	0.66	<QL
Fe_2O_3	11.51	0.27	<QL
MgO	10.01	<QL	0.97
Na_2O	2.85	<QL	<QL
K_2O	0.91	<QL	<QL
SO_3	0.05	<QL	0.18
CO_2	<QL	<QL	1.47
P_2O_5	0.45	<QL	<QL

QL—Quantifiable limit.

The sand used for the tests came from Industrial Silicate Production Plant (Ludynia, Poland): H + H Silikaty Sp. z o. o. The plant extracts sand locally. The chemical composition of sand used in this study is given in Table 1.

According to PN-EN 1008:2004 [43], potable water, which is not subject to additional quality tests, is readily suitable for the production of silicate products. Such water was used.

The mineral powder used in the tests was a waste generated in the process of producing a mix of hard aggregate in the mixing plant, which was used for the production of MMA mineral-asphalt masses. This powder was obtained in the process of drying aggregates at a temperature of approx. 200 °C. It was captured in the fabric filter of the coating machine and then collected in a special tank. For the production of mineral-asphalt masses, hard mineral aggregate is used, which comes mainly from mines, which means that the resulting powder—treated as waste—has properties similar to the rocks from which it was formed. The mineral powder used in the research came from the crushing of basalt rocks.

Table 1 show chemical composition of basalt powder in comparison with sand and quicklime.

2.2. Production Technology

The production of sand-lime products can be divided into several stages, specifying the delivery and storage of raw materials, preparation of the lime-sand mixture, and the forming of products and autoclaving, which is the most important process in production. The mixing of sand and lime can be carried out before or during lime slaking. The lime-sand mixture, after the hydration process and possible enrichment with additives and/or admixtures, is fed into the molding of products in the press, usually at a pressure of 20 MPa. The task of autoclaving is to ensure optimal thermal and humidity conditions, in which a chemical reaction between the binder (lime) and the aggregate (sand) takes place. The product of this process is the formation of low-basic hydrated calcium silicates, which determine the final quality of the material. The prefabricated elements introduced into the autoclave are initially heated with saturated steam until the required hydrothermal conditions are reached. The actual autoclaving process is carried out at a temperature above 180 °C to 203 °C, using steam supplied from the boiler with a pressure of 0.8 to 1.6 MPa, respectively. The hydrothermal treatment time is longer the lower the working pressure is used, and ranges from 8 h for autoclaving at 1.6 MPa to about 12 h at 0.8 MPa steam pressure.

2.3. Planning of Experiment

There are many methods of planning experiences and sometimes it is difficult to choose the right one. This is due to the presence of many factors and variables affecting the production process and the selection of statistically significant ones for the analysis. In the case of silicates, the technological factors affecting the results of the research conducted include autoclaving time of individual stages, pressing conditions, quality of ingredients, autoclaving conditions and the share of individual ingredients in the raw material mass.

After analyzing the different experimental design methods available in the STATISTICA software (version 10.0, StatSoft Polska Sp. z o.o., Cracow, Poland) (bivalent fractional designs, trivalent fractional designs, central compositional designs, mixture designs, etc.), designs for limited mixture areas were selected. This is a special group of plans that allow you to analyze mixtures of compositions that add up to a constant value and take into account the quantitative limitations of the individual factors. A commonly used way of presenting the shares of mixtures, most often composed of no more than three components, are graphs in triangular coordinates. For example, consider a mixture of three components A, B, and C. Any mixture of three components can be uniquely specified by specifying a point in a triangular coordinate system defined by three variables.

The sum for each mixture is 1, so the values of individual components can be interpreted as contributions of the component (Figure 1). If the above data is represented as points on a three-dimensional graph, it turns out that these points formed a triangle in three-dimensional space. Valid mixtures are only the points inside the triangle where the sum of the component values is equal to 1. Therefore, you can limit yourself to drawing a triangle in order to clearly determine the values (shares) of individual components for each mixture.

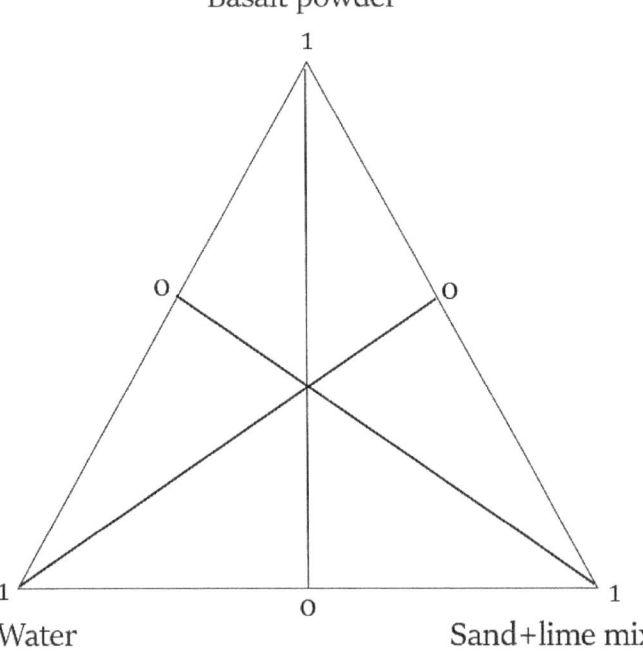

Figure 1. Model design of a mixture experiment with three independent variables.

The basic plans for mixtures assume that it is possible to design a mixture in which 100% of the share will be one of the ingredients and the others will have 0% shares. In the case of tests on sand-lime products that were prepared, this condition could not be met. Individual components have minimum and/or maximum share limits in the raw material mass, which enables appropriate reactions between the components during their mixing and autoclaving. For the selected plan for mixtures with restrictions, identical technological and material parameters were guaranteed during the tests, and only the composition of the silicate mixture was variable. As a result, 3 factors were obtained as input quantities—individual components of the mass: basalt dust (x_1), water (x_2), and a mixture of sand and lime (x_3). Limitations for individual components are presented in Table 2.

Table 2. Restrictions on the content of individual components in the raw material mass.

Component	Limitation
Basalt powder	$x_1 - 10 \geq 0$
(x_1)	$-x_1 + 30 \geq 0$
Water	$x_2 - 5 \geq 0$
(x_2)	$-x_2 + 15 \geq 0$
Sand-lime mix	$x_3 - 55 \geq 0$
(x_3)	$-x_3 + 95 \geq 0$

Basalt powder was used in the proportion of 10–30% and water 5–15%. The use of a minimum 5% mass share of water in the composition of the mixture is necessary due to the chemical reactions that take place in the products at the production stage. On the other hand, the use of a proportion of water greater than 15%, although it is possible at the stage of mixing the raw masses, is unjustified, because during the pressing of the bricks, excess water flows out of the molds. Due to the limitations of basalt powder and water, the amounts of sand and lime was properly calculated to make the sum of the components equal to 100%.

The imposition of multiple constraints in a linear form for all independent variables (x_1, x_2, x_3) in order to determine their appropriate combinations is shown in Figure 2.

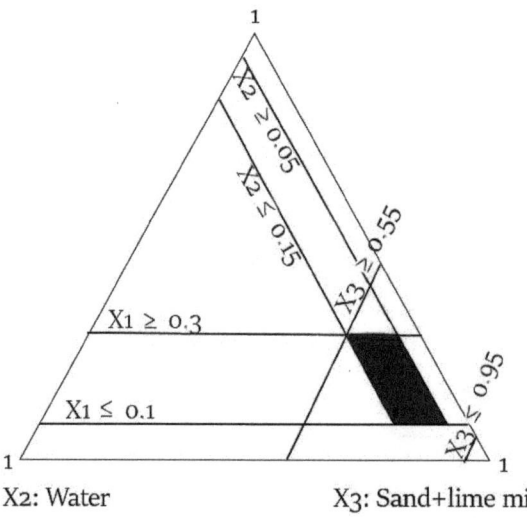

Figure 2. Model of a mixture experiment design with constraints for three independent variables.

The dark area marked in Figure 2 determines the possible values of individual factors and their combinations under the imposition of the given restrictions of these factors. Piepel's [44] and Snee's [45] algorithms were used to determine the vertices and centers of gravity of the selected area. The results of the calculations are presented in Figure 3. Table 3 shows that 9 experimental systems were generated.

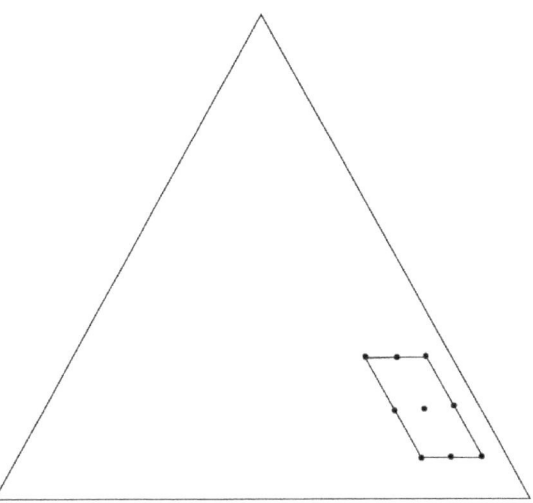

Figure 3. Experiment plan of mixtures with constraints—vertices and centroids generated for the tested range.

Table 3. Composition of individual samples.

No.	Sample	Ingredients		
		Sand-lime Mix [%]	Water [%]	Basalt Powder [%]
1	BP10-W5	85	5	10
2	BP10-W10	80	10	10
3	BP10-W15	75	15	10
4	BP20-W5	75	5	20
5	BP20-W10	70	10	20
6	BP20-W15	65	15	20
7	BP30-W5	65	5	30
8	BP30-W10	60	10	30
9	BP30-W15	55	15	30

The "BP" marking symbolizes a percentage by weight of basalt powder in samples and "W"—the percentage of water in the raw material mass. For example, the name "BP10-W5" means a sample made of traditional raw material with the addition 10% of basalt powder and 5% of water.

2.4. Tests

The research was carried out in the laboratories of the Kielce University of Technology, in the H + H Silikaty Sp. z o.o. and in cooperation with the Polish Ceramic Society. The analysis of the research results was carried out using the STATISTICA software and Design Expert software (version 11, Stat-Ease, Inc., Minneapolis, MN, USA).

Experimental studies of individual compositions of sand-lime products were carried out on samples with dimensions of 40 × 40 × 160 mm. Before testing, the samples were stored for a minimum of 14 days in the temperature and air humidity conditions of ≥15 °C and ≤65%, respectively. Six samples of each type of product with the same composition were tested.

Compressive strength is the greatest stress that a sample of the test material can withstand when compressed. The numerical value of the compressive strength is the quotient of the compressive force that caused the destruction of the material structure and the surface on which the compressive force acts. Samples with a face surface of 40 × 160 mm were placed on the plate of the testing machine coaxially with the center of the pressure plate with dimensions of 62.5 × 40.0 mm, and then compressed uniformly until the sample was destroyed. The arithmetic mean of six measurement results on samples of identical composition was taken as the determination result. The compressive strength was determined on a compressive strength testing machine, in accordance with the guidelines of PN-EN 772-1: 11 + A1: 2015-10 [46]. Using a Tecnotest KC 300 compressive strength testing machine (Tecnotest, Treviolo, Italy, Figure 4a), the samples were compressed in a uniform manner until they were destroyed.

(a) (b)

Figure 4. Equipment used during the tests: (**a**) compressive strength testing machine; (**b**) scanning microscope.

The density of the samples was tested on the basis of the PN-EN 772-13: 2001 standard [47]. Six samples of each product type were tested. Based on these measurements, the arithmetic mean of the density of each type of product was calculated. The density test was carried out using an ULTRAPYC 1200e helium pycnometer. The principle of operation of the apparatus is based on the use of gas to precisely determine the volume of the sample. The volume of the solid is that part of the previously calibrated measuring chamber that has not been occupied by gas. The mass of the sample is determined by weighing on an analytical balance and entered into the program operating the device.

The water absorption test was carried out in accordance with the guidelines set out in PN EN 772-11: 2011 [48] and PN-EN 772-21: 2011 [49]. Six samples of each type of product were dried to constant weight in a circulating air oven at 105 °C ± 5 °C, and then, after cooling, placed in a tank with water and left for 24 h. After this time, the samples were removed from the water and the tests were carried out.

Observations of the structure with optical microscopy were performed with the use of the Motic SMZ-168TP optical stereoscopic microscope (MOTIC, Xiamen, China). The maximum magnification of the microscope was 50×. Using the Moticam-3 camera compatible with Motic SMZ-168TP microscope digital photos of the samples were taken. Optical studies of the structure using an optical microscope were carried out for all samples. The material for the tests was obtained from the elements that remained after the destruction of the samples in the compressive strength tests. In order for the observations to be as accurate as possible, elements with the flattest surface were selected. The insides of the samples were analyzed, not the sides that were in contact with the press during brick formation.

The analysis of the microstructure of the obtained products was possible thanks to the use of the FEI Quanta FEG 250 scanning electron microscope (FEI, Brno, Czech Republic, Figure 4b), equipped with a detector measuring the energy of X-rays, EDS for short (Energy Dispersive Spectrometry, FEI, Hillsboro, OR, USA) or EDX (Energy Dispersive X-ray analysis, FEI, Hillsboro, OR, USA). Measurements were carried out in low vacuum conditions (water vapor pressure equaled 30 Pa) on non-sputtered samples. Magnifications from 150 to 20,000× were used. The tests were performed on fractures (shards) of the samples.

The phase composition of the obtained samples was determined using a PANalytical X-ray diffractometer, model Empyrean (Panalytical, Almelo, Netherlands). The share of individual phases was determined by the Rietveld method. Measurements were made using monochromatic radiation with a wavelength corresponding to the emission line $K_{\alpha 1}$ of copper, in the angular range of 5–90° on a 2θ. The results were based on the ICDD database (The International Center for Diffraction Data). The analysis was carried out using the Deby—Scherrer—Hull (DSH) powder method on samples taken from samples whose compressive strength was closest to the average compressive strength value.

3. Results and Discussion

3.1. Compressive Strength

Compressive strength is a particularly important property of sand-lime products, due to the fact that they are mainly used in the form of bricks for the construction of wall building elements. During strength tests, the samples were destroyed, but partially retained their structure. Visible during the tests was the characteristic hourglass shape appearing on the samples during the progress of loading (Figure 5). Although silicate is considered a brittle material, the samples after the end of the destructive tests still formed one element.

Figure 5. Cross-section of the sample as a result of compression.

Spatial and contour plots (Figures 6 and 7) were used to analyze the results of compressive strength tests. The models considered are presented in the form of graphs in a triangular coordinate system. Figure 6 shows the spatial distribution of compressive strength depending on the number of individual components in the raw material mixture.

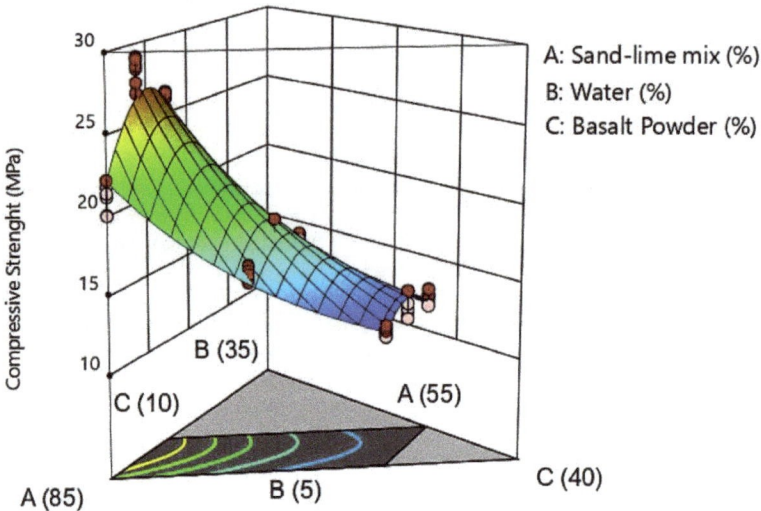

Figure 6. Compressive strength test sample.

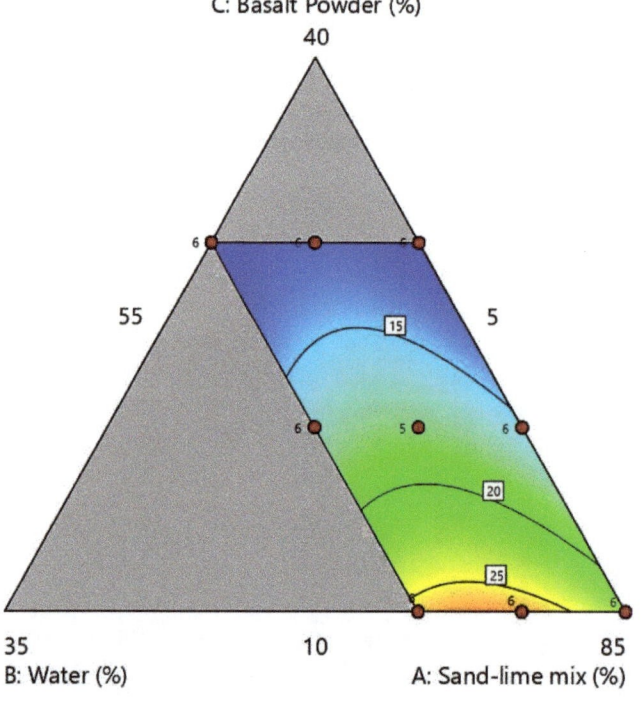

Figure 7. Ternary graph of compressive strength.

On the triangular contour plot (Figure 7), the vertices and centers of gravity calculated in the experiment plan, which correspond to the compositions of the research recipes, were marked. With the increase in the share of basalt powder in the raw material mixture, the compressive strength decreased. The best results were obtained for the addition of basalt powder in the proportion of about 10% by weight. The BP10-W10 sample obtained a strength of slightly over 28 MPa in the tests. The amount of water in the mixture is also important. The graphs show that increasing the share of water in the silicate mass has a positive effect on the tested parameter.

Sand consists mainly of the sand fraction (about 95% are grains with a size of 100–1000 μm, and half of the basalt meal consists of powder and sand fractions. The addition of basalt meal increases the amount of powder and clay fractions in the raw material mass. Samples with basalt powder, thanks to the addition of fine grains, obtain a compact structure, which increases the density of the samples, and thus increases the compressive strength.

Although too much basalt powder causes a decrease in compressive strength. Nevertheless, all analyzed samples reached a strength higher than 11 MPa. The standards allow the use of bricks whose compressive strength is greater than 5 MPa, and standard silicates have strengths between 10 MPa and 15 MPa.

3.2. Water Absorption

The test results of water absorption are shown in Figures 8 and 9. Standard sand-lime products achieve water adsorption of 16%. The results obtained in the tests are similar to standard silicates. Water adsorption of 17.5% was noted in the BP30-W5 sample.

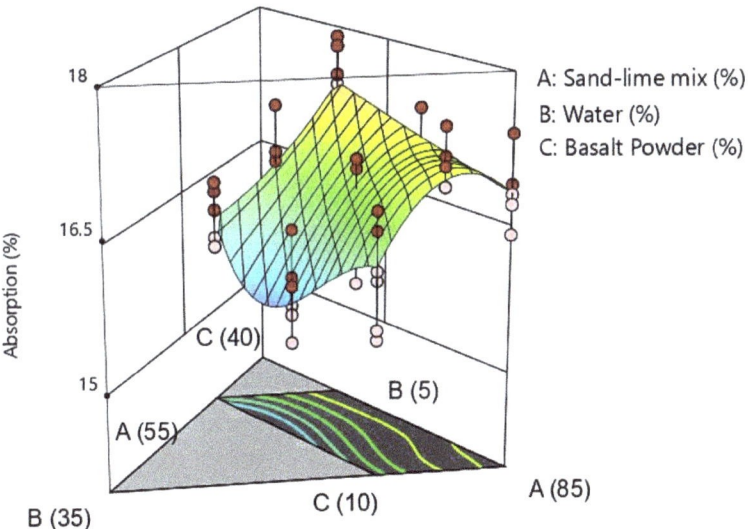

Figure 8. Water absorption of samples.

3.3. Density

The results of the density tests are presented in the Figures 10 and 11. The results of the density tests show a relationship between the share of basalt powder and the increase in density. As previously mentioned, basalt powder, due to its granulometry, fills free spaces in the mass and the products have a higher density. There was no significant difference in the products related to the water content of 10% and 15%. This may be due to the loss of some water during the sample compression stage. It was also a signal that a further increase in the water content above 15% in the samples is not possible with the production technology used in this case.

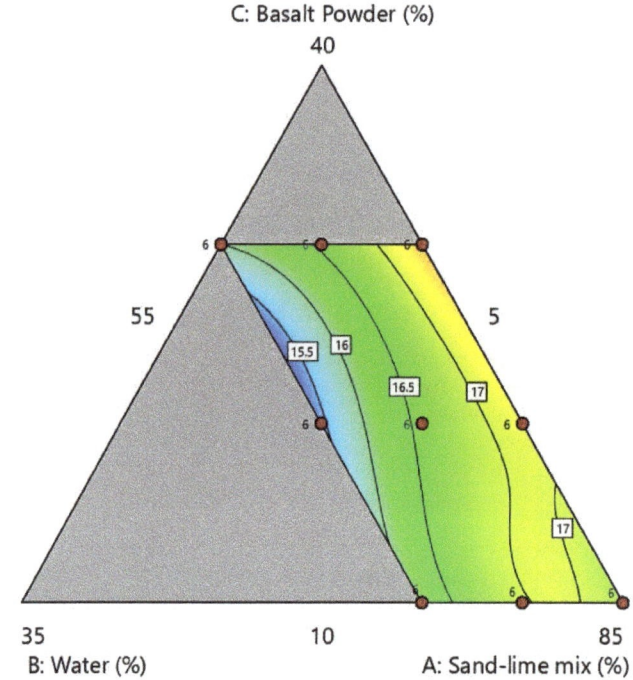

Figure 9. Ternary graph of water absorption.

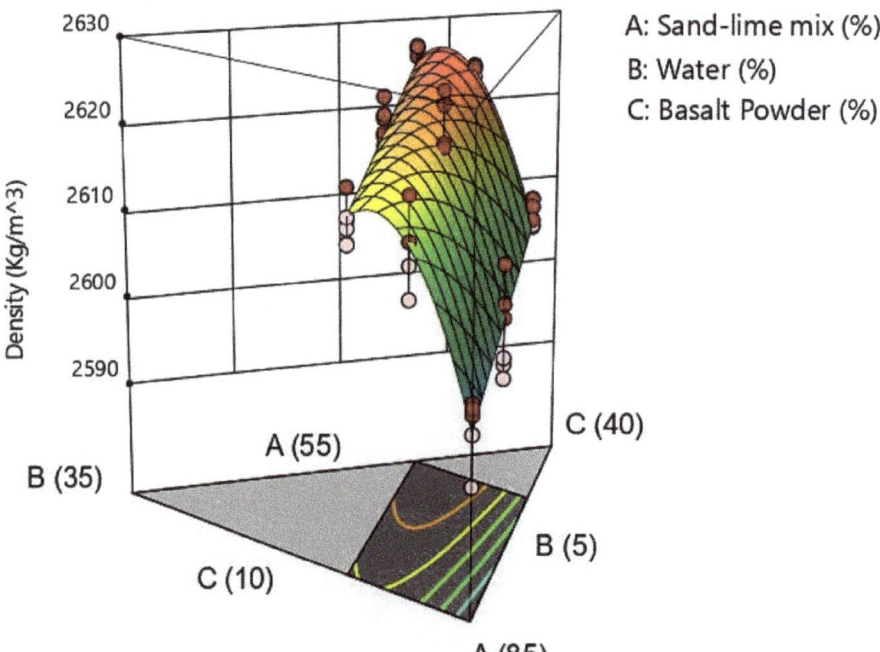

Figure 10. Density of samples.

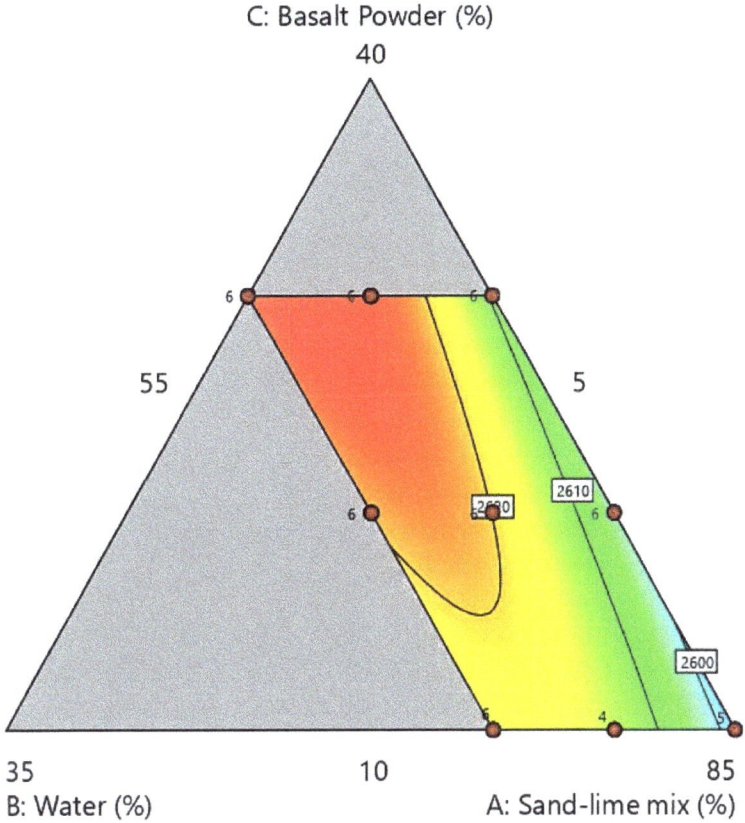

Figure 11. Ternary graph of density.

The increase in the density of the material in the case of silicates is a positive phenomenon, because it is usually associated with better thermal and acoustic insulation. Higher density is associated with an increase in the weight of the bricks. It cannot be unequivocally stated that this is a product defect, because it depends on the intended use of the product. Wall building materials with high density are used in residential construction due to increased acoustic insulation and in special construction, where high insulation of the wall is required, which will protect against, for example, X-rays diffraction (hospitals and laboratories).

3.4. Optical Microscopy

Figure 12 shows a photo of the BP10-W10 sample structure. There were no significant differences between individual samples. Higher magnification is necessary to better analyze the structure of the samples.

In Figure 12, grains of sand of various sizes are visible. The grains are surrounded to varying degrees by lime and because of this they form an inseparable and hard element. Lime is visible in the photo in the form of light (white) structures. In Figure 12, there is a fragment of the sample, which shows various cases of enveloping of sand grains by lime. In the upper right part, the grains are covered to a lesser extent than in the rest of the tested sample. This is the result of insufficient homogenization of the mixture or an optical illusion caused by the camera and the darkening of this part of the photo.

Figure 12. Optical microscope photo of the BP10-W10 sample.

3.5. SEM Scanning Microscopy

During autoclaving, basic chemical reactions occur, which determine the phase composition and microstructure of the autoclaved calcium silicate materials [50]. OH ions, after passing into the liquid phase, react with silicate ions, derived from the dissolution of SiO_2 [23], resulting in the formation of hydrated calcium silicate, called C-S-H (calcium silicate hydrate) with different ratios of CaO, SiO_2 and H_2O. They are characterized by a different degree of structure order—from amorphous (the so-called "C-S-H phase") to crystalline (tobermorite, xonotlite) [51,52].

There are many types of hydrated calcium silicates caused by the properties of the calcium atom, which with oxygen can form an octahedral coordination system with varying degrees of disorder. Therefore, these compounds are still the subject of numerous scientific studies. It is believed that hydrated calcium silicates with an ordered structure (crystalline phases) are formed from the amorphous C-S-H phase in hydrothermal conditions (at a temperature above 100 °C and a pressure above 0.1 MPa) [23,50–52]. The sequence of formation of individual phases is not accidental and is affected by the reactivity of the mixture components [51–53]. Hydrated calcium silicates are formed primarily on the surface of sand grains. Their nuclei are believed to form on the surface of the sand and then grow around the grains, filling the space occupied by water and dissolved anhydrous material as they grow. As a result, the free spaces between the grains are filled and the grains are bonded together to form a monolithic material with high strength [54]. It is estimated that with the increase in the number of synthesis products, especially the C-S-H phase and 1.1 nm tobermorite, the strength of the finished product increases [23,51,52], and the highest strengths are obtained when the reaction involves products with a molar ratio of C/S = 0.9–1.0 [52]. Figures 13–15 show SEM images.

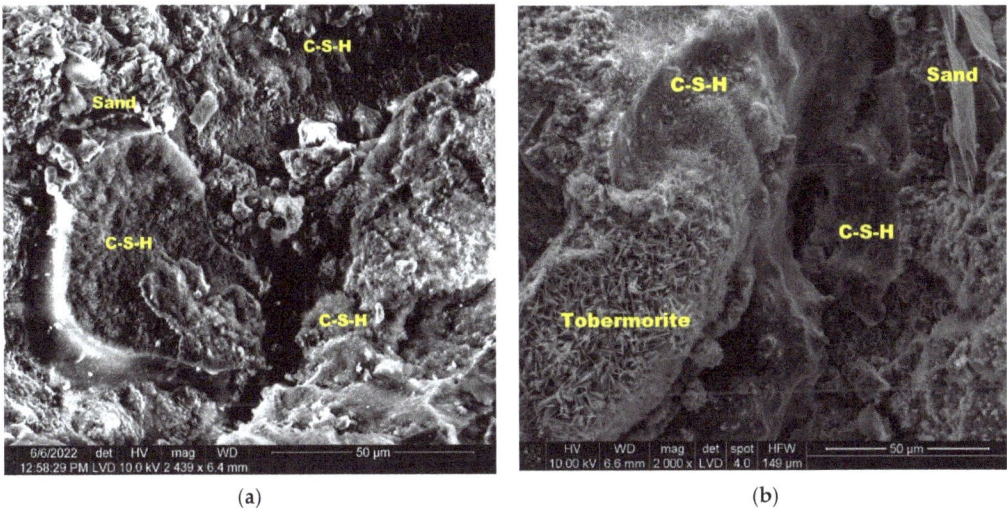

Figure 13. Microstructure of traditional sand-lime bricks; (**a**) industrial production; (**b**) laboratory production.

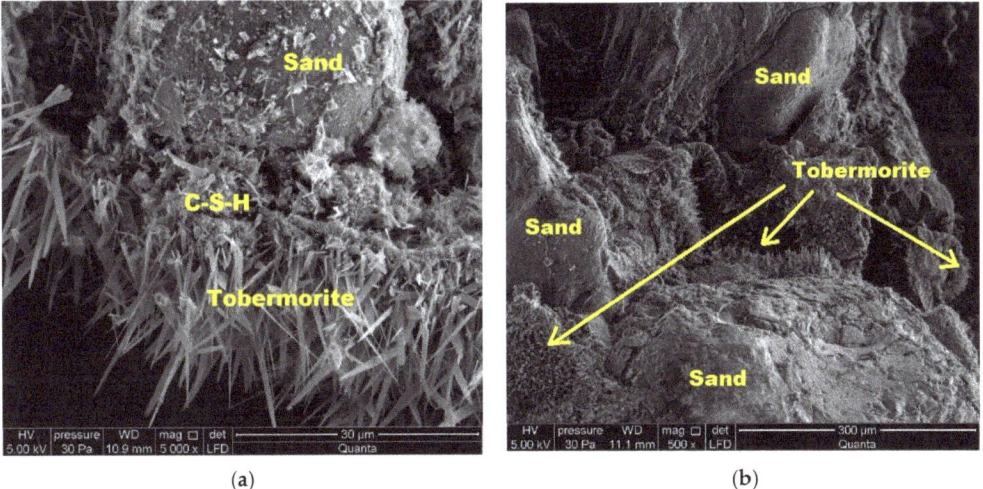

Figure 14. Microstructure of a BP10-W10 sample: (**a**) mag 5000×; (**b**) mag 500×.

In traditional samples (from both industrial and laboratory production), the C-S-H phase and tobermorite were observed (Figure 13). In the BP10-W10 sample (Figure 14), xonotlite crystals are well developed and cover a large part of the sample. In the sample discussed, an increase in strength was observed compared to the other samples tested, which is also confirmed by the presence and structure of xonotlite.

Figure 15 presents SEM photos of the BP30-W10 sample, where much smaller amounts of xonotlite were observed. The dominant phase here is the C-S-H phase, which covers the sand grains. In the sample discussed, lower strengths were also noted compared to the BP10-W10 sample.

Figure 15. Microstructure of a BP30-W10 sample: (**a**) mag 5000×; (**b**) mag 500×.

3.6. XRD Diffraction

Traditional silicates, due to the production process (high temperature and pressure above 1 Ba), consist mainly of crystalline phases. The dominant phase is the tobermorite phase. Depending on the type of modifier, however, crystallization may take different directions, which is caused by the chemical and mineralogical composition of the additive introduced to the raw material mass. In basalt powder, after the basic oxides such as SiO_2, CaO, Al_2O_3, Fe_2O_3 and K_2O, MgO, Na_2O and SO_3 are also present. Depending on the conditions and the amount of oxides, crystallization can proceed in the direction of natrolite, gyrolite, grucite, or the MSH phase. In the case of silicates, crystallization with such an amount of magnesium oxides proceed in the direction of brucite with a small share of the amorphous CSH phase. Depending on the course of crystallization of the CSH phase, it can transform into the MSH phase or the brucite phase, which is influenced by the content of magnesium oxide. The chemical composition of sand and basalt powder differs particularly in terms of the magnesium oxide content. The amount of magnesium oxide in the sand was below the measurable range. However, the chemical composition of the basalt powder (Table 1) shows that magnesium oxide constitutes even 10% of the composition of this raw material. The magnesium content in basalt can cause swelling of the aggregates, which is particularly noticeable in a humid environment (such conditions are also present in an autoclave). However, it depends on the amount of magnesium compounds in the material, the time of exposure to adverse weather conditions, or the quality of hydrothermal conditions (saturation with water vapor in an autoclave). In the tests presented and with the amount of the addition of basalt powder in the proportion of 10–30% by weight of the mass share, no process of swelling of the material was observed. Ph for such modified material is 10.7.

Figure 16 shows the XRD analysis, while Table 4 standard molar thermodynamic properties of minerals in sand-lime bricks modified with basalt powder.

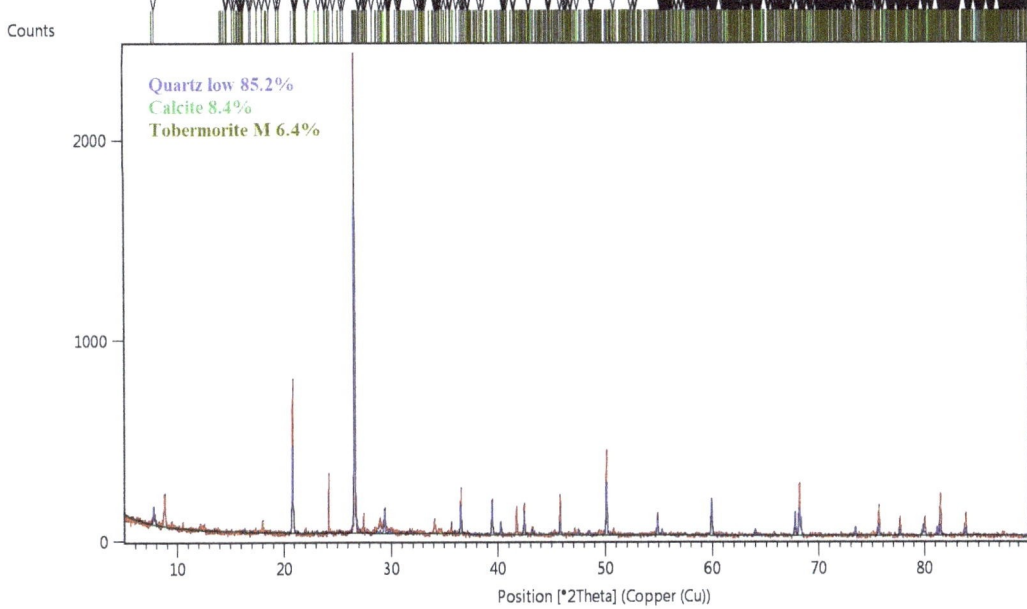

Figure 16. XRD analysis of sample BP10-W5.

Table 4. Standard molar thermodynamic properties of minerals in silicate bricks [55,56].

Mineral Name	Formula	$\Delta_fG°$ [kJ/mol]	$\Delta_fH°$ [kJ/mol]	$S°$ [J/K·mol]	$C°p$ [J/K·mol]	V^0 [cm³/mol]	M [g/mol]
Quartz	$Ca_{0.8}SiO_{2.8}·1.54H_2O$	−1769.00	−1945.13	107.85	138.30	59.20	132.60
Tobermorite 11A	$Ca_5Si_6H_{11}O_{22.5}$	−9889.30	−10,680.00	692.50	764.90	286.10	739.90
Brucite	$Mg(OH)_2$	−831.90	−924.10	59.43	77.27	24.63	58.32
Kaolinite	$Al_2Si_2O_5(OH)_4$	−3793.90	−4115.30	200.90	243.37	99.34	258.16

$\Delta_fG°$—standard molar Gibbs free energy of formation at T_o = 298 K; $\Delta_fH°$—standard molar enthalpy at T_o = 298 K; $S°$—standard molar entropy at T_o = 298 K; $C°p$—heat capacity at T_o = 298 K; $V°$—molar volume; M—molar mass.

Admixtures and additives in the form of dusts and powders (in this case, basalt powder), due to their chemical properties and reactivity (related to the direction of crystallization), are of particular importance in the modification of building materials.

In order to determine these relationships, it is necessary to carry out a long-term analysis of the modified material and to determine the chemical reactions and phase transformations that occur over time. As a standard, in reference silicate materials, the dominant phase is tobermorite 11A.

The system of hydrated calcium silicates is also influenced by the concentration of carbon dioxide. According to the indicated studies, carbon dioxide gradually develops from adsorption to desorption on the surface of tobermorite 9 Å (001). These findings provide further insights into understanding the carbonation of hydrated calcium silicates.

3.7. Profiles for Predicted Values and Desirability

The paper presents the approximation and utility profiles for compressive strength (Figure 17), water absorption (Figure 18) and density (Figure 19). Once the appropriate actual input values have been selected, the approximation profile can be checked to see which input values result in the most desirable approximate output response.

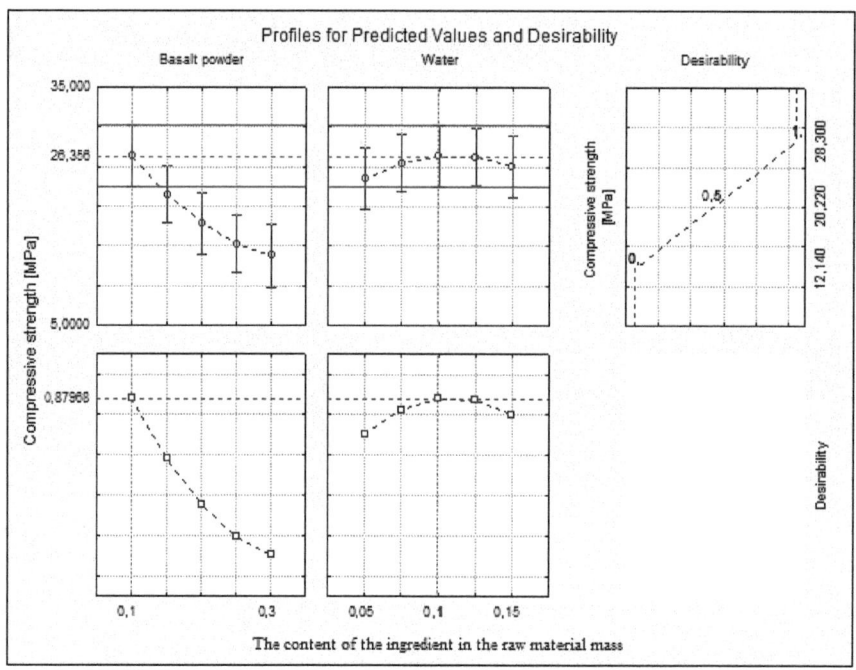

Figure 17. Profiles for predicted value and desirability for compressive strength.

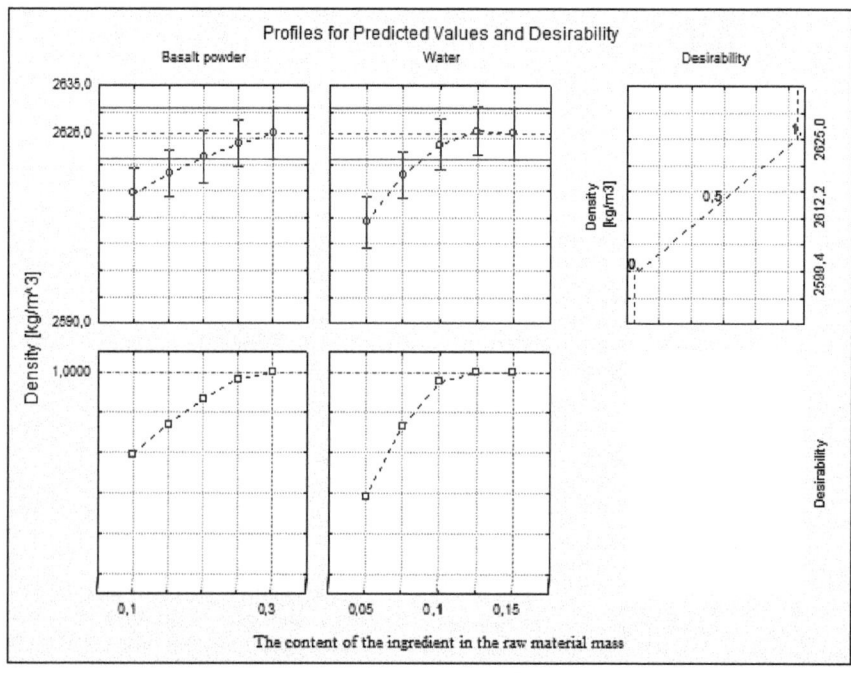

Figure 18. Profiles for predicted value and desirability for density.

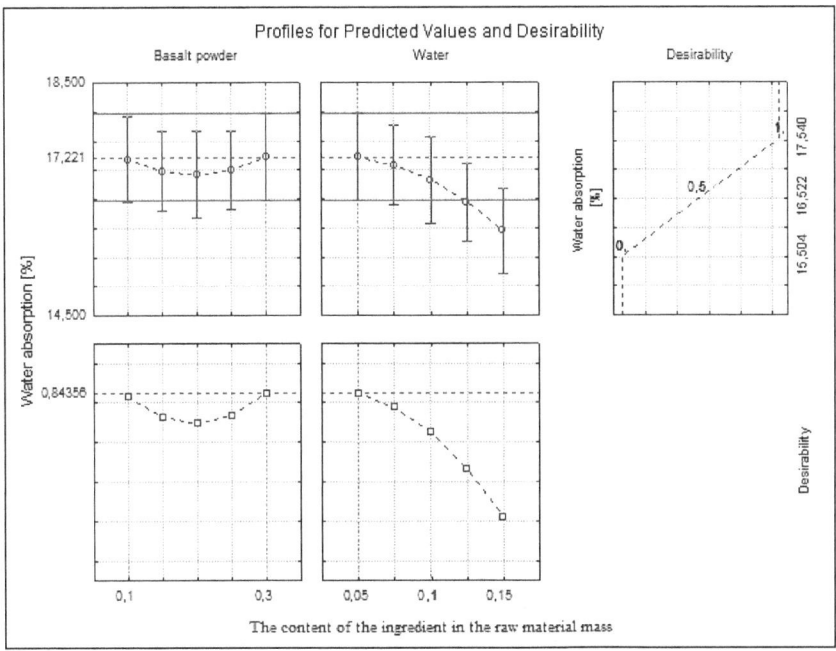

Figure 19. Profiles for predicted value and desirability for water absorption.

The utility function determines the relative utility of different values of the output quantity. In order to determine the utility profile, we first determined the utility functions for each output quantity (dependent variable). For each of the optimized quantities, a response utility profile was created, where values from the range 0–1 were assigned in three points, where the digit 0 represents the least desirable (low) value, the digit 0.5 represents the average value, while the most favorable values of the examined variable were marked 1.0 (high).

The profiles of approximated values and their usefulness were created using the STATISTICA 10.0 computer program. The inputs were set at their expected optimal output values. The total utility was calculated as the geometric mean of the individual utilities for the predicted output values. Utility profiles consist of a series of plots (one for each output) of the total utility for various values of one output (independent variable), with the other inputs constant. By analyzing the response profiles, it is possible to determine which values of the input quantities lead to the desired responses of the output quantities.

The desired values of the output variables were determined, taking into account the design requirements of the modified silicate products. In Section 3.1, it was mentioned that silicate products with a compressive strength above 5.0 MPa, a bulk density above 2000 kg/m^3 and an absorption of about 16% are sought. These values were the exponent to obtain such values of the output variables, in the utility profile for the approximated value.

Analyzing the profiles of approximated values and the utility in Figure 17, where the independent variables basalt powder and water were compared, the best results were obtained with the participation of 10% (0.1) by weight of basalt powder and 10% (0.1) by weight of water. The value of the dependent variable, compressive strength, is 26.36 MPa at the utility point of 0.88, i.e., in the range close to the most favorable value. Figure 18 shows the results for the dependent variable (bulk density) with the same values of the independent variables. The optimum value of about 2600 kg/m^3 was obtained, i.e., significantly higher than the expected value at the utility point of 0.5. Such a value qualifies silicate products for use in order to increase the sound insulation of, e.g., walls. Figure 19 shows the dependent variable (absorption), with the same values of the independent variables as above. The absorption value is around 17%, i.e., the limit value, at the utility point of 0.63 (above the average value).

The research conducted in this work is aimed at the development of sand-lime products and the management of waste basalt powder. The product development process typically involves finding a set of conditions or input values that produce the most desirable product in terms of its properties or output response. Occasionally, quantities that maximize one dependent variable for another independent variable may have the opposite effect (Figure 20).

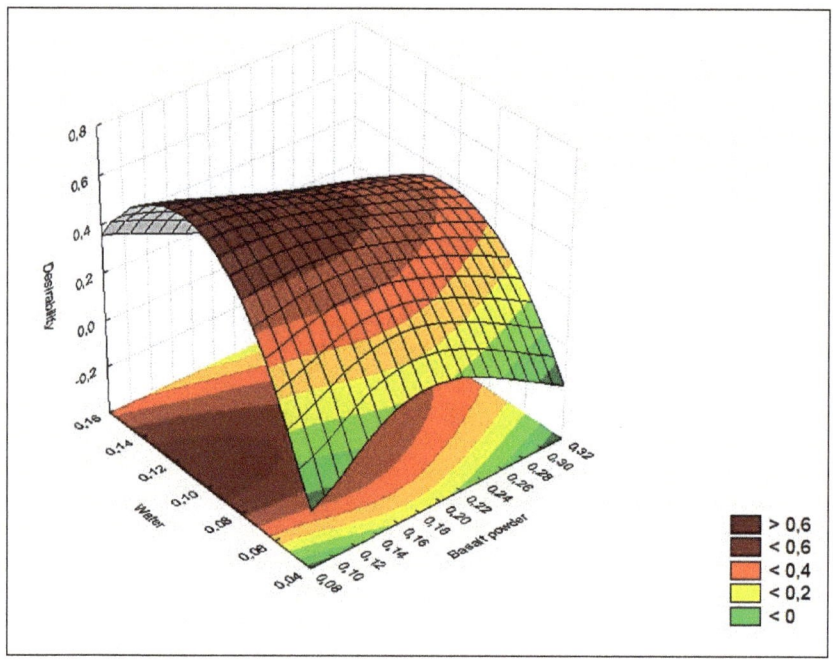

Figure 20. Desirability surface.

4. Conclusions

The paper analyzes the possibility of using waste in the form of basalt powder as a filler in autoclaved basalt materials, and its impact on the parameters of the modified product. Research clearly confirms that basalt powder can be used in the form of an additive to sand-lime products. In addition to ecology and protection of natural deposits, the advantage of this application is the economic aspect—reducing the cost of brick production resulting from the use of cheap post-production waste.

In order to eliminate the differences between the samples related to the technological conditions and the setting of the autoclave, the samples were made on the premises of the industrial silicate plant in Ludynia from one ready-made sand-lime mass, to which basalt powder and water were dosed. The best results were obtained for samples with a 10% mass share of basalt powder in the raw material mat and 10% water content (the remaining 80% by weight is a mix of sand and lime). For these samples, a compressive strength of 28 MPa was obtained, which is almost twice the typical values of silicates. Along with the increase in the use of basalt powder, a decrease in performance parameters was noted, but the products still met the conditions set for traditional silicate bricks. The deterioration of the properties of bricks with a higher content of basalt powder results from the decrease in the amount of binder—lime, the amount of which decreased in the mass inversely proportional to the amount of basalt powder. Powder was used as a filler, further research should be carried out to replace the appropriate fractions of sand with fractions of basalt powder and with different lime content. Such tests will allow the influence of basalt powder on sand-lime products to be fully determined.

The analysis of the microstructure of the samples indicates that the share of basalt powder affects the form of the tobermorite phase—according to the XRD analysis. In addition to the mechanical parameters of the autoclaving process (temperature, pressure), the quality and structure of the phases may be affected by the exploitation of the material and the concentration of CO_2 in the atmosphere. On the BP10-W10 samples, tobermorite crystals formed—a phase considered to accompany the increase in strength in direct proportion to the increase in its amount in the sample tested. The results of compressive strength and material structure tests are correlated.

Further research will focus on the replacement of sand with basalt dust and the study of changes in samples related to this. The next stage of the research will also include an XRF (X-ray Powder Diffraction) analysis for thermodynamic simulation.

Author Contributions: Conceptualization, P.K.-D.; Methodology, P.K.-D.; Software, P.K.-D. and R.B.d.S.; Validation, P.K.-D. and R.B.d.S.; Formal analysis, P.K.-D. and A.S.; Investigation, P.K.-D.; Data curation, R.D. and A.S.; Writing—original draft, P.K.-D.; Writing—review & editing, P.K.-D.; Visualization, P.K.-D. and R.B.d.S.; Supervision, R.D.; Project administration, P.K.-D.; Funding acquisition, R.D. All authors have read and agreed to the published version of the manuscript.

Funding: This research received no external funding. The APC was funded by Kielce University of Technology: 02.0.19.00/1.02.001 SUBB.BKOB.22.001.

Institutional Review Board Statement: Not applicable.

Informed Consent Statement: Not applicable.

Data Availability Statement: The data presented in this study are available in Kostrzewa-Demczuk, P.; Stepien, A.; Dachowski, R.; Krugiełka, A. The use of basalt powder in autoclaved brick as a method of production waste management. *J. Clean. Prod.* **2021**, *320*, 128900, https://doi.org/10.1016/j.jclepro.2021.128900.

Conflicts of Interest: The authors declare no conflict of interest.

Abbreviations

BP10-W5	sand-lime-basalt brick with 10% basalt powder and 5% water content
BP10-W10	sand-lime-basalt brick with 10% basalt powder and 10% water content
BP10-W15	sand-lime-basalt brick with 10% basalt powder and 15% water content
BP20-W5	sand-lime-basalt brick with 20% basalt powder and 5% water content
BP20-W10	sand-lime-basalt brick with 20% basalt powder and 10% water content
BP20-W15	sand-lime-basalt brick with 20% basalt powder and 15% water content
BP30-W5	sand-lime-basalt brick with 30% basalt powder and 5% water content
BP30-W10	sand-lime-basalt brick with 30% basalt powder and 10% water content
BP30-W15	sand-lime-basalt brick with 30% basalt powder and 15% water content
AAC	Autoclaved Aerated Concrete
SEM	Scanning Electron Microscope
XRD	X-ray Powder Diffraction [keV]
V	molar volume [cm3·mol^{-1}]
M	molar mass [g·mol^{-1}]
Log10K	solubility product at T = 298 K (25 °C)(this information is the basis for determining the pH of the material)
$\Delta fG°$	standard Gibbs free energy of formation [J·mol^{-1}]
$\Delta fH°$	standard enthalpy of formation, is the change in enthalpy when one mole of a substance is formed from its elements under a standard pressure of 1 atm/1 bar [J·mol^{-1}]
Cp°	specific heat [J·mol^{-1}·K^{-1}]
S°	standard entropy of formation [J·mol^{-1}·K^{-1}]

References

1. Dobiszewska, M. Zastosowanie pyłu bazaltowego, jako substytutu piasku w zaprawie i betonie cementowym. *Bud. I Archit.* **2016**, *15*, 75–85. [CrossRef]
2. Zhang, M.; Qian, H.; Zhang, Y.; Wang, S. Autoclaved Hollow Lime Sand. Brick. Patent CN104003662, 2014.
3. Barbosa da Silva, R.; Matoski, A.; Neves Junior, A.; Kostrzewa-Demczuk, P. Study of compressive strength of sand-lime bricks produced with coal tailings using mixture design. *Constr. Build. Mater.* **2022**, *344*, 127986. [CrossRef]
4. PN-EN 459-1:2003; Wapno Budowlane. Część 1: Definicje, Wymgania i Kryteria Zgodności. Polski Komitet Normalizacyjny: Warsaw, Poland, 2003.
5. Pytel, Z. Modyfication of the phase composition and microstructure of autoclaved sand-lime bricks. *Ceramics* **2014**, *116*, 14–18.
6. Rashad, A. Cementitious materials and agricultural wastes as natural fine aggregate replacement in conventional mortar and concrete. *J. Build. Eng.* **2016**, *5*, 119–141. [CrossRef]
7. Wzrost Cen Materiałów Budowlanych w 2022 r. Coraz Drożej jak Podaje GUS—Murator.pl. Available online: https://muratordom.pl/przed-budowa/organizacja-budowy/cena-materialow-budowlanych-w-gore-co-zdrozalo-aa-Rcg5-w3KP-q1LR.html (accessed on 31 December 2022).
8. Alyamac, K.E.; Aydin, A.B. Concrete properties containing fine aggregate marble powder. *KSCE J. Civ. Eng.* **2015**, *19*, 2208–2216. [CrossRef]
9. Dhanalaxmi, C.; Nirmalkumar, D.K. Study on the properties of concrete with various mineral admixtures—Limestone powder and marble powder (Reviev Paper). *Int. J. Innov. Res. Sci. Eng. Technol.* **2015**, *4*, 18511–18515.
10. Rahhal, V.; Bonavetti, V.; Trusilewicz, L.; Pedrajas, C.; Talero, R. Role of the filler on Portland cement hydration at early ages. *Construcuction Build. Mater.* **2012**, *27*, 82–90. [CrossRef]
11. Laibao, L.; Yunsheng, Z.; Wenkua, Z.; Zhiyong, L.; Lihua, Z. Investigating the influence of basalt as mineral admixture on hydration and microstructure formation mechanism of cement. *Constr. Build. Mater.* **2013**, *48*, 434–440. [CrossRef]
12. Uncik, S.; Kmecova, V. The effect of basalt powder on the properties of cement composites. *Concr. Concr. Struct. Conf. Procedia Eng.* **2013**, *65*, 51–56.
13. Kmecova, V.; Stefunkova, Z. Effect of basalt powder on workability and initial strength of cement mortar. *J. Civ. Eng. Archit. Res.* **2014**, *1*, 260–267.
14. Saraya, M.E.I. Study physico-chemical properties of blended cements containing fixe amount of silica fume, blast furnace slag, basalt and limestone, a comparative study. *Constr. Build. Mater.* **2014**, *72*, 104–112. [CrossRef]
15. Arivumangai, A.; Felixkala, T. Strength and durability properties of granite powder concrete. *J. Civ. Eng. Res.* **2014**, *4*, 1–6.
16. Arslan, M.E.; Aykanat, B.; Subaşı, S.; Maraşlı, M. Cyclic behavior of autoclaved aerated concrete block infill walls strengthened by basalt and glass fiber composites. *Eng. Struct.* **2021**, *240*, 112431. [CrossRef]
17. Stepien, A.; Kostrzewa, P. The Impact of Basalt Components on the Structure of Bricks Formed as a Result of Hydrothermal Treatment. *Buildings* **2019**, *9*, 192. [CrossRef]
18. Li, Z.; Chen, L.; Fang, Q.; Chen, W.; Hao, H.; Zhang, Y. Experimental and numerical study of basalt fiber reinforced polymer strip strengthened autoclaved aerated concrete masonry walls under vented gas explosions. *Eng. Struct.* **2017**, *152*, 901–919. [CrossRef]
19. Grzeszczyk, S.; Matuszek-Chmurowska, A.; Vejmelková, E.; Černý, R. Reactive Powder Concrete Containing Basalt Fibers: Strength, Abrasion and Porosity. *Materials* **2020**, *13*, 2948. [CrossRef] [PubMed]
20. Kostrzewa, P.; Stepien, A. Sand-lime composites with basalt fibers. In Proceedings of the MATEC Web of Conferences, 3rd Scientific Conference Environmental Challenges in Civil Engineering (ECCE 2018), Opole, Poland, 23–25 April 2018.
21. Kostrzewa-Demczuk, P.; Stepien, A.; Dachowski, R.; Krugiełka, A. The use of basalt powder in autoclaved brick as a method of production waste management. *J. Clean. Prod.* **2021**, *320*, 128900. [CrossRef]
22. Kostrzewa, P. Sand-lime bricks with addition of basalt powder. *Biul. WAT* **2019**, *68*, 147–157. [CrossRef]
23. Małolepszy, J. *Podstawy Technologii Materiałów Budowlanych i Metody Badań*; Wyd. AGH: Kraków, Poland, 2013.
24. Pytel, Z. Badanie wpływu zawartości i uziarnienia mułków chalcedonitowych na jakość tworzyw wapienno-piaskowych. In Proceedings of the V Konferencja Naukowo-Techniczna, Zagadnienia Materiałowe w Inżynierii Lądowej MATBUD'2007", Kraków, Poland, 20–22 June 2007.
25. Dachowski, R.; Kostrzewa, P.; Brelak, S. Autoclaved materials with chalcedonite addition. In Proceedings of the 3rd Scientific Conference Environmental Challenges in Civil Engineering (Ecce 2018), Opole, Poland, 23–25 April 2018; Curran Associates, Inc.: New York, NY, USA, 2018; Volume 174.
26. Pytel, Z. Wpływ mielonego wapienia na właściwości autoklawizowanych tworzyw wapienno-piaskowych. *Ceramics* **2005**, *20*.
27. Vancea, C.; Lazau, I. Glass foam from windows panels and bottle glass wastes. *Cent. Eur. J. Chem.* **2014**, *12*, 803–811.
28. Pytel, Z.; Małolepszy, J. Masa do Wytwarzania Autoklawizowanego Materiału Budowlanego. Patent PL18 5500B1, 2003.
29. Stepien, A.; Kostrzewa, P.; Dachowski, R. Influence of barium and lithium compounds on silica autoclaved materials properties and on the microstructure. *J. Clean. Prod.* **2019**, *236*, 117507. [CrossRef]
30. Owsiak, Z.; Kostrzewa, P. Efects of Bentonite Additives on Autoclaved Sand-Lime Product Properties. In *IOP Conference Series: Materials Science and Engineering*; IOP Publishing: Bristol, UK, 2017; Volume 251, p. 012021.
31. Pytel, Z. Wpływ aktywowanego termicznie łupku karbońskiego na jakość autoklawizowanych tworzyw wapienno-piaskowych. *Mater. Ceram.* **2014**, *66*, 412–420.

32. Nevrivova, L.; Kotucek, M.; Lang, K. Possibilities of Reducing the Apparent Porosity of Silica Bricks for the Coke Batteries. *Advenced Mater. Res.* **2014**, *897*, 121–124. [CrossRef]
33. Szudek, W.; Gołek, Ł.; Malata, G.; Pytel, Z. Influence of Waste Glass Powder Addition on the Microstructure and Mechanical Properties of Autoclaved Building Materials. *Materials* **2022**, *15*, 434. [CrossRef]
34. Stepien, A. Analysis of Porous Structure in Autoclaved Materials Modified by Glass Sand. *Crystals* **2021**, *11*, 408. [CrossRef]
35. Stepien, A.; Leśniak, M.; Sitarz, M. A Sustainable Autoclaved Material Made of Glass Sand. *Buildings* **2019**, *9*, 232. [CrossRef]
36. Jasińska, I.; Dachowski, R.; Jaworska-Wędzińska, M. Thermal Conductivity of Sand-Lime Products Modified with Foam Glass Granulate. *Materials* **2021**, *14*, 5678. [CrossRef] [PubMed]
37. Stepien, A.; Potrzeszcz-Sut, B.; Kostrzewa, P. Influence and application of glass cullet in autoclaved materials. In *IOP Conference Series: Materials Science and Engineering*; IOP Publishing: Bristol, UK, 2018.
38. Dachowski, R.; Stępień, A. The impact of various additives on the microstructure of silicate products. *Procedia Eng.* **2011**, *21*, 1173–1178. [CrossRef]
39. Dachowski, R.; Stępień, A. Zastosowanie Barytu Jako Dodatku w Wyrobach z Masy Silikatowej. Politechnika Świętokrzyska. Opis Patentowy PL216687, 2014.
40. Stępień, A. The impact of barium sulfate on the microstructural and mechanical properties of autoclaved silicate products. In Proceedings of the 9th International Conference Environmental Engineering, Vilnius, Lithuania, 22–23 May 2014.
41. Dachowski, R.; Stępień, A. Mass for the Production of Silicate Products with Higher Compressive Strength. Politechnika Świętokrzyska. Opis Patentowy PL217354, 2011.
42. Schober, G. Porosity in autoclave aerated concrete (AAC): A review on pore structure, types of porosity, measurement methods and effects of porosity on properties. In Proceedings of the 5th International Conference on Autoclaved Aerated Concrete, Bydgoszcz, Poland, 14–17 September 2011.
43. PN-EN 1008:2004; Woda Zarobowa do Betonu—Specyfikacja Pobierania Próbek, Badanie i Ocena Przydatności Wody Zarobowej do Betonu, w tym Wody Odzyskanej z Procesów Produkcji Betonu. Polish Committee for Standarization: Warsaw, Poland, 2004.
44. Piepel, G.F. Programs for generating extreme vertices and centroids of linearly constrained experimental regions. *J. Qual. Technol.* **1988**, *20*, 125–139. [CrossRef]
45. Snee, R.D. Experimental designs for mixture systems with multi-component constraints. *Commun. Stat. Theory Methods* **1979**, *A8*, 303–326. [CrossRef]
46. CEN. PN-EN 772-1, 2011a; 11+A1:2015-10 Methods of Test for Masonry Units. Part 1: Determination of Compressive Strength. CEN: Brussels, Belgium, 2001.
47. CEN. PN-EN 772-13, 2001; Methods of Test for Masonry Units-Part 13: Determination of the Density of the Net and Gross Density of Masonry in the Dry State (Except for Natural Stone). CEN: Brussels, Belgium, 2001.
48. CEN. PN EN 772-11: 2011; Methods of Test for Masonry Units-Part 11: Determination of Water Absorption of Aggregate Concrete, Autoclaved Aerated Concrete, Manufactured Stone and Natural Stone Masonry Units Due to Capillary Action and the Initial Rate of Water Absorption of Clay Masonry Units. CEN: Brussels, Belgium, 2001.
49. CEN. PN-EN 772-21, 2010; E Methods of Test for Masonry Units. Part 21: Determination of Water Absorption of Ceramic and Silicate Masonry Elements by Cold Water Absorption. CEN: Brussels, Belgium, 2001.
50. Nocuń-Wczelik, W. Struktura i właściwości uwodnionych krzemianów wapniowych. *Ceramika* **1999**, *59*, 18.
51. Kurdowski, W. *Chemia Cementu i Betonu*; Wyd. Naukowe PWN: Warszawa, Poland, 2010.
52. Zapotoczna-Sytek, G.; Balkovic, S. *Autoklawizowany Beton Komórkowy*; Wyd. Naukowe PWN: Warszawa, Poland, 2013.
53. Pimraksa, K.; Chindaprasirt, P. Lightweight bricks made of diatomaceous earth, lime and gypsum. *Ceram. Int.* **2009**, *35*, 471–478. [CrossRef]
54. Nonat, A. C-S-H i właściwości betonu. *Cem. Wapno Beton* **2010**, *15*, 315–326.
55. Stepien, A.; Potrzeszcz-Sut, B.; Prentice, D.P.; Oey, T.J.; Balonis, M. The role of glass compounds in autoclaved bricks. *Buildings* **2020**, *10*, 41. [CrossRef]
56. Available online: https://thermoddem.brgm.fr/species/kaolinite (accessed on 31 December 2022).

Disclaimer/Publisher's Note: The statements, opinions and data contained in all publications are solely those of the individual author(s) and contributor(s) and not of MDPI and/or the editor(s). MDPI and/or the editor(s) disclaim responsibility for any injury to people or property resulting from any ideas, methods, instructions or products referred to in the content.

Article

Cement Bypass Dust as an Ecological Binder Substitute in Autoclaved Silica–Lime Products

Katarzyna Borek *, Przemysław Czapik and Ryszard Dachowski

Faculty of Civil Engineering and Architecture, Kielce University of Technology, Al. Tysiąclecia Państwa Polskiego 7, 25-314 Kielce, Poland
* Correspondence: k.komisarczyk@tu.kielce.pl

Abstract: The cement industry is one of the most developed industries in the world. However, it consumes excessive amounts of natural resources and can negatively impact the environment through its by-products: carbon dioxide (CO_2), cement clinker dust (CKD) and cement bypass dust (CBPD). The amount of dust generated in the cement clinker production process depends largely on the technology used. It typically ranges from 0 to 25% by weight of the clinker, and a single cement plant is capable of producing 1000 tons of CBPD per day. Despite practical applications in many areas, such as soil stabilisation, concrete mix production, chemical processing or ceramic and brick production, the dust is still stored in heaps. This poses an environmental challenge, so new ways of managing it are being sought. Due to the significant content of free lime (>30%) in CBPD, this paper uses cement bypass dust as a binder replacement in autoclaved silica–lime products. Indeed, the basic composition of silicate bricks includes 92% sand, 8% lime and water. The investigation shows that it is possible to completely replace the binder with CBPD dust in the autoclaved products. The obtained results showed that all properties of produced bricks were satisfactory. The study concluded that many benefits could be achieved by using cement bypass dust in the production of bricks, including economic bricks for building, reducing the dependency on natural resources, reducing pollution and reducing negative impacts on the environment.

Keywords: CBPD; free CaO; autoclaved products; lime binder replacement; sand–lime bricks

Citation: Borek, K.; Czapik, P.; Dachowski, R. Cement Bypass Dust as an Ecological Binder Substitute in Autoclaved Silica–Lime Products. *Materials* **2023**, *16*, 316. https://doi.org/10.3390/ma16010316

Academic Editor: Malgorzata Ulewicz

Received: 21 November 2022
Revised: 16 December 2022
Accepted: 22 December 2022
Published: 29 December 2022

Copyright: © 2022 by the authors. Licensee MDPI, Basel, Switzerland. This article is an open access article distributed under the terms and conditions of the Creative Commons Attribution (CC BY) license (https:// creativecommons.org/licenses/by/ 4.0/).

1. Introduction

The most popular materials for building walls in Poland and Europe are autoclaved cellular concrete and sand–lime products [1]. These are products obtained in the process of hydrothermal treatment in autoclaves. They are manufactured using binders (cement, lime and gypsum), aggregates (sand, fly ash and blast furnace slag), water and, in the case of cellular concrete, an additional blowing agent and additives. In silicates, the lime binder is 8% of the mass of the entire mixture. Unfortunately, during the production of quicklime, significant amounts of CO_2 are generated into the atmosphere, which has a negative impact on the environment, polluting the air and exacerbating the greenhouse effect [2]. In the era of growing demand for construction materials, the use of a lime substitute for waste material could reduce CO_2 emissions. Such a solution would be in line with the idea of sustainable development in construction [3–5].

Most Portland cement is used to make concrete, mortars or stuccos, and competes in the construction sector with concrete substitutes, such as aluminium, asphalt, clay brick, fiberglass, glass, gypsum (plaster), steel, stone and wood. In 2021, the total world production of cement was 4.4 billion tons. In 2021, China cement production was an estimated 2.5 billion tons; in Vietnam, 810 million tons; in India, 330 million tons; and in the US, 92 million tons [6]. According to the Polish Cement Association, cement production in Poland in 2021 reached almost 19.3 million tons, the third highest in Europe [7]. The cement industry is one of the largest emitters of mineral dust, which generates problems with its management and disposal. In Portland cement production, about 90% of emissions are direct emissions from the combustion of fossil fuels, the decomposition of calcium carbonate (mainly limestone) and energy consumption [8,9]. Despite the reuse of significant amounts of raw materials, intermediates or final products, there are some dusts that, due to their chemical composition, cannot be reintroduced into the cement production cycle. These include alkali-, chlorine- and sulphur-rich cement kiln dust (CKD) and cement bypass dust (CBPD) [10]. CBPD can be divided into two types depending on the place from which they are collected: dust collected immediately at the cold end of the furnace, and dust collected after adding limestone powder to prevent clogging of the ducts [11,12]. The quality of the bypass dust depends on the clinker burning technology, the type of raw materials and fuels used and the method for removing dust. It is usually between 0 and 25% of the clinker mass, as reported by researchers [13,14]. These dust types differ in their chemical and mineral composition and physical properties [15].

According to the Polish Cement Association, the amount of dust produced during cement manufacturing is decreasing. Their latest report found that the annual quantity of dusts from cement kilns in Poland was about 1200 tons [16]. Compared to 25,000 tons in Oman, 2.7–3.5 million tons in Egypt, 8 million tons in the UK, and 2.5–12 million tons in the US [13,17–19], the amount of 1200 tons appears minor. However, as a single cement plant is capable of producing 1000 tons of CBPD daily [20], the reported dust emission rate is not the same as the total dust quantity generated in Polish cement plants and does not include the CKD and CBPD that are recycled back into the kiln system. The amount of dust so used in Poland is much higher and ranges from 9000 to 25,000 tons a year [21]. In 2016, 15,071 tons of CKD and CBPD were reused in the cement production process [22]. However, as recycled dusts lower the cement quality, research is being conducted into new opportunities for dust management [13].

Because of its high alkali content, CBPD is used to produce binders in which it would act as an activator for components with latent binding properties. Binders formed by mixing CBPD, cement and mineral additives in the form of lime, granulated blast furnace slag or fly ash are used in geotechnical soil stabilisation [23–26].

The use of CBPD for the production of alkali-activated binders (so-called "geopolymers"), where the high content of water-soluble alkali as well as the presence of active CaO contribute to the binding of aluminosilicates in the presence of alkali-metal silicate, is known in the literature [27,28].

Many researchers have used waste dust as a partial replacement for cement binder in concrete materials [29–31]. Al-Harthy et al. [32] investigated the effect of CBPD on mortar–concrete mixtures for the partial replacement of Portland cement in concrete. The results indicated that the overall compressive strength, compared to the control mixture, decreased with increasing content of bypass dust in the tested materials. However, the researchers concluded that a mere 5% exchange of dust to cement does not adversely affect the tested strength parameter.

The overall decrease in compressive strength of concrete with CBPD compared to the control mixture was confirmed by Udoeyo and Hyee [33]. However, based on experimental research, it was noted from the experimental study that a minimal decrease in compressive strength occurred when replacing 20% of Portland cement with CBPD.

Aydin et al. [34], based on their research, concluded that cement kiln dust could be used as a source of CaO in the production of ceramic wall tiles. Mahrous and Yang [35] successfully applied CKD dust to clay bricks. Abdulkareem and Eyada used two types of CKD with sand and cement to produce pressed building brick [36]. Abdel-Gawwad et al. [37] utilized CKD with the waste of red clay bricks and silica fume as the main ingredients to produce unfired building bricks. Based on the above, it can be concluded that the introduction of CKD and CBPD, therefore, has a beneficial effect on the properties of wall materials. In addition, since cement bypass dust consists mainly of calcium oxide (CaO), it becomes highly probable that it can be used as a substitute for lime binder in materials in which lime is one of the basic ingredients.

Lime-based wall materials include sand–lime products (92% quartz sand + 8% lime + water) [4,5] and autoclaved aerated concrete [38]. During hydrothermal treatment in an autoclave, chemical processes take place between the components of individual mixtures, and are responsible for the physical and mechanical properties of the finished products. In sand–lime products, crystalline silica, derived from quartz sand, reacts with lime at an elevated temperature of 180 °C and a pressure 10 bar to form hydrated calcium silicates, such as amorphous C–S–H phase and crystalline tobermorite [39]. The use of a substitute for quicklime in autoclaved products would allow for economic advantages the production process stage, and would additionally have a positive impact on the environment, as demonstrated by Vojvodikova, Prochazka and Bohancova [40,41].

Tkaczewska [42] conducted research on the use of small amounts of CBPD (0.05–5.00% by weight of the binder) with free CaO content <10% for the production of autoclaved materials. Autoclaved mortar samples showed a greater increase in compressive strength. After two days of autoclaving, the strength of mortar containing 5% bypass dust increased to 55.60 MPa, while the strength of Portland cement mortar was only 50.05 MPa. After 28 days of autoclaving, the strength values of these mortars were 64.85 MPa and 69.85 MPa, respectively. The increase in compressive strength of the mortar after autoclaving is attributed to hydrothermal treatment, which allows the formation of the C–S–H phase with a higher degree of crystallization and tobermorite in the form of well-formed needles, allowing the cement mortar to achieve higher strength [42].

The use of waste dust as a substitute for natural lime, a component of the classic sand–lime mix, is also well-known. In patent application EP 3 705 462 A1 [43], the use of dust is limited to a range of 5–60% of one or more quicklime sources. The complete replacement of lime by cement kiln dust in the tests carried out resulted in silicate bricks with a significantly reduced compressive strength compared to products with a typical formulation.

In the literature, it was found that limited amounts of waste dust have been used as cement or lime binder. However, no studies have been carried out to determine conclusively whether lime binder can be completely replaced by CBPD dust. In addition, few data indicate the use of waste dust in lime–silica products. The dusts used to date have contained less than 30% free CaO in their composition [43].

The main objective of this study is, therefore, to see whether the extent to which CBPD can be used in autoclaved products (sand–lime products) can be increased. If this is a possibility, a further question arises as to whether the proportion of bypass dust in the mixture of autoclaved products depends on its phase composition and, in particular, on the free CaO content of the dust. Furthermore, the additional question arises whether those dusts with a high free lime content (>30%) should be treated as lime binders.

2. Materials and Methods

2.1. Material Characterization

Quartz sand, quicklime and CBPD were used to prepare the mass for autoclaved sand–lime samples. The sand applied in this study was used in the production of silicate bricks at one of the production plants in the Świętokrzyskie Voivodship (Ludynia) in Poland. The lime was sourced from the local Trzuskawica plant (Sitkówka, Poland) and was intended for silica products. The CBPD dust used was taken from a local cement plant. The physical and chemical properties of these raw materials are shown in Table 1.

Research indicates that the main component of CBPD (Figure 1) is free lime. Free CaO is highly reactive and reacts rapidly with water. The second largest component is belite, which is the clinker phase formed at the lowest temperature [44]. Chlorides present in the CBPD are condensed in the form of potassium chloride (sylvine). The secondary phases, calcite and quartz, come from unprocessed raw materials.

Table 1. Physical and chemical properties of the materials used.

Quartz sand	
Grain size [mm]	2 sieve = 100%, 0.2 sieve ≥ 97% by weight, 0.09 sieve ≥ 97% by weight,
Chemical composition [%]	CaO + MgO ≥ 91 MgO ≤ 2.0 CO_2 ≤ 3.0 SO_3 ≤ 0.50
Bulk density [kg/m^3]	790
Ground quicklime	
Reactivity	60 °C ≤ 2.0 min
Grain size [mm]	2.5–0.5 = 19% 0.5–0.05 = 81%
Chemical composition	Si and O, Al, Mg, Fe, K and C
Density [kg/m^3]	2650
Cement bypass dust (CBPD)	
Grain size [µm]	od 0.20–15 µm

Chemical composition [%]	SiO_2	Al_2O_3	Fe_2O_3	CaO	MgO	Na_2O	K_2O	Na_2O_e	Cl	SO_3	LOI
	15.44	3.42	1.77	52.17	1.31	0.26	6.03	4.22	3.53	1.65	14.40

Phase composition [%]	Free lime	Sylvine	C_2S (belite)	Calcite	Quartz	Arcanite	Portlandite
	38.1	16.2	36.9	4.0	2.5	0.9	1.5

Density [kg/m^3]	3010
Specific surface area [cm^2/g]	5480

Figure 1. Cement bypass dust (CBPD).

2.2. Preparation of Samples Treated with CBPD

A variable amount of CBPD was introduced into a basic mix consisting of quartz sand (92%), quicklime (8%) and water. Four series of six samples each were performed. The first series included the reference samples and the next three series of samples were modified with CBPD. CaO derived from the quicklime was replaced by free CaO from the bypass dust, dosed at 33, 66 and 100% of the CaO from the lime in the mixture (Figure 2).

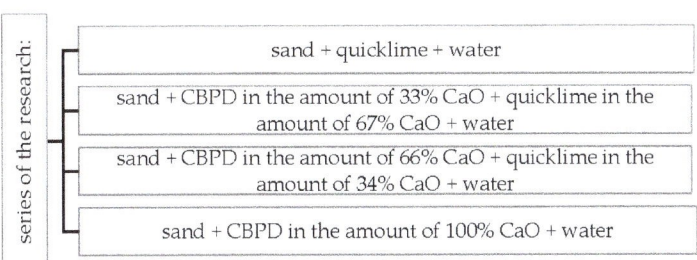

Figure 2. Scheme of the research carried out.

Mixtures were prepared with the compositions given in Table 2. After mixing the components and adding water, the mixtures were placed in a sealed glass vessel and dried in an oven at 65 °C for 1 h. Cylindrical samples (with a moisture content of 6–8%) with a diameter and height of 25 mm were then formed by two-stage pressing at pressures of 10 MPa and 20 MPa. The last stage, autoclaving, was divided into three phases shown in Figure 3. In the first stage, the samples were heated in an autoclave to a temperature of 180 °C for 2.5 h. When the assumed temperature and steam pressure were obtained, the second stage began, in which the samples were autoclaved at 180 °C under a saturated steam pressure of 1.002 MPa for 8 h. The third stage lasted 12 h and involved cooling the samples to ambient temperature.

Table 2. Composition of mixtures with CBPD (at the moisture content of the mixture of 6–8%).

No.	Sample	Proportion of Quartz Sand		Proportion of CBPD		Proportion of Quicklime	
		[%]	[g]	[%]	[g]	[%]	[g]
1	R	92.00	230.00	0.00	0.00	8.00	18.40
2	CBPD33	90.04	225.11	4.68	11.69	5.28	13.20
3	CBPD66	88.04	220.11	9.28	23.19	2.68	6.70
4	CBPD100	85.89	214.73	14.11	35.27	0.00	0.00

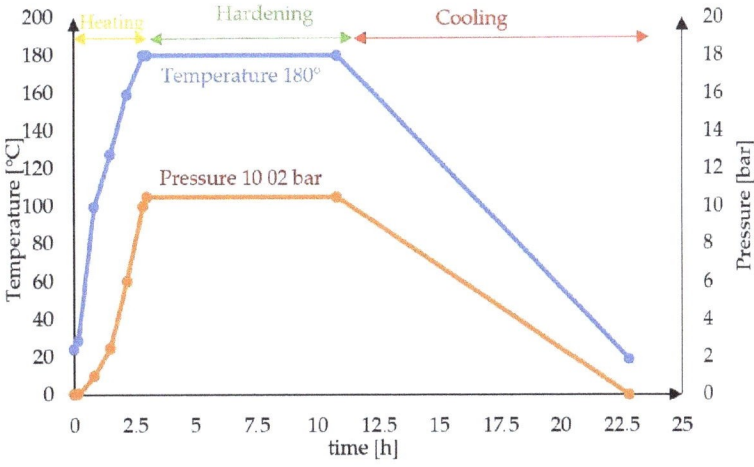

Figure 3. Autoclaving process as a function of time.

2.3. Testing Methods

The paper examines the physical and mechanical properties of samples with the use of cement bypass dust. These tests are necessary to verify the suitability of CBPD dust in sand–lime products used for building walls. The compressive strength was tested 21 days after autoclaving on six cylindrical samples with heights and diameters of 2.5 cm, according to PN-EN 772-1+A1:2015-10 [45]. A Controls 50-C9030 (Controls, Manchester, Barcelona) compression device was used. Prior to testing, the samples were stored in laboratory conditions at room temperature.

The bulk density was determined by the hydrostatic method. Cylindrical samples of autoclaved limestone and sand materials were used for the test, following the scheme below:

1. The limestone and sand samples were completely immersed in water. The water-saturated samples were weighed (to the nearest 0.01 g);
2. One by one, the tested samples were suspended on a thin wire on one of the balance arms and immersed in such a way that the samples and the weighing pan did not come into contact with the vessel, liquid or the table on which the vessel stood;
3. Samples were weighed in water and the measurement results were recorded (to the nearest 0.01 g);
4. The temperature and density of the water in the vessel were determined;
5. The samples were then dried at 103 °C to constant weight;
6. After cooling down, the dry samples were weighed and the measurement result was recorded;
7. The calculations were performed using the Equation (1):

$$\rho = \frac{m_d}{V_0} \; [kg/m^3] \tag{1}$$

where:

m_d —the weight of the dry sample,
V_0 —the volume of the sample including pores calculated according to Equation (2).

$$V_0 = \frac{m_1 - m_2}{\rho_w} \tag{2}$$

where:

m_1—the weight of the saturated sample, weighed in air, [g]
m_2—the weight of the saturated sample, weighed in water, [g]
ρ_w—water density, [g/cm^3]

In each case, the tests were conducted on six samples of the same batch. The arithmetic mean of the obtained values was assumed as the result of the determination.

Water absorption was determined in accordance with the PN-EN 772-21: 2011 [46] standard. The test was carried out on 6 samples from each batch, using those used for the bulk density analysis. The water absorption was calculated using Equation (3):

$$w = (m_1 - m_d)/m_d \times 100 \; [\%] \tag{3}$$

where:

m_d—weight of the sample after drying,
m_1—weight of the sample after soaking.

Microstructure and phase composition studies were performed as explanatory tests for the parameters of the physical and mechanical characteristics. Microstructure analysis of silica–lime samples with CBPD was carried out using a Quanta FEG 250 (FEI, Brno, Czech Republic) scanning electron microscope. Measurements were taken in a low vacuum at 5 kV. The test material for the silica–lime products was taken from the damaged samples after the compressive strength test. The tests were, therefore, performed on undusted crumbs of limestone and silica products. To analyse the microstructure of individual samples, magnifications from 1.000 to 20.000× were used and subjected to EDS point analysis.

3. Results and Discussion

3.1. Compressive Strength

Based on the test results obtained (Figure 4), it can be concluded that modifying the limestone–silica samples with CBPD has a negligible effect on the change in compressive strength. Replacing 100% of free lime in the silica–lime mixture with CBPD dust increases the compressive strength by ~5% (4.00 MPa), compared to the R samples (3.81 MPa). The average compressive strength values of CBPD66 and CBPD33 samples are 0.5% and 4.2% lower, respectively, compared to products with a typical formulation. Temmermans, Tromp and Santamaria noticed a decrease in the examined parameter with an increase in the share of CBPD in autoclaved products [43]. In their paper, they presented that the decrease in compressive strength was more than 50% (28.1 MPa for the reference sample, 12.9 MPa for the sample with 100% dust). In their study, a reduction in the results of the parameter tested was noted for samples with 33 and 66% dust. However, the complete replacement of the lime binder with dust from cement kiln dusting improves the compressive strength. The difference between the compared test results may result from a different share of free CaO in CBPD. In their research, [39] the authors of the patent used dust containing 23.6% of free CaO, which is a lower value compared to the amount of CBPD used in this study.

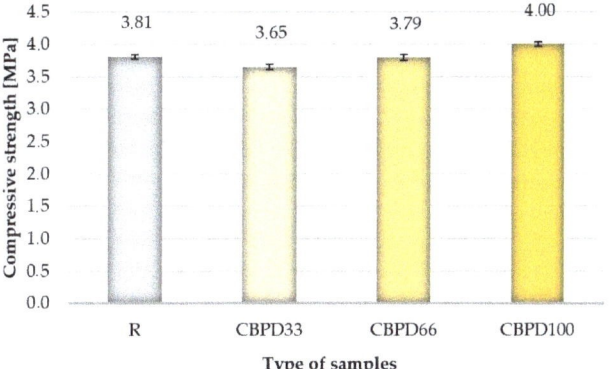

Figure 4. Compressive strength results.

Tkaczewska [42] used bypass dust as a cement replacement in autoclaved products. Mortar samples with CBPD subjected to hydrothermal treatment showed a greater increase in compressive strength in relation to the samples of typical composition. Tkaczewska explained this phenomenon by the formation of substantial amounts of highly crystallized hydrated calcium silicates, such as tobermorite, resulting from autoclaving.

3.2. Bulk Density

Modifying the samples with bypass dust, regardless of its share in the mixture, reduces the volume density in relation to the R samples (Figure 5). The use of dust at 33% relative to the free CaO in the mixture reduces the bulk density to the greatest extent relative to the R samples (about 9.7%). The average bulk density values of the CBPD66 and CBPD100 samples are 8.3% and 5.3% lower in relation to the reference products. The average bulk density of the modified product increases with increasing CBPD content of the sample. A similar phenomenon is illustrated by the specific density of the samples (CBPD33 = 2583 kg/m^3, CBPD66 = 2604 kg/m^3, CBPD100 = 2625 kg/m^3). It is concluded that this may be due to the higher specific density of cement kiln dust, used as a binder replacement, compared to that of quicklime. However, the bulk density of the samples with 100% CBPD is lower compared to the R samples, which may be related to the increase in the open porosity of the modified products.

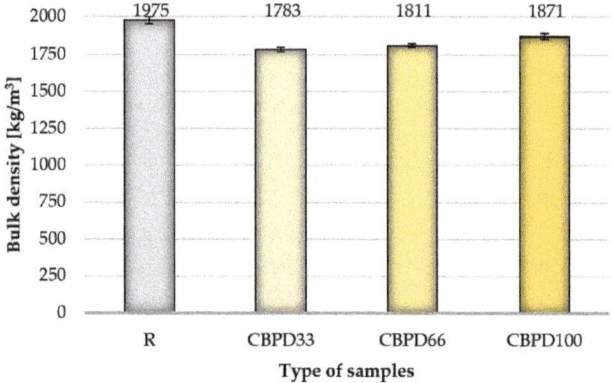

Figure 5. Bulk density results.

3.3. Water Absorption

Modifying the samples with bypass dust, regardless of its share in the mixture, increases the water absorption in relation to the R samples (Figure 6). The average water absorption values for CBPD100 and CBPD66 samples are 40.4% and 50.4% higher, respectively, compared to the R samples. The use of dust at 33% relative to free CaO in the mixture increases water absorption to the greatest extent in relation to the R samples (by about 61.3%).

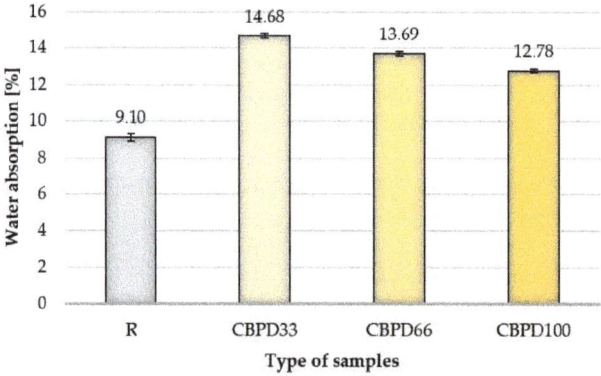

Figure 6. Water absorption results.

The complete replacement of quicklime with dust from the cement kiln dedusting increases the packing density; as a consequence, a denser microstructure is created. The spaces between the aggregate could be sealed as a result of the hydration of the dust grains that previously filled them.

3.4. Phase Composition and Microstructure

Diffractograms of silica–lime products are shown in Figure 7. The R sample revealed intense reflections of quartz and calcite. The phase compositions of the products with CBPD and of the R sample are different. All the modified products are dominated by quartz and calcite reflections. The introduction of CBPD resulted in the presence of portlandite and sylvine peaks. As the CBPD content in the products increases, the intensity of quartz (50 2θ°) and portlandite reflections decreases (to show this, a clearer separation of the diffraction patterns would be required). The presence of portlandite indicates that the active lime (contained in free CaO and belite) has neither carbonated nor reacted to form hydrated calcium silicates. Increasing the share of bypass dust increases the intensity of the reflections characteristic of sylvine, particularly visible in the 28 ÷ 30 2θ° angle range. This is justified by the fact that CBPD contains significant amounts of potassium chloride (sylvine) [10]. In the CBPD100 samples, a peak appeared in the 5 ÷ 10 2θ° angle range, characteristic of the crystalline phase, tobermorite. Based on the X-ray analysis, it can be concluded that the formation of hydrated calcium silicates in autoclaved products is possible with the complete replacement of lime with CBPD.

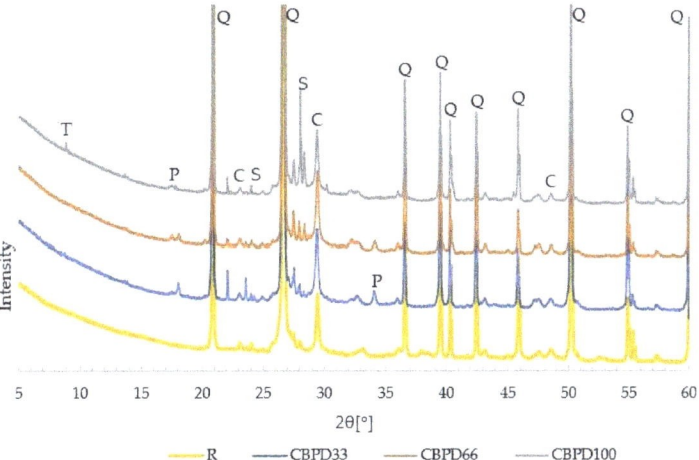

Figure 7. Diffraction pattern of autoclaved sand–lime products with 33%, 66% and 100% CBPD in relation to free CaO, compared to a reference sample. Designations: C—calcite, P—portlandite, S—sylvine, T—tobermorite, Q—quartz.

The SEM microscopic image of autoclaved silica–lime products with 33% CBPD in the binder mass shown in Figure 8a indicates the formation of phases near the aggregate surface, characteristic of silicate products [47]. In the upper part, a sand grain covered by residual C–S–H phase is visible, as indicated by the result of the EDS analysis (Figure 8b), in which, in addition to an intense quartz signal, a clear calcium signal was obtained. In Point 2, in a small area near the surface of the sand grain, a phase with a higher degree of crystallization—tobermorite [48]—was formed, which is indicated by its lamellar structure and a Ca/Si ratio of 0.8 (Figure 8c). At Point 3, according to EDS analysis (Figure 8d), a spongy C–S–H phase is visible, largely covering the analysed surface of the sample.

The SEM microscopic image of limestone–silica samples with 100% CBPD seen in Figure 9a shows a different morphology compared to samples modified with a lower dust content. A microstructure created by very densely packed hydrated calcium silicates in the form of well-developed needles was observed over the entire surface of the samples. The short grassy forms develop into long networks of threads filling the areas between the aggregate grains. This can lead to an increase in the tightness of the product, resulting in improved mechanical and physical properties. The formation of hydrated calcium silicates with a higher degree of crystallisation in the structure of the CBPD-modified material is confirmed by the results obtained in the following papers: [42,49]. The EDS analysis (Figure 9b) showed that the resulting structure was characterised by significantly elevated K and Cl readings, which is justified by the proportion of CBPD in the analysed sample.

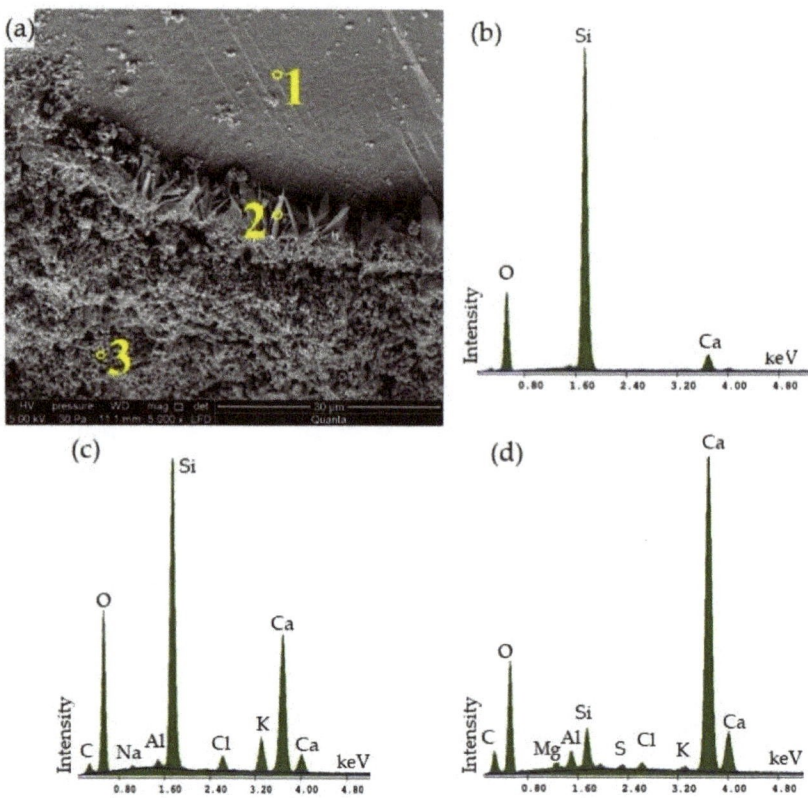

Figure 8. Microscopic SEM image of calcium–silica samples with 33% CBPD at (**a**) 5000× magnification and X-ray microanalysis of its elemental composition at (**b**) 1, (**c**) 2 and (**d**) 3.

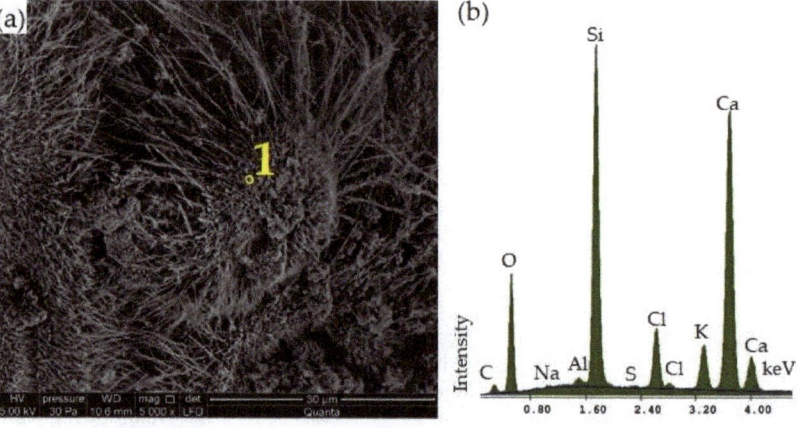

Figure 9. Microscopic SEM image of calcium–silica samples with 100% CBPD at (**a**) 5000× magnification and micro-X-ray analysis of its elemental composition in Point (**b**) 1.

4. Conclusions

The following general conclusions can be drawn from the study:

1. The amount of CBPD dust used in autoclaved products depends on the dust's chemical composition, primarily on its free CaO content;
2. The use of CBPD dust with >30% CaO content makes it possible to obtain full-value silicate bricks.

The results of this research allow the following specific conclusions to be reached:

1. The complete replacement of quicklime with CBPD contributes to a 5% higher compressive strength with a 5.3% reduction in bulk density. This effect can be considered beneficial from a logistical point of view and the structural properties of the construction material. Lighter construction materials with lower density allow slimmer structures to be erected;
2. Replacing quicklime with CBPD to a lesser extent (33 and 66%) contributes to a greater decrease in density; however, this is accompanied by a reduction in the strength of the sand–lime product;
3. The decrease in the density of sand–lime products due to the addition of CBPD is accompanied by a corresponding increase in water absorption. Thus, it can be concluded that the decrease in density is related to an increase in porosity that allows water to be absorbed;
4. Based on the compressive strength, density, and water absorption tests, it can be concluded that even a small addition of CBPD can significantly affect the properties of the sand–lime product, and this effect can be eliminated by increasing the proportion of dust in the mixture. This may be related to the fact that replacing free CaO from quicklime with that derived from CBPD increases the total amount of binder used. In addition, in the CBPD, not only free lime but also belite shows binding properties, which can positively influence the properties of the material obtained;
5. The modification of the traditional silica–lime mixture with bypass dust alters its phase composition and introduces new phases—portlandite and sylvine—into the system. The increase in modifier content decreases the intensity of the quartz and portlandite reflections and, in the CBPD100 sample, increases the intensity of the peak characteristic of tobermorite;
6. The use of CBPD from cement kilns as a lime substitute does not inhibit the formation of hydrated lime silicates, characteristic of autoclaved products. On the contrary, increasing the proportion of waste dust in the products generates the formation of denser structures with a higher degree of crystallisation, i.e., tobermorite. The key is the content of free CaO in the dust.

The research presented in this article opens up further areas of research related to the elucidation of the microstructural differences and the investigation of practically useful material properties, such as thermal conductivity, water vapour permeability, hygroscopic properties and frost durability. Given the economic and environmental benefits of replacing lime binder in silicate products with waste dust, a study in which the aggregate of products is also replaced should be considered. The substitute for quartz sand could be a waste material containing a high proportion of silica. This would make it possible to produce a product containing only waste materials.

Author Contributions: Conceptualization, K.B., R.D. and P.C.; methodology, K.B.; software, K.B.; formal analysis, K.B. and P.C.; investigation, K.B. and P.C. (XRD); resources, K.B.; data curation, K.B. and P.C.; writing—original draft preparation, K.B.; writing—review and editing, K.B., R.D. and P.C. All authors have read and agreed to the published version of the manuscript.

Funding: The work is supported by the Kielce University of Technology 02.0.19.00/1.02.001 SUBB.BK OB.22.001.

Institutional Review Board Statement: Not applicable.

Informed Consent Statement: Not applicable.

Data Availability Statement: Not applicable.

Conflicts of Interest: The authors declare no conflict of interest.

References

1. Statistics Poland. *Statistical Analyses. Market of Construction Products Manufactures in 2015–2018*; Zakład Wydawnictw Statystycznych: Warsaw, Poland, 2020; pp. 45–67.
2. The Science in Poland. Available online: https://naukawpolsce.pl/aktualnosci (accessed on 14 October 2022).
3. Ajith, S.; Arumugaprabu, V.; Szóstak, M. A framework for systematic assessment of human error in construction sites—A sustainable approach. *Civil Eng. Arch.* **2022**, *10*, 1725–1737. [CrossRef]
4. Borek, K.; Czapik, P. Utilization of waste glass in autoclaved silica–lime materials. *Materials* **2022**, *15*, 549. [CrossRef] [PubMed]
5. Borek, K.; Czapik, P.; Dachowski, R. Recycled glass as a substitute for quartz sand in silicate products. *Materials* **2020**, *13*, 1030. [CrossRef] [PubMed]
6. Hatfield, A.K.U.S. Geological Survey, Mineral Commodity Summaries. 2022. Available online: https://pubs.usgs.gov/periodicals/mcs2022/mcs2022.pdf (accessed on 14 December 2022).
7. Polish Cement Association. Available online: https://www.polskicement.pl/emisje/ (accessed on 14 October 2022).
8. Hanein, T.; Hayashi, Y.; Utton, C.; Nyberg, M.; Martinez, J.C.; Quintero-Mora, N.I.; Kinoshita, H. Pyro processing cement kiln bypass dust: Enhancing clinker phase formation. *Constr. Build. Mater.* **2020**, *259*, 120420. [CrossRef]
9. Zhang, C.Y.; Han, R.; Yu, B.; Wei, Y.M. Accounting process-related CO_2 emissions from global cement production under Shared Socioeconomic Pathways. *J. Clean. Prod.* **2018**, *184*, 451–465. [CrossRef]
10. Owsiak, Z.; Czapik, P.; Zapała-Sławeta, J. Properties of a three-component mineral road binder for deep-cold recycling technology. *Materials* **2020**, *13*, 3585. [CrossRef]
11. Al-Jabri, K.S.; Taha, R.A.; Al-Hashmi, A.; Al-Harthy, A.S. Effect of copper slag and cement by-pass dust addition on mechanical properties of concrete. *Constr. Build. Mater.* **2006**, *20*, 322–331. [CrossRef]
12. Al-Jabri, K.S.; Taha, R.A.; Al-Ghassani, M. Use of copper slag and cement by-pass dust as cementitious materials. *Cem. Concr. Aggreg.* **2002**, *24*, 7–12. [CrossRef]
13. Adaska, W.S.; Taubert, D.H. Beneficial uses of cement kiln dust. In Proceedings of the 2008 IEEE Cement Industry Technical Conference Record, Miami, FL, USA, 18–28 May 2008. [CrossRef]
14. Sreekrishnavilasam, A.; Santagata, M.C. *Report No. FHWA/IN/JTRP-2005/10 Development of Criteria for the Utilization of Cement kiln Dust (CKD) in Highway Infrastructures*; Joint Transportation Research Program; Purdue University: West Lafayette, IN, USA, 2006.
15. Czapik, P.; Zapała-Sławeta, J.; Owsiak, Z.; Stępień, P. Hydration of cement by-pass dust. *Constr. Build. Mater.* **2020**, *231*, 117139. [CrossRef]
16. 2012—INFORMATOR SPC—Przemysł Cementowy w liczbach", Polish Cement Association Reports. Available online: https://www.polskicement.pl/2012-informator-spc-przemysl-cementowy-w-liczbach/ (accessed on 14 December 2022).
17. Abdel-Ghani, N.T.; El-Sayed, H.A.; El-Habak, A.A. Utilization of by-pass cement kiln dust and air-cooled blast-furnace steel slag in the production of some "green" cement products. *HBRC J.* **2018**, *14*, 408–414. [CrossRef]
18. Darweesh, H.H.M. A review article on the influence of the electrostatic precipitator cement kiln dust waste on the environment and public health. *Am. J. Biol. Environ. Stat.* **2017**, *3*, 36–43. [CrossRef]
19. Taha, R.; Al-Rawas, A.; Al-Harthy, A.; Qatan, A. Use of cement bypass dust as filler in asphalt concrete mixture. *J. Mater. Civil Eng.* **2002**, *14*, 338–343. [CrossRef]
20. Khater, G.A. Use of bypass cement dust for production of glass ceramic materials. *Adv. Appl. Ceram.* **2006**, *105*, 107–111. [CrossRef]
21. Uliasz-Bocheńczyk, A. Chemical characteristics of dust from cement kilns. *Gospod. Surowcami Miner.* **2019**, *35*, 87–102. [CrossRef]
22. 2019—Informator SPC—Przemysł Cementowy w iczbach, Polish Cement Association Reports. Available online: https://www.polskicement.pl/2019-informator-spc-przemysl-cementowy-w-liczbach/ (accessed on 14 December 2022).
23. Yoobanpot, N.; Jamsawang, P.; Horpibulsuk, S. Strength behavior and microstructural characteristics of soft clay stabilized with cement kiln dust and fly ash residue. *Appl. Clay Sci.* **2017**, *141*, 146–156. [CrossRef]
24. Peethamparan, S.; Olek, J.; Lovell, J. Influence of chemical and physical characteristics of cement kiln dusts (CKDs) on their hydration behavior and potential suitability for soil stabilization. *Cem. Concr. Res.* **2008**, *38*, 803–815. [CrossRef]
25. Buczyński, P.; Iwański, M. The influence of hydrated lime, portland cement and cement dust on rheological properties of recycled cold mixes with foamed bitumen. In Proceedings of the "Environmental Engineering" 10th International Conference, Vilnus, Lithuania, 27–28 April 2017.
26. Al-Aghbari, M.Y.; Mohamedzein, Y.E.-A.; Taha, R. Stabilisation of desert sands using cement and cement dust. *Proc. Inst. Civ. Eng. -Ground Improv.* **2009**, *162*, 145–151. [CrossRef]
27. Sultan, M.E.; Abo-El-Enein, S.A.; Sayed, A.Z.; EL-Sokkary, T.M.; Hammad, H.A. Incorporation of cement bypass flue dust in fly ash and blast furnace slag-based geopolymer. *Case Stud. Constr. Mater.* **2018**, *8*, 315–322. [CrossRef]
28. Heikal, M.; Zaki ME, A.; Ibrahim, S.M. Preparation, physico-mechanical characteristics and durability of eco-alkali-activated binder from blast-furnace slag, cement kiln-by-pass dust and microsilica ternary system. *Constr. Build. Mater.* **2020**, *260*, 119947. [CrossRef]

29. Nocuń-Wczelik, W.; Stolarska, K. Calorimetry in the studies of by-pass cement kiln dust as an additive to the calcium aluminate cement. *J. Therm. Anal. Calorim.* **2019**, *138*, 4561–4569. [CrossRef]
30. Abdel-Gawwad, H.A.; Heikal, M.; Mohammed, M.S.; Abd El-Aleem, S.; Hassan, H.S.; García, S.V.; Alomayri, T. Sustainable disposal of cement kiln dust in the production of cementitious materials. *J. Clean. Prod.* **2019**, *232*, 1218–1229. [CrossRef]
31. Wojtacha-Rychter, K.; Król, M.; Gołaszewska, M.; Całus-Moszko, J.; Magdziarczyk, M.; Smoliński, A. Dust from chlorine bypass installation as cementitious materials replacement in concrete making. *J. Build. Eng.* **2022**, *51*. [CrossRef]
32. Al-Harthy, A.S.; Taha, R.; Al-Maamary, F. Effect of cement klin dust (CKD) on mortar and conrete mixtures. *Constr. Build. Mater.* **2003**, *17*, 353–360. [CrossRef]
33. Udoeyo, F.F.; Hyee, A. Strengths of cement kiln dust concrete. *J. Mater. Civ. Eng.* **2002**, *14*, 524–526. [CrossRef]
34. Aydin, T.; Tarhan, M.; Tarhan, B. Addition of cement kiln dust in ceramic wall tile bodies. *J. Therm. Anal. Calorim.* **2019**, *136*, 527–533. [CrossRef]
35. Mahrous, M.A.; Yang, H.S. Utilization of cement kiln dust in industry cement bricks. *Geosystem Eng.* **2011**, *14*, 29–34. [CrossRef]
36. Abdulkareem, A.H.; Eyada, S.O. Production of Building Bricks Using Cement Kiln Dust CKD Waste. In *Sustainable Civil Infrastructures*; Springer: Berlin/Heidelberg, Germany, 2018; pp. 102–113.
37. Abdel-Gawwad, H.A.; Rashad, A.M.; Mohammed, M.S.; Tawfik, T.A. The potential application of cement kiln dust-red clay brick waste-silica fume composites as unfired building bricks with outstanding properties and high ability to CO_2-capture. *J. Build. Eng.* **2021**, *42*, 102479. [CrossRef]
38. Qu, X.; Zhao, X. Previous and present investigations on the components, microstructure and main properties of autoclaved aerated concrete—A review. *Constr. Build. Mater.* **2017**, *135*, 505–516. [CrossRef]
39. Li, J.; Lv, Y.; Jiao, X.; Sun, P.; Li, J.; Wuri, L.; Zhang, T.C. Electrolytic manganese residue based autoclaved bricks with $Ca(OH)_2$ and thermal-mechanical activated K-feldspar additions. *Constr. Build. Mater.* **2020**, *230*. [CrossRef]
40. Vojvodikova, B.; Prochazka, L.; Bohacova, J. Posiibilities of application cement by-pass dust into the garden architecture elements. *Crystals* **2021**, *11*, 1033. [CrossRef]
41. Vojvodikova, B.; Prochazka, L.; Bohacova, J. X-ray diffraction of alkali-activated materials with cement by-pass dust. *Materials* **2021**, *11*, 782. [CrossRef]
42. Tkaczewska, E. The influence of cement bypass dust on the properties of cement curing under normal and autoclave conditions. *Struct. Environ.* **2019**, *11*, 5–22. [CrossRef]
43. Temmermans, F.; Tromp, O.; Santamaria Razo, D.A. A1 Method of Making Calcium Silicate Bricks. European Patent 3 705 462, 9 September 2020.
44. Kurdowski, W. *Cement and Concrete Chemistry*; Springer: Dordrecht, The Netherlands, 2014. [CrossRef]
45. *PN-EN 772-1+A1:2015-10*; Metody Badań Elementów Murowych- Część 1: Określenie Wytrzymałości na Ściskanie. Polish Committee for Standarization: Warsaw, Poland, 2015.
46. *PN-EN 772-21*; 2011 Określanie Absorpcji Wody Ceramicznych i Silikatowych Elementów Murowych Przez Absorpcję Zimnej Wody. Polish Committee for Standarization: Warsaw, Poland, 2011.
47. Stępień, A.; Leśniak, M.; Sitarz, M.A. Sustainable autoclaved material made of glass sand. *Buildings* **2019**, *9*, 232. [CrossRef]
48. Stępień, A.; Potrzeszcz-Sut, B.; Prentice, P.D.; Oey, T.; Balonis, M. The role of glass compounds in autoclaved bricks. *Buildings* **2020**, *10*, 41. [CrossRef]
49. Huang, Z.; Yuan, Y.; Chen, Z.; Wen, Z. Microstructure of autoclaved aerated concrete hydration products in different water-to-binder ratio and different autoclaved system. *Adv. Mater. Res.* **2012**, *602–604*, 1004–1009. [CrossRef]

Disclaimer/Publisher's Note: The statements, opinions and data contained in all publications are solely those of the individual author(s) and contributor(s) and not of MDPI and/or the editor(s). MDPI and/or the editor(s) disclaim responsibility for any injury to people or property resulting from any ideas, methods, instructions or products referred to in the content.

Article

Evaluation of Strength Properties of Sand Stabilized with Wood Fly Ash (WFA) and Cement

Sanja Dimter [1,*], Martina Zagvozda [1], Tea Tonc [1] and Miroslav Šimun [2]

1. Josip Juraj Strossmayer University of Osijek, Faculty of Civil Engineering and Architecture Osijek, 31000 Osijek, Croatia; mzagvozda@gfos.hr (M.Z.); ttonc15@gfos.hr (T.T.)
2. Zagreb University of Applied Sciences, Civil Engineering Department Zagreb, 10000 Zagreb, Croatia; msimun@tvz.hr
* Correspondence: sdimter@gfos.hr

Abstract: The article describes the laboratory evaluation of mixtures of sand modified with wood fly ash (WFA) and additionally stabilized with different amounts of cement. Laboratory research includes determining the California Bearing Ratio (CBR), compressive and indirect tensile strengths of the mixtures, and the resistance of mixtures to freezing/thawing cycles. The aim of the research is to determine if WFA, an alternative material, can improve sand bearing capacity and contribute to strength development while reducing necessary cement amounts and satisfying the technical regulation for use in pavement base courses. The test results obtained show that WFA has a considerable stabilization effect on the sand mixture and improves its load bearing capacity. By adding a small quantity of the cement, the hydraulic reaction in the stabilized mixture is more intense and results in greater strengths and an improved resistance to freezing. The test results show that, by replacement of part of the sand with WFA (in the quantity of 30%), greater strengths can be achieved in relation to the mixture of only sand and cement. Additionally, the content of cement necessary for the stabilization of sand (usually 8–12%) is considerably reduced, which enables cost savings in the construction of pavement structures.

Keywords: wood fly ash (WFA); sand; cement stabilized mixtures; mechanical properties

1. Introduction

Load-bearing layers stabilized by cement or other hydraulic binders (fly ash, slag) are widely applied in pavement structures of roads throughout Europe and the world [1]. These layers take the static and dynamic traffic loads from the pavement surface and transfer them to the substructure. Cement stabilized layers with their mechanical properties must meet the stresses of traffic load and durability requirements during the design period of construction. They are carried out in asphalt pavement structures, in which they act as one of the most important elements in terms of load-bearing capacity, as well as in concrete pavement structures in which, as a base under the concrete slab, they prevent the occurrence of material "pumping". The pavement structures that have cement stabilized layers in their composition are less sensitive to seasonal influences, their load bearing capacity is considerably more uniform, and they withstand heavy traffic loads.

Although there are several different definitions and names of mixtures for the construction of cement stabilized layers [1,2], the simplified one defines the cement stabilized mixture as a mixture of different types of granular material to which cement is added for the purpose of improving specific properties. For the production of cement-stabilized bearing layers, granular stone material (natural sandy gravel, crushed stone material, or their mixtures) or sand is used as the basic aggregate. The use of sandy gravel and the crushed stone aggregate also results in better mechanical properties of stabilization mixtures, while sandy materials produce mixtures with somewhat lower mechanical properties. Depending on the type of the aggregate and the desired mechanical properties of the stabilization mixture, cement is added to the stabilization mixture in the content of 3–12% [1–3]. By adding cement, the stabilized mixture increases in compressive and tensile strength, moduli of elasticity, the resistance to change of humidity, and the resistance to freezing and thawing cycles [1,3]. Thus, the improved properties and the increased load bearing capacity of the mixture make the cement stabilized mixture indispensable in the construction of pavement structures for heavy and very heavy traffic loads.

In addition to the mechanical properties of the material during the construction of pavement structures, the design and the selection of material for its execution and the cost-effectiveness, sustainability, and environmental impact [4] should also be taken into account. In designing cost-effective pavement structures, the accessibility and availability of local materials, of natural or industrial origin, are of great importance.

Thus, in the area of eastern Croatia, unlike other parts of Croatia, large amounts of river and dug sand are used in road construction. Sands are used for the construction of embankments, bedrock, and in the construction of unbound and cement-bound load-bearing layers of the pavement structure. The application of river and dug sands in eastern Croatia was imposed as a necessity in order to rationalize the costs of construction and was confirmed in the *Study of the possibilities of application of sand in the construction of roads of Slavonia and the Baranya region* [5]. Sand is applied in the natural state or stabilized with cement. The content of cement necessary for stabilization of sand is increased in relation to other granular stone materials and it amounts to 8–12%. Sand application as a local material enabled savings of as much as 50–60% in costs of transport of granular stone material. Through many years of application in road construction, sand has proven to be a quality and suitable local engineering material in the design and construction of cost-effective pavement structures and embankment layers [6].

Besides natural local materials, waste materials and industrial by-products also play a considerable role in the construction of cost-effective pavement structures [7,8]. These are materials that require complex depositing of large quantities at dump sites and are, as such, harmful for the environment, whilst their application has a significant economic justification and is clearly encouraged by sustainable development guidelines [4]. Ash from wood biomass (WA) is one of the industrial by-products with great potential for application in civil engineering, especially in road construction [4,8,9], where finer and coarser fractions of ash can be implemented in all layers of the pavement structure. WA is characterized as non-hazardous waste and is formed as a residue from the combustion of wood biomass for the production of electricity and heat. Based on the place of their creation and accumulation, WA are divided into three fractions: bottom ash (WBA) that is collected under the boiler grate and fly ashes (WFA) collected on a cyclone filter and electrostatic filter. The basic composition of WA depends on the type of biomass, and for the part that is being burned, it depends on ground type or climate [9].

2. State of the Art

One of the first waste materials that began to be used in construction was coal fly ash (CFA), produced by the combustion of coal in thermal power plants. Depending on the type, CFA found numerous and diverse applications in cement stabilized mixtures [4,10–24]. Unlike stabilization with cement, in which the basic granular material is stabilized instantly, stabilization by materials with pozzolanic properties, such as CFA, is based on the positive effects of delayed pozzolanic reaction.

Characteristic of stabilization mixtures with CFA, in comparison with mixtures with pure cement (CSM), is a gradual increase in strength which also continues for a longer period of time [1,7,11]. Namely, the increase in compressive strength of cement stabilized mixtures (CSM) is intensive during the first 28 days of curing, and it slows down after that. In the case of CFA stabilized mixtures, the development of compressive strength during the first 28 days of curing strength is slightly lower; however, with prolonged curing time (90 days or more), strength may be higher than CSM strength. This can be explained by the delayed pozzolanic activity of this material [1,7,10]. The overall developed heat of hydration in the mixture is, as a rule, reduced with the addition of CFA, which reduces the appearance of thermal cracks from shrinking in the constructed layer [11,12]. A unique phenomenon in stabilized layers is the ability of self-healing—meaning the ability of these layers to activate a mechanism of healing and rebinding of cracks, i.e., damages even after the cracking of a layer (whether the cracks arose due to the characteristics of a semi-rigid layer or whether the damage was caused by external force) [7]. This mechanism of self-healing is the result of the continuing pozzolanic reaction between the activator and CFA in the stabilized mixture. All the mentioned positive effects of CFA have made this industrial by-product the subject of multiannual and numerous studies in mixtures of cement stabilized sand [4,7,10–24].

With time, a possible application of other types of ash also started to be investigated in stabilized mixtures, such as ash from municipal waste incinerators [25,26], ash from paper mills, and particularly, bio ash of varying origin and of various fractions [27–40]. In the load-bearing layers of the pavement structure, to improve the load bearing capacity, finer (fly ash (WFA)) and coarser (bottom ash (WBA)) wood ash fractions can be used in two modes of action, depending on their properties. If inert and unable to react on its own, and given its fine granulation, it can be used as filler for mechanical stabilization. On the other hand, WA with a favorable physiochemical composition can be used as a binder if hydraulic or pozzolanic properties are present [41]. According to the research done by Sigvardsen et al. on WA [41–43], WA can show pozzolanic activity when high silicone, aluminum, and iron oxide content is present, but in more cases it has been recorded that, unlike the above mentioned CFA, WA are scarce in those pozzolanic minerals and have high calcium content. According to this oxide content, WA are more likely to show hydraulic properties, i.e., to be able to chemically react with water, set and harden, and retain strength and stability under water [43]. Keeping in mind this property and considering the application in roads, other research on this topic is described in more detail below. The stabilization effect of 10% and 20% WFA in mixtures with sand [27] and the unbound pavement of a forest road executed from crushed stone or gravel [28] was investigated by Škels et al. The results showed a multiple increase in load bearing capacity of the mixture expressed by the CBR bearing ratio in both stabilized mixtures, without the appearance of significant swelling. WFA was characterized by the authors as a good independent hydraulic binder for sand stabilization. Vestin et al. [29] investigated the possibility of the application of WFA from burning tree bark for stabilization of a base layer of gravel. Mixtures with 20% and 30% WFA had approximate initial strengths of 4.7, i.e., 4.4 MPa. An increase in compressive strength in the mixture with 20% ash amounted to 2.0 MPa, and in a mixture with 30% ash, it amounted to 5.3 MPa. The same authors reported about the stabilization of a test section with 30% WFA, which resulted in an increase in the load bearing capacity over time. The authors Bohrn and Stampfer [30] reported an increase in the modulus of elasticity E by 20.0 MPa by stabilization of the load-bearing layer of an existing forest road with WBA, in relation to the values obtained by mechanical stabilization of materials. Kaakkurivaara et al. [31] compared the effect of restructuring on test sections with the addition of a new surface layer of gravel (on the reference section) with sections on which, as well as gravel surfacing, the load-bearing layers from WFA or a combination of WFA and gravel were also constructed. An increase in the load bearing capacity was recorded in sections with WFA, but it was less than expected. The greatest improvement of the load bearing capacity was noticed on the sections that had the load-bearing layer made of a mixture of WFA and gravel. The authors attributed poorer results to insufficient compacting during execution, poor technology for mixing of the materials, and different times of storage of ash of particular sections. Sarkkinen et al. [32] ascertained that WA could also be used as a binder for stabilization of granular stone material from dolomite rock waste. The mixtures that consisted of ash from a wood fired power plant, peat and waste from a paper mill, and waste stone material in the proportion of 20:80 had a uniaxial compressive strength of 2.85 MPa after 7 days and 7.3 MPa after 28 days. This mixture, without additional additives, meets the strength properties necessary for installation in load-bearing layers in Finland. More recently, Cherian and Siddiqua [44] researched stabilization of silty sand with the addition of pulp mill (wood) fly ash (PFA). They showed that this type of ash in amounts of 20–30% acts as an effective stand-alone binder for weak subgrade stabilization. It improved strength, resulting in unconfined compressive strength of around 1.6 MPa at 28 days. However, no significant strength gain was recorded with prolonged curing at 60 and 90 days. Cabrera et al. [33] used WBA as a component in cement stabilized load-bearing layers from natural or recycled aggregate. They used a dose of WBA in the contents of 15% and 30%, while the content of cement amounted to 3% and 5%. The research showed that using an amount of WBA up to 15% increases the compressive strength, indirect tensile strength, and the modulus of elasticity of all the mixtures with this WBA in comparison to those that did not contain it, while using a dose of 30% WBA caused a reduction in the modulus of elasticity and indirect tensile strength.

In numerous studies of various types of WA carried out so far, the potential of its application in stabilization mixtures has been confirmed [36,37]. By applying WA, it is possible not only to improve the physical and mechanical properties of the stabilized mixture but also to reduce the proportion of cement in the mixture, which, along with the reduction in WA in landfills, is encouraged by the guidelines of sustainable development [38]. In addition to the numerous advantages of using WA, it should not be forgotten that any application of WA, given that it is a waste material of a variable chemical composition, should be confirmed by conducting complete laboratory tests [39,40].

3. Objective of This Study

According to Demirbas [45], biomass, as a renewable energy source, is considered one of the most diverse and valuable resources in the world. According to [46], producing energy from biomass during combustion results in a quantity of ash between 2.7% and 3.5%, on average, of the original weight of wood biomass.

Currently, numerous countries in the world (including Croatia) are faced with ever increasing quantities of bio ash at landfill, which will continue to be "produced" and disposed of, and the reuse/recycling of bio ash is strongly encouraged [9,37]. Unlike other countries, in Croatia, research into the application of WA started to be carried out only in recent years, initiated by large quantities of WA being produced in newly built power plants [47–53]. WA is produced in an amount of approximately 3.1% of the combusted biomass or about 25,414 tons per year in Croatia. According to the total power of plants which use biomass for energy production, the estimated amount of WA in Croatia is 38,461 tons per year [48]. The passing of the Energy Development Strategy of the Republic of Croatia intensified the construction of power plants on biomass fuels and the first biomass co-generation plant was commissioned in 2011. Almost half of the total number of commissioned power plants are in eastern Croatia, and the largest ones use wood biomass for energy production. The research on the possibilities of recovery of local WA is emphasized by the fact that the Osijek-Baranya and Vukovar-Srijem counties in eastern Croatia are part of the SRCplus project [47]. The European Commission initiated this project in order to produce sufficient quantities of wood chips to meet the national and European energy goals. The project planned the development of a sustainable supply of local power plants/co-generation plants with wood biomass that would come from short rotation, woody crops grown on the agricultural land of poorer quality.

However, even though WA is being progressively produced, its usage in road construction is still the area that requires more research before wider application. While there is some research on the application in asphalts or subgrade improvement, application in hydraulically bound base courses is sparse, considering WA varying properties. Its usage in combination with or as an improvement to local materials of lower quality (sand) for usage as base courses is even less known and therefore the subject of this research. The aim of the research described in this study is (1) to determine, by carrying out the planned tests and analysis of the results, the possibility of using WFA in the load-bearing layers of the pavement structure, while achieving the required and prescribed engineering properties of mixtures. Furthermore, the aim is (2) to enable the rationalization of costs of construction of the pavement structure by using local materials (natural; sand and waste; WFA) and ultimately, in accordance with the concept of sustainable development, (3) to contribute to the reduction in WFA in landfills.

4. Experimental Section

When using new and alternative materials in road construction, all aspects of their application should be determined. Along with the effect on mechanical properties in a given desired application, both the durability and the effect on the environment should be determined. However, determination of such mechanical properties is the first step to check if the application of the material complies with regulation and if further evaluation is justified. In technical regulations, the properties of cement stabilized mixtures are mostly defined by compressive and tensile strength. For cement stabilized materials (CSM), it is usual to test compressive strength after 7 and 28 days of curing of samples, while the period to achieve the required compressive strength may be even longer for materials with pozzolanic properties with a prolonged binding time. In addition to strengths, it is important to know the resistance of stabilized mixtures to freezing, since low environmental temperatures in combination with water can cause additional stress in the cement stabilized layer and also a loss in the load bearing capacity during thawing. In the text that follows, the article describes the tests of the mechanical properties of stabilized mixtures of sand, WFA, and cement: the compressive and indirect tensile strength and the resistance of mixtures to freezing, as is shown in Figure 1.

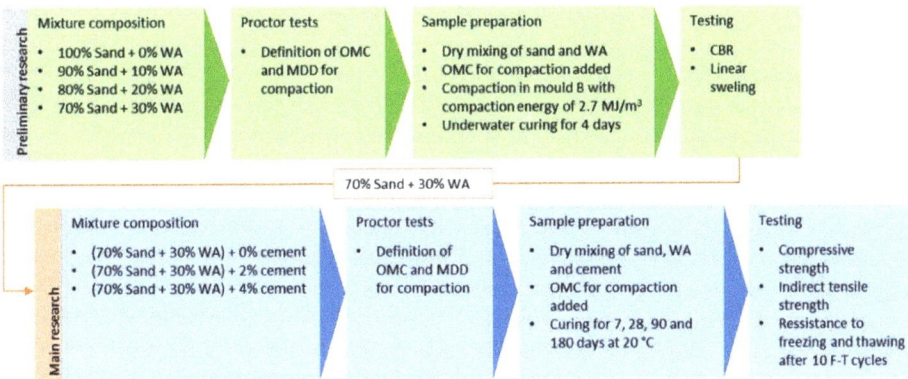

Figure 1. Flow chart of experimental research.

Laboratory research was carried out on mixtures composed of sand from the Drava River as the basic aggregate, WFA from a co-generation plant for the production of electricity, and heat based on the combustion of wood biomass in eastern Croatia, with cement CEM II/B-M (P-S) 32.5 N and water.

Sand from the Drava River is a material that has proven its application in the construction of embankments and load-bearing layers of pavement structures, although it is a uniformly graded material (SP) that is somewhat harder to compact due to its composition [5]. Determination of the granulometric composition of Drava sand was carried out on samples of 500 g in accordance with the standard HRN EN ISO 17892-4 [54] on a mechanical vibrating table with a set of sieves whose openings ranged from 31.5 to 0.063 mm. According to the sieving results, 96% of the sand particles have a diameter of less than 0.5 mm, and almost 70% of them have a diameter of less than 0.25 mm. The diameter of the largest grain is 1.50 mm, and the proportion of particles smaller than 0.063 mm is under 1% of the mass. Table 1 shows the physical and mineralogical properties of the sand and Figure 2 its particle size distribution. In the granulometric diagram, curves of the boundary area for the application of sandy material in load-bearing layers with stabilized cement have been added, as defined by the Croatian standard for the production of cement-stabilized load-bearing layers, HRN U.E9.024 [55].

Table 1. Physical and mineralogical [56] properties of sand from the Drava River.

Physical Properties	UCSC	Density (Mg/m^3)	Color	D_{10} (mm)	D_{30} (mm)	D_{60} (mm)	C_u	C_c
	SP	2.68	Grayish-Brown	0.13	0.16	0.22	1.68	0.895
Mineralogical properties	type of mineral	quartz	calcite	dolomite	feldspars	clay minerals		
	mass %	71	2	2	25	1.2		

Note: C_u = coefficient of uniformity; C_c = coefficient of curvature; D_X is the diameter of material particle below which X percent of materials are finer than this D_X size.

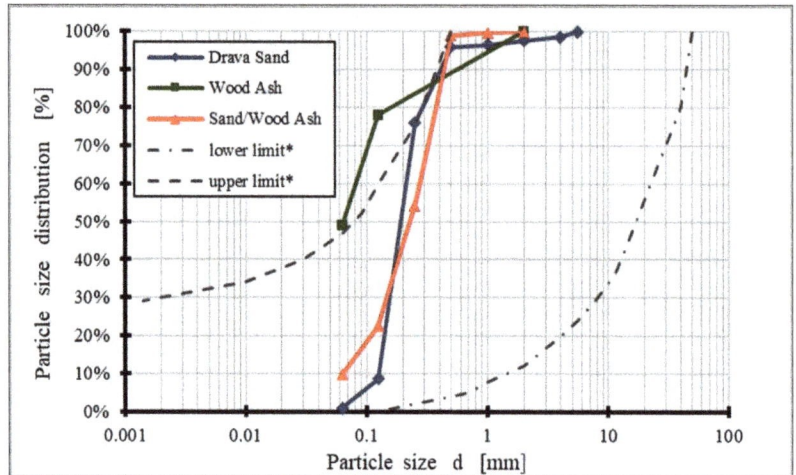

Figure 2. Particle size distribution curves, (* curves of the boundary area are defined according HRN U.E9.024 [55]).

Mineralogical composition of sand was determined by the qualitative and semi-quantitative X-ray diffraction analysis [56]. For XRD measurement, a PANalytical X'Pert Powder diffractometer was used. X-ray source was CuKα radiation with wavelength of λ = 1.54 Å. The sample was measured in step-scan mode in area between 4 and 66°2θ with step size of 0.02°2θ in sample time of 4 s. The obtained XRD measurements were evaluated by X'Pert HighScore Plus. The mineralogical composition of Drava sand is presented in Figure 3. Results show that quartz is the predominant mineral (71%) and there are high amounts of felspars (24%) in the form of plagioclase ad potassium feldspar. Along with small amounts of calcite and dolomite, there is also a small amount of clay minerals, probably in the form of muscovite and ilite. Even though the main component—quartz—is completely inert and should not contribute to strength or particle packing (uniform sand), components such as feldspar and clay minerals could potentially present a source of aluminosilicates and contribute to strength if activated by source of calcium and water.

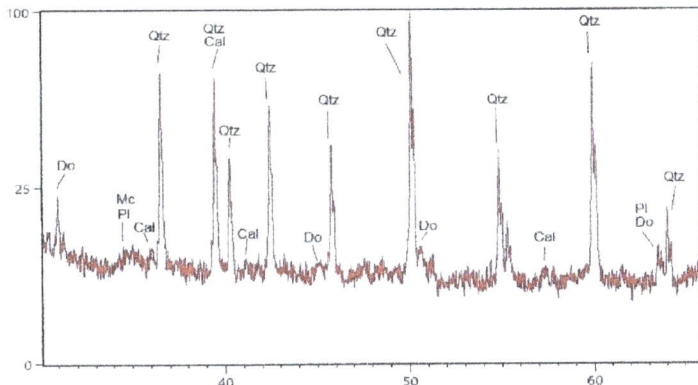

Figure 3. The results of the XRD analysis of Drava sand (Cal—calcite, Ch—chlorite, Do—dolomite, Kfs—potassium feldspar, Mc—mica minerals, Ph—plagioclase, Qtz—quartz) [56].

As mentioned in [57], both the shape and the texture of sand play an important role in the performance of sand as a pavement material. The shape of the sand particles has a considerable impact on compactness and stability, i.e., the engineering behavior of the sand mixture. Thus, during compacting, irregular sand particles with a rough surface are more favorable than smooth rounded particles and enable a stronger bond to be formed during binding with the cement [58]. Therefore, scanning electron microscopy (JEOL SEM JSM-IT200) was used for images of the shape and size of sand particles (Figure 4). SEM images showed that sand particles vary from oblate to equate with the highest amount of equant, and in terms of the degree of roundness they are angular to subrounded, according to classification in [57], and they have a rough surface.

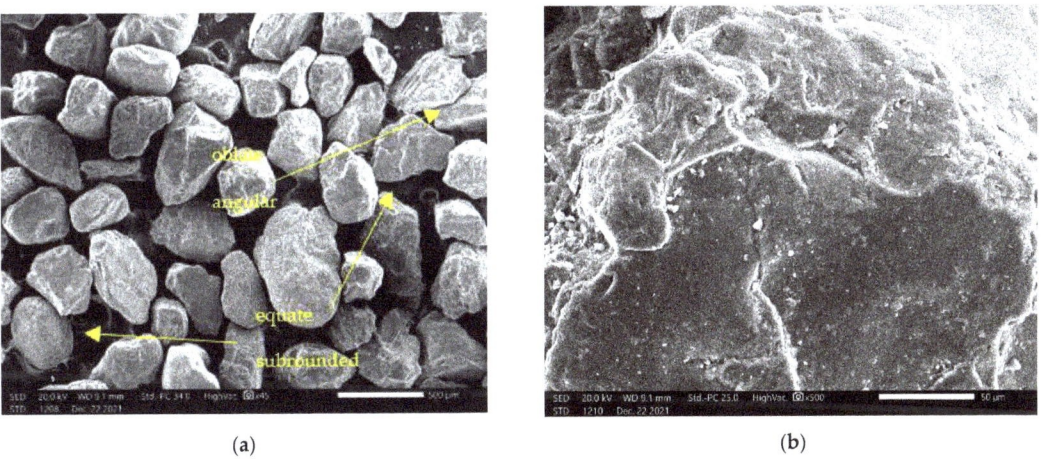

Figure 4. SEM microphotograph of sand: (**a**) magnification SEM_MAG = 45×; (**b**) magnification SEM_MAG = 500×.

WFA used in this study is collected on a cyclone filter. The granulometric composition of WFA was determined on sieve openings of 0.063 mm, 0.125 mm, and 2 mm according to the specification Fly Ash for Hydraulically Bound Mixtures (HRN EN 14227-4 [59]) and air jet sieving method was used (HRN EN 933-10 [60]). Granulometric composition of WFA is shown in Figure 2, and its chemical composition is shown in Table 2.

Table 2. Chemical composition of WFA.

Components	MgO	Al$_2$O$_3$	SiO$_2$	P$_2$O$_5$	SO$_3$	K$_2$O	CaO
mass %	3.06	0.44	4.05	2.90	1.59	2.82	46.9

As seen from the composition, the main component is calcium oxide, which indicates that WFA could have hydraulic properties. In turn, content of pozzolanic oxides is quite low and well below the set limit of 70% (EN 450-1 [61]) to be considered source of pozzolan. As a result of testing for the mineral composition of ash samples by the X-ray diffraction method (XRD) [62] (Figure 5), it was determined that the main components of WFA were calcite, quartz, and CaO, and, in smaller quantities, portlandite (CaOH)$_2$ and fairchildite (K$_2$Ca(CO$_3$)$_2$).

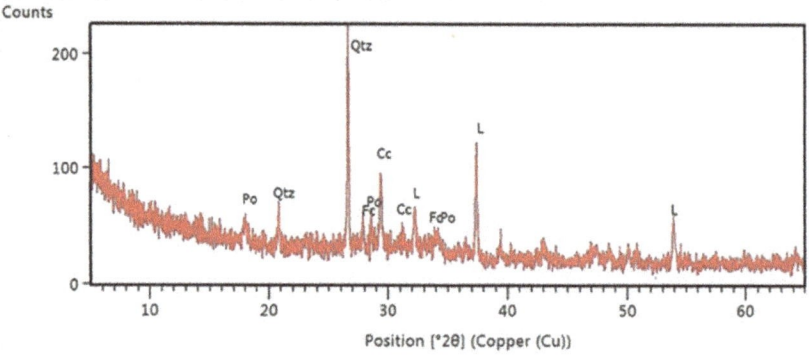

Figure 5. The results of the XRD analysis of WFA [62].

A morphological analysis was carried out using scanning electron microscopy (SEM) (Figure 6). SEM images show that WFA is composed of particles of different shapes and sizes. Ash particles are irregular in shape and have a rough, porous surface, and, in addition, particle sizes are very uneven. The overall structure of particles does not follow a specific pattern.

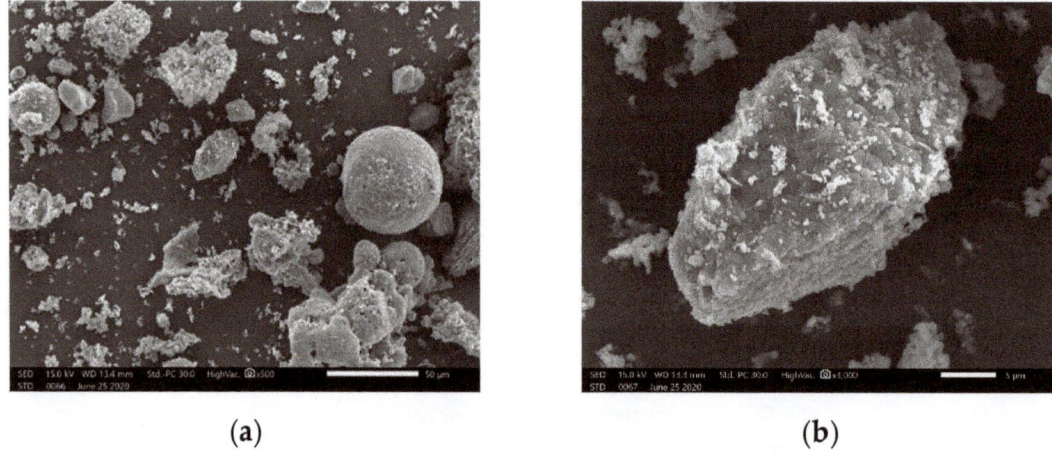

Figure 6. SEM microphotograph of WFA particles: (a) magnification SEM_MAG = 500×; (b) magnification SEM_MAG = 3000×.

Cement CEM II/B-M (P-S) 32,5 N [63] is a type of cement that is most often used for the construction of stabilized layers in eastern Croatia, and its properties are presented in Table 3. For the production of samples of the stabilized mixture, water from the city water supply was used.

Table 3. Properties of cement CEM II/B-M (P-S) 32,5 N according to HRN EN 197-1:2012 [63].

Initial Time Setting (min)	Stability of Volume According to Le Chatelier (mm)	Unconfined Compressive Strength (2, 28 Days) (MPa)	SO_3 (%)	Cl (%)
200	0.4	16.0; 42.0	3.20	0.009

4.1. Stabilized Mixture Composition

4.1.1. Determination of Optimum WFA Amount in Stabilized Mixtures

The composition of stabilized mixtures is defined, taking into account the fact that WFA can be used in mixtures in two ways: (1) as a binding component, when its addition initiates certain chemical reactions as a result of pozzolanic or hydraulic activity, and (2) as a filler, when it is necessary to improve the physical properties of mixtures by increasing the proportion of fine particles. When initiating chemical reactions in a mixture, the chemical composition of WFA is important, whilst its granulometric composition is also important for improving the physical properties of the mixture.

Since Drava sand is a uniformly graded material (SP), preliminary tests were conducted on mixtures of sand with varying amounts of WFA (0%, 10%, 20%, 30%) as means of mechanical stabilization (Figure 1). The tests included the determination of the optimal moisture content (OMC) and the maximum dry density (MDD), as well as the load bearing capacity of the mixture expressed in accordance with the California Bearing Ratio (CBR). Firstly, sand was dried at 105 °C and both sand and WFA were sieved on a 2 mm sieve opening to remove any larger materials such as shells, rocks, branches, bark, or any unburnt material in ash. No other conditioning was done on those materials. Prior to sample preparation for the test, first the dry sand and WFA were thoroughly mixed, and then the different water amounts were added and mixed until uniform mixture was achieved. Five samples of each mixture were than compacted with energy of modified Proctor test (2.70 MJ/m^3) in mold A (h = 120 mm, d = 100 mm). After that, samples were weighed and dried to determine the OMC and MDD according to EN 13286-2 [64]. Such determined OMC were then used to prepare samples of all four mixtures for CBR tests. Three samples of each mixture were prepared in a Proctor's cylindrical mold B with a diameter of 150 mm and a height of 120 mm. After compaction, the samples were submerged in water with a preload of 4.5 kg for 4 days, and the daily reading from the micrometer was recorded so that the increase in linear swelling could be seen. Determination of the CBR ratio on mixtures of sand and WFA was carried out in accordance with the 13286-47 [65] standard, where CBR 1 and CBR 2 are defined by measuring the forces at piston penetration of 2.5 mm and 5 mm and expressed as ratios of standard forces od 13.2 kN and 20 kN, respectively. CBR results are presented in Table 4.

Table 4. Results of the CBR ratio and linear swelling of sand/WFA mixtures.

Test		Amount of WFA in Mixtures			
		0%	10%	20%	30%
CBR 1	%	27.44	48.99	52.69	82.09
CBR 2	%	18.96	59.39	56.82	90.70
Linear swelling	%	0.02	0.08	0.22	0.64

The conducted tests proved that WFA has a direct impact on the increase in the load bearing capacity of the mixture. The CBR ratio increased considerably with the increase in the content of WFA (10%, 20%, 30%) in the mixtures, in relation to pure sand. The highest value of load bearing capacity was recorded for the mixture with 30% WFA and it amounted to 90.70%, which is three times more than bearing ratio of Drava sand (27.44%). This is the result of mechanical stabilization as added finer WFA acted as filler for pore in uniformly graduated sand. However, the samples with highest ash content were also quite cohesive and hardened, which proved that the bearing capacity increase is not a result of better compaction solely and that some form of hydraulic binding occurred. For this reason, in order to carry out further research, a mixture of sand and 30% WFA was chosen as the basic mixture to which cement is added in different proportions.

4.1.2. Designing the Composition of Stabilization Mixtures

The usual proportion of cement required to stabilize Drava sand and achieve the required values of compressive strength in previous tests was 8–12% [5,6]. Considering the WFA content in the mixture and its hydraulic activity potential, and based on preliminary tests, the cement content in the mix was significantly reduced compared to the usual percentage. The aim was to determine the minimum proportion of cement in the stabilized mixture that would ensure that the mixture had the prescribed strength values and adequate resistance to freezing. Mixture composition is visible in Table 5.

Table 5. Proctor elements OMC and MDD for sample preparation.

Mix No.	Mixture Composition	OMC [%]	MDD [g/cm^3]
1	(70% sand + 30% WA) + 0% cem	13.50	1.72
2	(70% sand + 30% WA) + 2% cem	12.69	1.73
3	(70% sand + 30% WA) + 4% cem	11.61	1.70
Drava Sand		14.20	1.64

A modified Proctor experiment was conducted in accordance with the HRN EN 13286-2 standard [64]. Five samples of each mixture with various moisture contents were prepared to determine OMC and MDD. For the purposes of the research, a Proctor's cylindrical mold A with a diameter of 100 mm and height of 120 mm was used. Five layers of samples were compacted with the appropriate energy (2.7 MJ/m^3) in an automatic Proctor device. The resulting values of the OMC and MDD (Table 5) were used to prepare the samples by the same method; they were wrapped in cling film and cured 7, 28, 90, and 180 days in climate chambers at 20 °C. Three samples of each mixture were used to determine compressive strength, indirect tensile strength, and the mixture's resistance to freezing and thawing.

5. Test Methods

5.1. Compressive Strength Test

Compressive strength is calculated as average stress of three samples of the same mixture that are exposed to uniaxial pressure at fracture force. The compressive strength of the samples is determined in accordance with the standard HRN EN 13286-41 [66], usually after 7 and 28 days of curing. Each sample is loaded evenly between two slabs, continuously and without jerking with constant increase in force until the samples break. Thus, the maximum load of the sample up to breaking point is recorded and the compressive strength of the sample is calculated according to the formula:

$$f_c = \frac{F}{A} \tag{1}$$

where: f_c = compressive strength (MPa); F = max compressive force (MN); A = sample cross section area (m^2).

The standard HRN EN 13286-14 [66] also prescribes that the breakage of the sample/appearance of cracks occurs from between 30 and 60 s from starting to apply the load. After breakage, the sample is taken out of the press and types of breakages are studied, which may be satisfactory or unsatisfactory. Testing of compressive strength was carried out by the device Shimadzu Autograph AG-X Series (Figure 7), and the results were processed by the computer program TRAPEZIUM MX.

(a) (b)

Figure 7. Compressive strength test: (**a**) before loading and (**b**) after loading.

Testing of compressive strength of the samples was carried out after curing samples at a temperature of 20 °C for 7, 28, 90, and 180 days for mixtures 2 and 3, while for mixture 1 (0% cement), testing of compressive strength was carried out after 7, 28, and 90 days. After curing, the samples were first unwrapped and testing was conducted at room temperature between 21 and 24 °C. Compressive strength was calculated as mean values of three samples for each mixture and curing length. The results of compressive strength testing are presented in Section 6.1.

5.2. Tensile Strength Test

Tensile strength was determined by the indirect test method in accordance with HRN EN 13286-42 [67], which is considered the most appropriate for practical testing of cement stabilized materials. Loading is applied on a cylindrical sample by a wood loading bar. Such a load causes a relatively uniform stress perpendicular to the diameter plane at which the pressure is applied. Indirect tensile strength was calculated according to:

$$f_t = \frac{2F}{\pi h d} \quad (2)$$

where: f_t = indirect tensile strength (MPa); F = max compressive force (MN); h = length of the sample (m); d = diameter of the sample (m).

Testing of indirect tensile strength of the samples was carried out after curing samples at the temperature of 20 °C for 7, 28, 90, and 180 days, except for the samples of mixture 1, on which testing was carried out after 7, 28, and 90 days. As in compressive strength, samples were first unwrapped and then tested at room temperature, while each indirect tensile strength result shown is calculated as mean of three tested samples. The equipment used in this testing is the same as that used for testing of compressive strength with the exception of the connection through which the load is applied (Figure 8). The results of indirect tensile strength testing are presented in Section 6.3.

(a) (b)

Figure 8. Indirect tensile strength tests: (**a**) before loading and (**b**) after loading.

5.3. Testing of Resistance to Freezing and Thawing

The resistance of samples to freezing was tested in accordance with the European technical specification HRS CEN/TS 13286-54 [68] according to which the resistance of mixtures to freezing and thawing is defined by measuring the compressive strength of the samples that have been subjected to freezing and thawing cycles and the control samples that were cured for the same period according to the prescribed curing regime. Two sets of samples, with the same dimensions and shape as for compressive strength testing, were prepared, wrapped in plastic adhesive foil, and cured for 28 days at a temperature of (20 ± 2) °C. The technical specification HRS CEN/TS 13286-54 [68] prescribes that the samples should be in this regime of curing for 28 days during which strengths develop, and then followed by a two day (secondary) curing of the samples under water, in a water saturated climate chamber or curing the same as in the first 28 days. In this research, the form of curing least favorable for the samples was used, i.e., two-day curing under water (Figure 9a). This method of curing enables direct contact of water with the sample and the action of water during freezing of the sample. After the two-day curing under water, Set A of each mixture was subjected to freezing and thawing cycles in the climate chamber (Figure 9b) while Set B—the control set—was left at the secondary curing during this period. Each cycle of freezing lasted 24 h and consisted of a period during which the temperature dropped from +20 °C to −18 °C, which was then maintained for a further 6 h, and then a period of thawing. After 10 freeze/thaw cycles, the samples were returned to the original cure regimen with a control set of samples for 24 h, and then both sets of samples were subjected to compressive strength testing (Figure 9c). Considering that significant quantities of salt are spread on road pavements in eastern Croatia during the winter period, an additional freeze/thaw test was carried out with the presence of salt, in accordance with EN 1367-1 [69] (Set C). The water, in the second stage of curing, was replaced with a solution of 1% sodium chloride (NaCl).

Retained compressive strength after freezing was calculated according to the formula:

$$\text{RFT} = \frac{M_A}{M_B} * 100 \tag{3}$$

where: RFT = retained strength factor, after freeze/thaw testing, M_A = the mean value of strength for Set A (MPa), M_B = the mean value of strength for Set B—control (MPa).

(a) (b) (c)

Figure 9. Testing of resistance to freezing and thawing: (**a**) two-day care of samples under water; (**b**) climate chamber; (**c**) compressive strength test after freezing/thawing cycles.

6. Results Analysis and Discussion

6.1. Analysis of Compressive Strength Results

The compressive strength results of stabilized mixtures are presented in Figures 10 and 11 and the development of compressive strength over time compared with the results from the study [5] is presented in Figure 12.

The stabilized mixture with 0% cement had the following compressive strength values: 1.4 MPa after 7 days, 2.52 MPa after 28 days, and 4.06 MPa after 90 days. By adding a very small quantity of the cement, the reaction is even more intensive and results in higher strength values. Thus, the mixture with 2% cement on average had 55.7% higher compressive strength than the mixture with 0% cement; the greatest increase in strength recorded between mixtures was after curing for 28 days. The highest values of compressive strength were, as expected, obtained by mixtures with 4% cement: 2.5 MPa after 7 days, 5.10 MPa after 28 days, 8.06 MPa after 90 days, and a value of compressive strength of 8.89 MPa after 180 days. The average increase in compressive strength of mixtures with 4% cement compared to mixtures with 0% cement is as much as 75.7%, and in this case the largest increase in compressive strength (102.4%) was recorded after curing for 28 days.

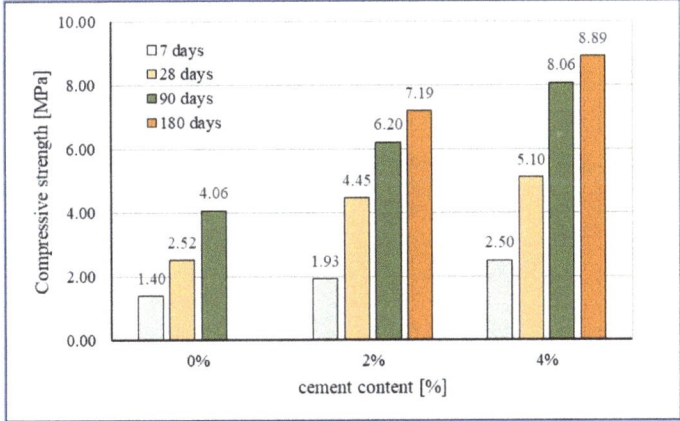

Figure 10. Effect of the cement content on the compressive strength.

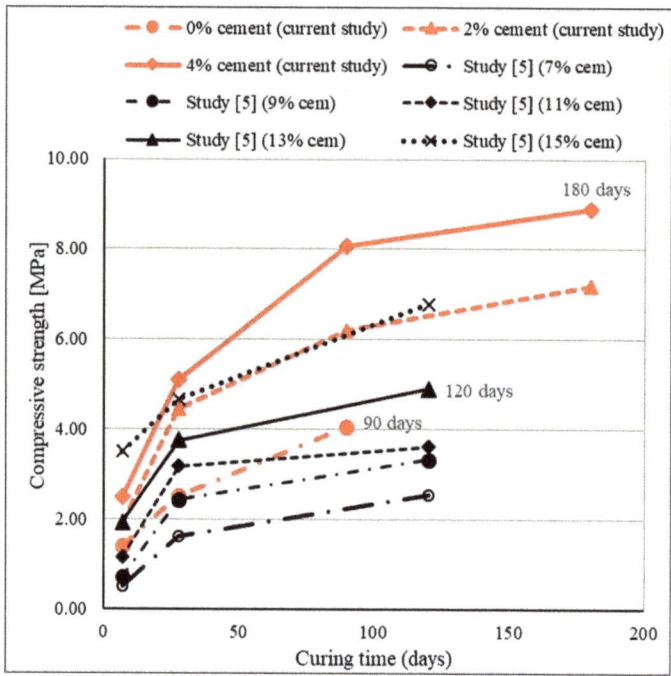

Figure 11. Effect of curing time on the compressive strength.

Although the bearing capacity (CBR) improvement can be attributed to the change in particle size distribution of sand by the addition of WFA, the compressive strength development in mixtures without cement points to some form of hydraulic hardening as well. According to some authors [43], this is sometimes due to WFA acting as a pozzolanic binder but more commonly it is due to WFA acting as a hydraulic binder. The WFA used here contains only a small amount of pozzolanic oxide but it has large amounts of calcium oxide. Even though total oxide content does not necessarily predict the actual performance and WFA has shown only low self-cementing properties (per ASTM D5239 [70]), recorded compressive strength on mixtures without cement is significant. As WFA could not on its own develop this compressive strength, results obtained on mixtures with no cement (0%) must be the product of the interaction of WFA and sand, i.e., their chemical and mineralogical compounds. This WFA that has high CaO content, in the mixtures with presence of water, creates a highly alkaline environment (pH of 12.7 [71]). As in the lime application, Ca+ and OH- ions could also initiate chemical reactions if aluminosilicates were available and this would result in the development of the compressive strength. Given that the major constituent of sand is quartz (71%), which is generally not reactive, an explanation could come from the joint effect of the CaO reaction with the small number of clay minerals in sand, the low SiO_2 content of WFA, and also the feldspar minerals in sand (25% of sand content). The pozzolanic reaction of lime with clay minerals that results in hydration products, such as in cement hydration (CSH and CAH), is well known [72]. However, at 1% content of clay minerals in sand, it can hardly be responsible for all the gained strength. However, feldspar is also an aluminosilicate with a chemical composition that includes Ca, Si, and Al cations and, therefore, pozzolanic properties could be expected [73]. Because around 80% of feldspar is made from SiO_2 and Al_2O_3, the authors in [74] believe that in their research it reacted directly with $Ca(OH)_2$ available from cement to form a hydration product such as CSH. According to [75], lime added to properly treated feldspar can create an agent that can act as an activator for the development of pozzolanic reactions in known silica sources such as slag and fly ash of silica fume. This should result in material with cementitious composition that can be used in filings, paving, or stabilizations.

In this research, sand containing feldspar minerals was mixed with WFA that had high CaO content, and this resulted in hydraulic activity-mixture hardening and strength development. It is reasonable to assume that the gained compressive strength of the mixture without cement can be attributed to the alkaline environment created by the WFA addition that then led to a hydraulic reaction with the available silica and the development of cementitious hydrates. However, even though measured mechanical properties point to this, further research into the developed mineralogic phases, which was not in the scope of this study, is necessary to fully understand these effects.

The addition of 2% and 4% of cement to mixtures further increases their strength. By increasing the cement content in the mixtures, the compressive strength value increases—mainly due to the hydraulic reactions between cement components (lime and aluminosilicates), which leads to the production of hydrated calcium silicate (CSH) and hydrated calcium aluminate (CAH).

The binding activity continued over time, which is evident by the increase in mechanical properties (compressive and tensile strengths) of the samples over time. However, since cements are formulated such that most of the compressive strength gain is achieved in the first 28 days, the significant long term strength gain recorded here (90 and 180 days) could be due to slower pozzolanic reaction. Pozzolanic reactions are well known to occur over longer time periods, such as months or years, in normal curing conditions. The mechanism of strength gain involves using available and unused $Ca(OH)_2$ in moist conditions (presence of water) to activate the source of pozzolans and to create further cementitious materials (CSH and CAH) through time. This mechanism is particularly noticeable in the diagram in Figure 11, where the contribution of WFA achieves higher values and the development of compressive strength of the mixture over time is clearly seen.

Increasing the curing time of the specimens had a positive effect on the increase in compressive strengths of the mixtures. From the results, it can be seen that the increase in compressive strength in mixtures with 2% and 4% cement, although greatest during the first 28 days of curing (130.57% for 2% cement and 104% for 4% cement), is significant and continuous for longer periods of curing. Namely, in addition to the relatively rapid increase in compressive strengths during the first 28 days of curing due to hydraulic reactions, long-term pozzolanic reactions also develop. A particularly significant increase in compressive strength over time was achieved by mixtures with 0% cement with WFA only, taking into account the properties of sand and WFA and necessary prolonged curing for the pozzolanic reaction. Mixtures with WFA achieved 80% higher compressive strength after 28 days and 190% higher strength after 90 days (compared to 7-days compressive strength) and thus confirmed the ability of WFA to act as a binder in the sand mixture. The test results show that, by replacement of part of the sand with WFA, in the quantity of 30%, much higher values of compressive strengths can be achieved than in mixtures of sand stabilized only with cement. This is particularly evident in Figure 11 where the obtained results are compared with the results from the study [5], in which compressive strengths of stabilized mixtures composed of Drava sand and cement were analyzed. Cement that was then used in testing had the same properties and class as the cement that was used in the research described in this article. The determination of compressive strength was conducted after 7, 28, and 120 days. The measured compressive strengths were 0.5 MPa and 1.62 MPa for the mixture with 7% cement; 0.71 MPa and 2.43 MPa for the mixture with 9% cement; 1.15 MPa and 3.18 MPa for the mixture with 11% cement; 1.925 MPa and 3.75 MPa for the mixture with 13% cement; and finally, 3.5 MPA and 4.67 MPa for the mixture with 15% cement, measured at 7 and 28-days curing, respectively.

Comparing the results obtained in this study with the results of the study [5], a significant difference in the values of compressive strengths can be seen, i.e., the significant contribution of WFA in the mixture to the increase in compressive strengths. Thus, a mixture of sand and WFA with 2% cement and mixture of pure sand and 13% cement had the same compressive strengths at 7-days curing, while higher compressive strength was measured at 28-days curing on the mixture with WFA and 2% cement (4.45 MPa vs. 3.75 MPa). The mixture of sand and WFA with 4% cement achieved a 7-day compressive strength (2.50 MPa) greater than the mixture of sand stabilized with (as much as) 13% cement (1.925 MPa). The most interesting result is certainly that of the mixture with 0% cement which (with only WFA) achieved a higher compressive strength already after 7 days (1.40 MPa) compared to mixtures of sand stabilized with 7%, 9%, and 11% cement. In addition, for this mixture without cement, an intensive increase in compressive strength is noticed even after 28 days, which was less prominent for mixtures of sand and lower cement content from the study [5]. The stabilized mixture with 2% cement achieved higher values of compressive strengths from all the mixtures of sand and cement, except the mixture with 15% cement. The stabilized mixture with 4% cement had a lower value of compressive strength (2.50 MPa) after 7 days than the mixture of sand and 15% cement (3.50 MPa); however, over time, already after 28 days, the mixture with 4% cement achieved a compressive strength of 5.10 MPa and the strength continued to increase with the duration of curing. Clearly evident in the diagram is the exceptional contribution of WFA in the mixture of sand to the results of compressive strength, as well as the continuous development of compressive strength over time. Additionally, the results confirm the conclusions of previous research [11,16,18,33] that the compressive strength in mixtures with fly ash (with a notable content of CaO) and cement is considerably higher than the compressive strength of mixtures stabilized with cement only.

The Geotechnical Classification of NRC materials, suggested by the author in [76] (NRC: New construction based on recycled materials) classifies mixtures into five groups (A–E) according to the minimum compressive strength. The materials, i.e., mixtures that achieve a minimum compressive strength of 4.5 MPa belong to Group A (very strong), while Group E is for materials in which the development of strength does not occur (no strength development). According to the proposed classification, stabilization mixtures with 0% and 2% cement belong to Group B: strong (min. 1.5 MPa) while the mixture with 4% cement belongs to Group A: very strong (min. 4.5 MPa); therefore, they could be used as bearing layers.

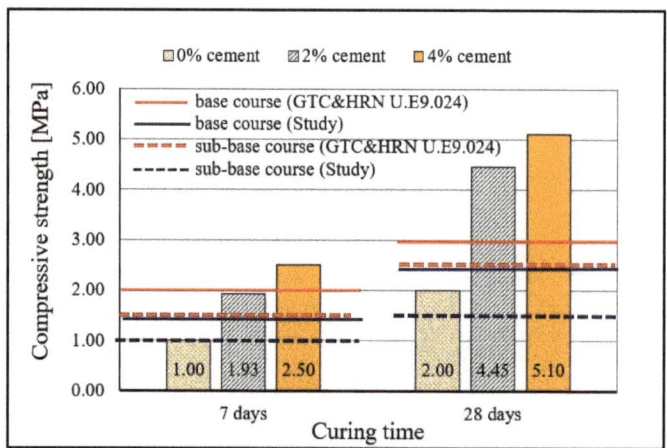

Figure 12. Comparison of compressive strengths of mixtures with the prescribed criteria of the Croatian technical regulation [5,55,77].

6.2. The Comparison of Values of Compressive Strengths of Stabilized Mixtures with the Criteria Prescribed by the Applicable Croatian Technical Regulations

In order to determine a possible application of stabilized mixtures in the construction of stabilized load-bearing layers, the results of compressive strength after 7 and 28 days must meet the criteria prescribed by the standard or technical conditions. There are still no precisely defined requirements regarding quality and the load bearing capacity of the layers stabilized with fly ash (or any other pozzolan) in the Republic of Croatia, and the comparison of the results obtained will be carried out on the basis of the criteria applicable for mixtures stabilized with cement.

As already stated in the introductory part of the study, for the needs of construction of pavement structures with sand in load-bearing and stabilized layers, the *Study of the possibilities of the application of sand in the construction of roads of Slavonia and Baranya region* [5] was prepared. The study [5] prescribed somewhat lower criteria of compressive strength in relation to the criteria defined by the HRN U.E9.024 standard [55]. The standard does not differentiate between the type of aggregate in the load-bearing layer (sand, gravel, or crushed stone) and the values of compressive strengths are defined in relation to the road category (i.e., the expected traffic load) and the position of the layer in the pavement structure. Therefore, the conditions for the application of mixtures of sand are stricter. Non-standard binders such as fly ash or slag are mentioned only in the General Technical Requirements for Road Work (GTR) [77]. The GTR stipulates that when using binders other than cement (fly ash, slag), the compressive strength limits of the mixture remain the same as for cement but with a prolonged duration of curing that needs to be determined on the basis of laboratory tests. The defined minimum compressive strength criteria and the results of the comparison are shown in Figure 12.

From the diagram in Figure 12, it is evident that the mixture with 4% cement meets all the criteria for application, regardless of the position of the layer in the pavement structure. These results are particularly good when we take into account that the criteria for application are defined for mixtures of crushed stone or gravel. The mixture with 2% cement meets all the criteria of application according to the study and it meets the (stricter) criterion for application in lower base layers, according to the standard. Although, after 28 days, the compressive strength of this mixture is more than sufficient for application in upper base layers, the value of the strength of the mixture with 2% cement is somewhat lower than required after 7 days. The mixture with 0% cement, strictly speaking, does not meet the criteria for application in upper base layers and may be used in lower base layers of the pavement structure. However, the achieved results of compressive strength are good, bearing in mind the type of binder in the mixture. Namely, in the case of fly ash, a longer time of curing for the development of compressive strength should be defined, whilst here the results for curing of 7 and 28 days are analyzed, according to the criterion for mixtures stabilized with cement. In the results for curing for 28 days, an improvement of 80% in compressive strength of the mixture with 0% cement is already evident.

6.3. Analysis of Indirect Tensile Strength Results

The results of the indirect tensile strength tests are presented in Figure 13. The trend of the development of indirect tensile strength for all stabilization mixtures and all durations of curing is very similar to that for compressive strengths and it is attributed to the same binding effects. The stabilized mixture with 0% cement gave the following values of indirect tensile strength: 0.2 MPa after 7 days, 0.64 MPa after 28 days, and 0.86 MPa after 90 days, which confirms the activity of these ashes as binders in the mixture of sand of this mineral composition. By adding a very small quantity of cement, the reaction was more intense, and the strength results were considerably higher. Thus, the mixture with 2% cement achieved on average (for curing durations of 7–90 days) 49.9% higher tensile strengths than the mixture with 0% cement, with the greatest increase in tensile strength recorded during the curing of 28 days (56.25%). The greatest values of tensile strength were, as expected, achieved by the mixture with 4% cement: 0.48 MPa after 7 days, 1.0 MPa after 28 days, 1.56 MPa after 90 days and, after 180 days, the value of tensile strength was 1.93 MPa. The average increase in indirect tensile strength of mixtures with 4% cement in relation to the mixtures with 0% cement, for curing of 7 to 90 days, amounts to 75.55%, almost the same as in the case of the compressive strengths of the mixture.

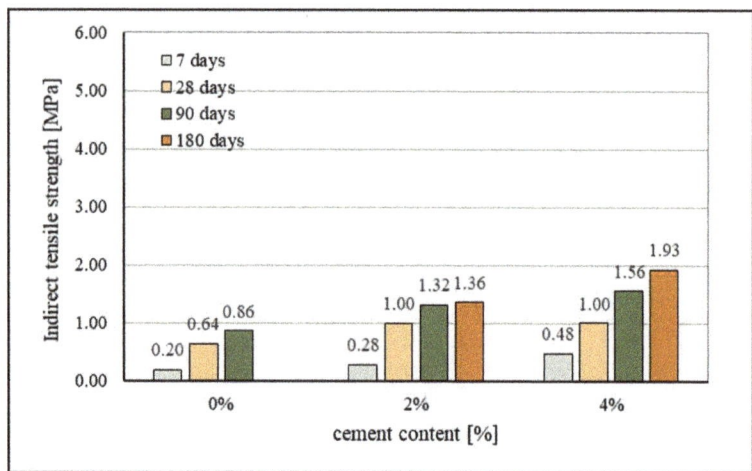

Figure 13. Effect of cement content and curing on indirect tensile strength.

With prolonged curing, a considerable increase in indirect tensile strength can also be seen. Thus, for mixtures with 0% cement, the increase after 90 days, in comparison with an initial strength after 7 days, amounts to 330%. Mixtures with 2% cement even achieved an increase in strength of 371.4% after curing of 90 days, whilst the average increase in indirect tensile strength for all durations of curing (7–180 days) was 245.0%. Mixtures with 4% cement achieved a 302% increase in strength after 180 days in comparison to 7 days, while, after 90 days, they had a strength of over 225% in comparison to the 7-day strength. All the obtained results of indirect tensile strengths, as was the case with compressive strengths, are the result of the hydraulic and pozzolanic activity between constituents of the mixture, further improved by the addition of cement, and also somewhat due to the favorable shape and roughness of the surface of the grains of sand [78].

6.4. Correlations between Compressive and Indirect Tensile Strength

Regression analysis of the test results enabled the establishment of connections (correlations) between compressive and indirect tensile strengths of stabilization mixtures and the determination of the shape and strength of these connections. The regression analysis conducted was between compressive and indirect tensile strength of the samples of the same mixture and same curing length, for example, between f_c and f_t of mixture 1 at 28 days of curing. This type of analysis is useful as it enables the prediction of indirect tensile strength of mixtures based on their compressive strength with a high degree of confidence and also shows possible variability of different design mixtures. For each mixture group and all mixtures combined, a linear correlation model in the form of

$$f_t = a * f_c + b \tag{4}$$

was analyzed and coefficient of determination, R^2, was calculated.

The results of the regression analysis are shown in Table 6 and in the diagram in Figure 14.

Table 6. The results of the regression analysis.

Mix No.	Days of Curing	Model	Coefficient of Determination R^2
1	7, 28, 90	a = 0.2428 b = 0.0777	0.9078
2	7, 28, 90, 180	a = 0.2089 b = 0.0411	0.9227
3	7, 28, 90, 180	a = 0.2075 b = 0.0323	0.9435
All mixtures		a = 0.2052 b = 0.0087	0.9470

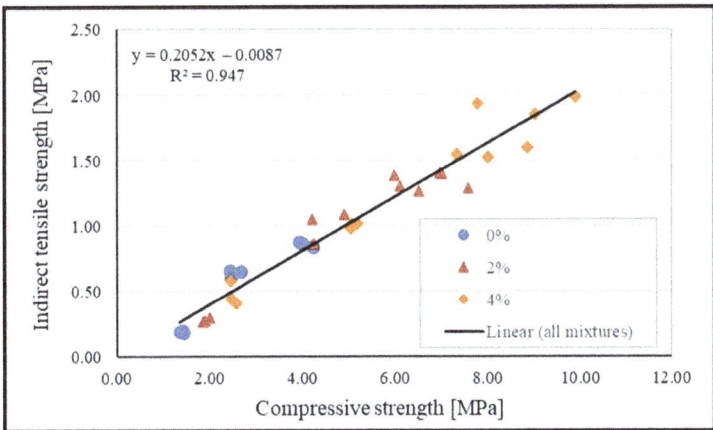

Figure 14. Diagram of the correlation of strengths of stabilized mixtures.

The linear correlation between the indirect tensile strength and compressive strength of stabilizing mixtures is very strong as the coefficients of determination R^2 for all models were between 0.90 and 0.947. The indirect tensile strength of stabilization mixtures with cement amounts to 20% of the value of compressive strength for both contents of cement, and the mixture without cement to 24% of compressive strength. This difference in values of mixtures with and without cement can be explained by the difference in the composition but also by the fact that for the mixture with 0% cement, strengths for 180 days of curing were not determined and therefore not included in the analysis. The average correlation of indirect tensile strength in relation to compressive strength for all the tested mixtures and all the durations of curing amounts to 20%. This mutual strength relationship is higher and differs from the strength results of stabilized mixtures of similar composition [1,7] which range from 10% to 17%. The mentioned differences in strength ratios can be explained by differences in the type of fly ash used as well as differences in the composition of mixtures. This study points out that WA should be further researched as a conclusion on the conventional hydraulically bound mixtures or those with CFA cannot fully be applied on WA.

6.5. The Results of Resistance to Freezing and Thawing

The results of testing the resistance of mixtures with 2% and 4% cement to freezing are presented in Figure 15. Due to the lower value of initial strengths, testing of resistance to freezing of the mixture with 0% cement was not carried out. The control mixture with 2% cement achieved a compressive strength of 4.83 MPa, while the control mixture with 4% cement achieved a compressive strength value of 5.58 MPa. It can be seen that the increase in content of cement in the control mixtures results in a 13% higher value of compressive strength and strengths here are following those measured at 28 days of curing for compressive strengths tests (4.45 MPa and 5.1 MPa, respectively). In mixtures that were subjected to freezing, the difference in the achieved strengths for different contents of cement is greater.

The mixtures with 2% cement achieved a compressive strength of 2.65 MPa after freezing, which represents a considerable reduction in comparison with the control mixture for which the compressive strength amounted to 4.83 MPa. The retained compressive strength of the mixture after freezing (RFT 1) amounts to 55% of the compressive strength of the mixture before the freezing process. The compressive strength after freezing of the mixture with 4% cement amounted to 3.63 MPa, while compressive strength of the control mixture with 4% cement was 5.58 MPa. For stabilization mixtures with 4% cement, the retained compressive strength after freezing amounts to 65%.

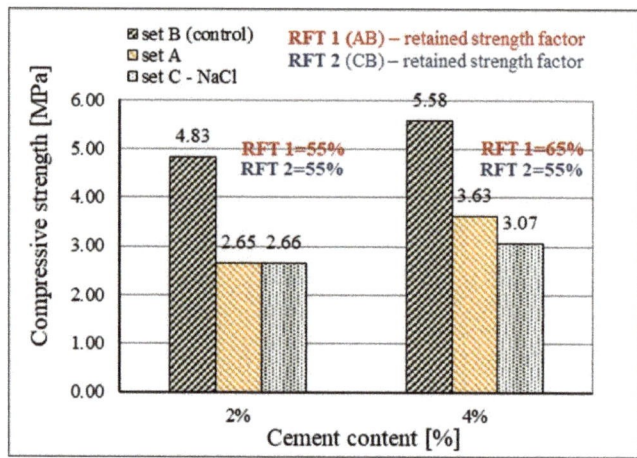

Figure 15. Results of resistance to freezing and thawing of stabilized mixtures.

The results of additional testing of resistance to freezing with the presence of salts (RFT 2) showed that the value of retained strength after freezing for both mixtures is identical, regardless of the content of cement in them, and it amounts to RFT = 55%.

The standard [68] does not define the criteria for the satisfactory resistance of a mixture to freezing, but the American standard ASTM C593 [79] precisely defines the resistance of the mixture with fly ash to freezing, prescribing a minimum compressive strength of mixtures in the quantity of 2.8 MPa. Furthermore, Lahtinen [76] states that for the application of stabilization mixtures with fly ash in the base layers of roads with low traffic (low volume roads), the greatest reduction in compressive strength after freezing can be 40%. Besides the prescribed reduction in strength, the same author states that the samples must remain firm and undamaged after freezing.

It should be pointed out that testing of the freezing/thawing of samples simulates conditions that are much more complex than the actual conditions "in situ" and that rarely occur in the field. Stabilized layers are usually in the middle of the pavement structure and are protected by an asphalt or concrete pavement from above, and there are unbound base layers constructed under them, which prevent capillary penetration of water or the rising of the underground water level. In addition, the reduced strength of the mixture after curing in water with a content of salt can be seen in a similar way. Although a smaller quantity of NaCl can appear through cracks in the pavement in stabilized base layers during the winter period, it is not expected that such a quantity of salt will jeopardize the load bearing capacity and durability of the layer.

On the basis of the above, it can be concluded that the mixture with 4% cement has an adequate resistance to freezing, and that it can be used for the construction of base layers of the pavement structure in a region with a continental climate, such as the climate of eastern Croatia. The mixtures with 2% cement (although very similar to the mixture with 4% cement according to the results obtained) did not meet the criteria for satisfactory resistance to freezing.

7. Conclusions

In this study, tests on mixtures of Drava sand, WFA, and cement were carried out, which included determining the CBR bearing ratio, the compressive and indirect tensile strength of mixtures, and their resistance to freezing/thawing. The test results obtained show that local materials of lower quality, WA and sand, can be used in hydraulically bound mixtures and they satisfy the conditions given for road base courses of higher quality materials. The following can be concluded:

(1) WFA with a considerable content of CaO (46.91 mass %) has a stabilization effect on the sand mixture and improves the load bearing capacity of the mixture expressed by the CBR index. The CBR index increases considerably with an increase in the content of ash in the mixture (10%, 20%, 30%) both due to the filler effect and through hydraulic binding between minerals in sand and WFA. The greatest value of CBR (90.70%) was achieved by mixtures of sand and 30% WFA, which is three times more than the bearing ratio of pure sand (CBR = 27.44%).

(2) WFA in the mixture of sand improves compressive strength, so that the mixtures with 0% cement achieved compressive strength values of 1.4 MPa (7 days) to 4.06 MPa (90 days). These results indicate the development of hydraulic and pozzolanic reactions between sand and WFA, but further research is necessary to clearly define the reactions and minerals that occur. By adding a small quantity of the cement (2% and 4%), the hydraulic reaction is even more intensive and results in higher values of compressive strength. According to the proposed classification based on a minimum compressive strength for new construction based on recycled materials [76], stabilizing mixtures with 0% and 2% cement belong to Group B: strong (min 1.5 MPa) whilst the mixture with 4% cement belongs to Group A: very strong (min 4.5 MPa).

(3) The comparison of the values of compressive strengths after 7 and 28 days with the criteria of the Croatian technical regulations (prescribed for mixtures stabilized exclusively with cement) showed that the mixtures with 4% cement meet the conditions for application in the upper and bottom base layers of the pavement structure. The mixtures with 2% cement can be used for the construction of the subbase layers, as well as the mixtures without cement.

(4) The trend of the indirect tensile strength development for all stabilized mixtures and durations of curing is very similar to that of compressive strengths. The stabilized mixture with 0% cement achieved the values of indirect tensile strength from 0.2 MPa (7 days) to 0.86 MPa (90 days). The average increase in indirect tensile strength of the mixtures with 4% cement in comparison with the mixtures with 0% cement, for curing of 7 to 90 days, amounts to 75.55%, almost the same value as in the relation of compressive strengths of the mixture.

(5) For compressive and indirect tensile strengths of mixtures, the model of the correlation $f_t = a * f_c + b$ was analyzed. The results obtained show that the linear correlation between indirect tensile strength and compressive strength of stabilization mixtures is very strong, and coefficients of determination are very high (R^2 = 0.90–0.947). The average correlation of indirect tensile strength in relation to compressive strength for all the tested mixtures and all the durations of curing amounts to 20%.

(6) The mixture with 4% cement has a satisfactory resistance to freezing, and it can be used for the construction of load-bearing layers of pavement structures in a region with a continental climate, such as the area of eastern Croatia.

(7) In addition to a significant contribution to achieving the required mechanical properties of mixtures, it is necessary to emphasize that the environmental and economic benefits of WFA in load-bearing layers of a pavement structure are reflected in:

   (I) The rationalization of costs of stabilized load-bearing layers through savings in cement and sand amount. WFA is a cost-free material, and with the replacement of 30% of sand with WFA and the decrease in cement content from the standard amount of 8% to 4%, the construction costs of pavement structure are reduced.

   (II) A reduction in the quantity of this waste material at landfills. The daily quantities of WFA that are produced are significant and they are mostly deposited in landfills. Depositing in landfills is demanding and expensive and takes up valuable space, so reuse/recycling is strongly encouraged.

   (III) Protection of natural aggregates. Clear requirements for conservation and protection of non-renewable sources of natural aggregates is stated in the sustainable development guidelines and the usage of WFA means a lower requirement for the exploitation of natural aggregates.

In highlighting the advantages of WFA usage, one should not neglect the fact that WFA is a waste material with variable chemical and mineralogical compositions, and as such, it could have environmental impacts. However, prior to their wider usage in the road construction industry, the durability and environmental effect of such materials should be further investigated.

Author Contributions: Conceptualization, methodology, supervision S.D.; validation S.D. and M.Z.; investigation M.Z. and T.T.; writing—original draft preparation, S.D.; writing—review and editing, S.D.; M.Z. and M.Š. All authors have read and agreed to the published version of the manuscript.

Funding: This research was funded by the Faculty of Civil Engineering and Architecture Osijek, Croatia (scientific-research project IZIP-GrAFOS-2018) entitled "The application of ash from wood biomass in layers of pavement structure-Bio PAV".

Acknowledgments: The authors would like to acknowledge Goran Vrhovac, Assistant Director at Institut IGH d.d. Zagreb for his help in providing part of the laboratory testing and Robert Bušić, Ph.D. in Civ.Eng. from the Faculty of Civil Engineering and Architecture Osijek for SEM pictures of sand and WFA.

Conflicts of Interest: The authors declare no conflict of interest. The funders had no role in the design of the study; in the collection, analyses, or interpretation of data; in the writing of the manuscript, or in the decision to publish the results.

References

1. European Commission. *COST 333 Developement of New Bituminous Pavement Design Method. Final Report of the Action*; European Commission: Brussels, Belgium, 1999; pp. 130–137. ISBN 92-828-6796-X.
2. Barišić, I.; Rukavina, T.; Dimter, S. Cement stabilizations—characterization of materials and design criteria. *Građevinar* **2011**, *63*, 135–142.
3. Little, D.N.; Scullion, T.; Kota, P.B.V.S.; Bhuiyan, J. *Guidelines for Mixture Design and Tickness Design for Stabilized Bases and subgrades*; Texas Transportation Institute: College Station, TX, USA, 1995.
4. Dimter, S.; Rukavina, T.; Barišić, I. Alternative, environmentally acceptable materials in road construction. In *Handbook of Research on Advancements in Environmental Engineering*; Gaurina-Međimurec, N., Ed.; IGI Global: Hershey, PA, USA, 2014; pp. 557–583.
5. Institute of Civil Engineering. *Study of The Possibilities of Application of Sand in Construction of Roadso Slavonia and Baranja Region*; Institute of Civil Engineering, Faculty of Civil Engineering in Osijek: Osijek, Croatia, 1990; p. 105. (In Croatian)
6. Bošnjak, J. Constructing of the roads with sand. In Proceedings of the 1st Croatian Congress of Roads, Croatian Road Society—Via Vita, Opatija, Croatia, 23–25 October 1995; pp. 536–543, (In Croatian with English Summary).
7. Dimter, S. Properties of Stabilized Mixtures for Road Construction. Ph.D. Thesis, University of Zagreb, Faculty of Civil Engineering, Zagreb, Croatia, 2005; 156p. (In Croatian with English Summary).
8. Dimter, S.; Zagvozda, M.; Tonc, T.; Šimun, M. Construction of economical pavement structures with wood ash. In Proceedings of the 30th International Baltic Road Conference, Riga, Latvia, 23–24 August 2021; pp. 1–7.
9. Zagvozda, M.; Dimter, S.; Rukavina, T.; Grubeša, I.N. Possibilities for using bioash in road construction. *Građevinar* **2018**, *70*, 393–402. [CrossRef]
10. Dimter, S.; Rukavina, T.; Dragčević, V. Strength Properties of Fly Ash Stabilized Mixtures. *Road Mater. Pavement Des.* **2011**, *12*, 687–697. [CrossRef]
11. Ksaibati, K.; Conklin, T. *Field Performance Evaluation of Cement-Treated Bases with and Without Fly Ash*; Transportation Research Bord: Washington, DC, USA, 1994; Record No. 1440.
12. Hadi Shirazi, P. *Field and Lab Evaluation of the Use of Lime Fly Ash to Replace Soil Cement as a Base Course*; Report No. 303; Louisiana Transportation Research Center: Baton Rouge, LA, USA, 1997. [CrossRef]
13. Dimter, S.; Rukavina, T.; Minažek, K. Estimation of elastic properties of fly ash—Stabilized mixtures using nondestructive evaluation methods. *Constr. Build. Mater.* **2016**, *102*, 505–514. [CrossRef]
14. Dimter, S.; Rukavina, T.; Barišić, I. Application of the ultrasonic method in evaluation of properties of stabilized mixture. *Baltic J. Road Bridge Eng.* **2011**, *6*, 177–184. [CrossRef]
15. Sear, L.K.A. *Properties and Use of Coal Fly Ash. A Valuable Industrial by Product*; Thomas Telford Publishing: London, UK, 2001; p. 261. ISBN 0727730150.
16. Maher, M.H.; Balaguru, P.N. Properties of flowable high-volume fly ash-cement composite. *J. Mater. Civil Eng.* **1993**, *5*, 212–225. [CrossRef]
17. Zenieris, P.; Laguros, J.G. Fly ash as a binder in aggregate base course. *Mater. Res. Soc. Simp. Proc.* **1988**, *113*, 231–241. [CrossRef]
18. Sobhan, K.; Jesick, M.R.; Dedominicis, E.J.; McFadden, J.P.; Cooper, K.A.; Roe, J.R. A Soil-Cement-Fly Ash Pavement Base Course Reinforced with Recycled Plastic Fibers. In Proceedings of the Transportation Research Board, 78th Annual Meeting, Washington, DC, USA, 10–14 January 1999; pp. 1–12.

19. Simatupang, M.; Mangalla, L.K.; Edwin, R.S.; Putra, A.A.; Azikin, M.T.; Aswad, N.H.; Mustika, W. The Mechanical Properties of Fly-Ash-Stabilized Sands. *Geosciences* **2020**, *10*, 132. [CrossRef]
20. Mahvash, S.; López-Querol, S.; Bahadori-Jahromi, A. Effect of class F fly ash on fine sand compaction through soil stabilization. *Heliyon* **2017**, *3*, e00274, ISSN 2405-8440. [CrossRef]
21. Mahvash, S.; López-Querol, S.; Bahadori-Jahromi, A. Effect of fly ash on the bearing capacity of stabilised fine sand. *Proc. Inst. Civ. Eng.-Ground Improv.* **2018**, *171*, 82–95. [CrossRef]
22. Chenari, R.J.; Fatahi, B.; Ghorbani, A.; Alamoti, M.N. Evaluation of strength properties of cement stabilized sand mixed with EPS beads and fly ash. *Geomech. Eng.* **2018**, *14*, 533–544. [CrossRef]
23. Lav, A.H.; Lav, M.A.; Goktepe, A.B. Analysis and design of stabilized fly ash as pavement base material. *Fuel* **2006**, *85*, 2359–2370. [CrossRef]
24. Lav, M.A.; Lav, A.H. Effects of stabilization on resilient characteristics of fly ash as pavement material. *Constr. Build. Mater.* **2014**, *54*, 10–16. [CrossRef]
25. Forteza, R.; Far, M.; Seguı, C.; Cerda, V. Characterization of bottom ash in municipal solid waste incinerators for its use in road base. *Waste Manag.* **2004**, *24*, 899–909. [CrossRef]
26. Marieta, C.; Guerrero, A.; Cascante, I.L. Municipal solid waste incineration fly ash to produce eco-friendly binders for sustainable building construction. *Waste Manag.* **2021**, *120*, 114–124. [CrossRef]
27. Škēls, P.; Bondars, K.; Plonis, R.; Haritonovs, V.; Paeglītis, A. Usage of Wood Fly Ash in Stabilization of Unbound Pavement Layers and Soils. In Proceedings of the 13th Baltic Sea Geotechnical Conference, Vilnius, Lithuania, 22–24 September 2016; pp. 122–125. Available online: https://www.researchgate.net/publication/309015822_Usage_of_Wood_Fly_Ash_in_Stabilization_of_Unbound_Pavement_Layers_and_Soils (accessed on 2 December 2021). [CrossRef]
28. Škēls, P.; Bondars, K.; Haritonovs, V. Wood fly ash stabilization of unbound pavement layers. In Proceedings of the 19th International Conference on Soil Mechanics and Geotechnical Engineering (ICSMGE 2017), Seoul, Korea, 17–21 September 2017; pp. 2014–2017.
29. Vestin, J.; Arm, M.; Nordmark, D.; Lagerkvist, A.; Hallgren, P.; Lind, B. Fly ash as a road construction material. In Proceedings of the WASCON 2012, Gothenburg, Sweden, 30 May–1 June 2012; pp. 1–8.
30. Bohrn, G.; Stampfer, K. Untreated Wood Ash as a Structural Stabilizing Material in Forest Roads. *Croat. J. For. Eng.* **2014**, *35*, 81–90.
31. Kaakkurivaara, T.; Kolisoja, P.; Uusitalo, J.; Vuorimies, N. Fly Ash in Forest Road Rehabilitation. *Croat. J. For. Eng.* **2009**, *37*, 119–130.
32. Sarkkinen, M.; Luukkonen, T.; Kemppainen, K. A wasterock and bioash mixture as a road stabilization product. In Proceedings of the 3rd Edition of the International Conference on Wastes: Solution, Treatments and Opportunities, Viana do Castelo, Portugal, 14–16 September 2015; pp. 283–288.
33. Cabrera, M.; Agrela, F.; Ayuso, J.; Galvin, A.P.; Rosales, J. Feasible use of biomass bottom ash in the manufacture of cement treated recycled materials. *Mater. Struct.* **2015**, *49*, 3227–3238. [CrossRef]
34. Cabrera, M.; Rosales, J.; Ayuso, J.; Estaire, J.; Agrela, F. Feasibility of using olive biomass bottom ash in the sub-bases of roads and rural paths. *Constr. Build. Mater.* **2018**, *181*, 266–275. [CrossRef]
35. Behak, L.; Musso, M. Performance of Low-Volume Roads with Wearing Course of Silty Sand Modified with Rice Husk Ash and Lime. *Transp. Res. Procedia* **2016**, *18*, 93–99, ISSN 2352-1465. [CrossRef]
36. Cabrera, M.; Díaz-López, J.L.; Agrela, F.; Rosales, J. Eco-Efficient Cement-Based Materials Using Biomass Bottom Ash: A Review. *Appl. Sci.* **2020**, *10*, 8026. [CrossRef]
37. Sáez del Bosque, I.F.; Sánchez de Rojas, M.I.; Asensio, E.; Frías, M.; Medina, C. Industrial waste from biomass-fired electric power plants as alternative pozzolanic material. In *Waste and Byproducts in Cement-Based Materials. Innovative Sustainable Materials for a Circular Economy*; de Brito, J., Thomas, C., Medina, C., Agrela, F., Eds.; Elsevier: Amsterdam, The Netherlands, 2021; pp. 243–283. ISBN 9780128208953/9780128205495. [CrossRef]
38. Zagvozda, M.; Rukavina, T.; Dimter, S. Wood bioash effect as lime replacement in the stabilisation of different clay subgrades. *Int. J. Pavement Eng.* **2020**, 1–11. [CrossRef]
39. Hinojosa, M.J.R.; Galvín, A.P.; Agrela, F.; Perianes, M.; Barbudo, A. Potential use of biomass bottom ash as alternative construction material: Conflictive chemical parameters according to technical regulations. *Fuel* **2014**, *128*, 248–259. [CrossRef]
40. Zagvozda, M.; Dimter, S.; Rukavina, T.; Estokova, A. Ecological aspect of bioashes as road building material. In Proceedings of the 5th International Conference on Road and Rail Infrastructures-CETRA 2018, Zadar, Croatia, 17–19 May 2018. [CrossRef]
41. Sigvardsen, N.M.; Kirkelund, G.M.; Jensen, P.E.; Geiker, M.R.; Ottosen, L.M. Impact of production parameters on physiochemical characteristics of wood ash for possible utilisation in cement-based materials. *Resour. Conserv. Recycl.* **2019**, *145*, 230–240. [CrossRef]
42. Sigvardsen, N.M.; Geiker, M.R.; Ottosen, L.M. Phase development and mechanical response of low-level cement replacements with wood ash and washed wood ash. *Constr. Build. Mater.* **2020**, *269*, 121734. [CrossRef]
43. Sigvardsen, N.M.; Geiker, M.R.; Ottosen, L.M. Reaction mechanisms of wood ash for use as a partial cement replacement. *Constr. Build. Mater.* **2021**, *286*, 122889. [CrossRef]
44. Cherian, C.; Siddidua, S. Engineering and environmental evaluation for utilization of recycled pulp mill fly ash as binder in sustainable road construction. *J. Clean. Prod.* **2021**, *298*, 126758. [CrossRef]

45. Demirbas, A. Potential applications of renewable energy sources, biomass combustion problems in boiler power systems and combustion related environmental issues. *Prog. Energy Combust. Sci.* **2005**, *31*, 171–192. [CrossRef]
46. Directorate-General for Energy (European Commission), Navigant. *Technical Assistance in Realisation of the 5th Report on Progress of Renewable Energy in the EU. Analysis of Bioenergy Supply and Demand in the EU (Task 3): Final Report*; Publications Office of the European Union: Luxembourg, 2020.
47. Zagvozda, M.; Dimter, S.; Rukavina, T. The potential of bioash for utilization in road construction in eastern Croatia. *Sel. Sci. Pap. J. Civ. Eng.* **2017**, *12*, 121–127. [CrossRef]
48. Milovanović, B.; Štirmer, N.; Carević, I.; Baričević, A. Wood biomass ash as a raw material in concrete industry. *Građevinar* **2019**, *71*, 504–514. [CrossRef]
49. Dimter, S.; Zagvozda, M.; Milovanović, B.; Šimun, M. Usage of Wood Ash in Stabilization of Unbound Pavement Layers. Road and Rail Infrastructure VI. In Proceedings of the 6th International Conference on Road and Rail Infrasturcture, Pula, Croatia, 20–22 May 2020; pp. 981–987. [CrossRef]
50. Carević, I.; Baričević, A.; Štirmer, N.; Šantek Bajto, J. Correlation between physical and chemical properties of wood biomass ash and cement composites performances. *Constr. Build. Mater.* **2020**, *256*, 119450. [CrossRef]
51. Carević, I.; Štirmer, N.; Serdar, M.; Ukrainczyk, N. Effect of wood biomass ash storage on the properties of cement composites. *Materials* **2021**, *14*, 1632. [CrossRef] [PubMed]
52. Dimter, S.; Šimun, M.; Zagvozda, M.; Rukavina, T. Laboratory evaluation of the properties of asphalt mixture with wood ash filler. *Materials* **2021**, *3*, 575. [CrossRef] [PubMed]
53. Gabrijel, I.; Jelčić Rukavina, M.; Štirmer, N. Influence of Wood Fly Ash on Concrete Properties through Filling Effect Mechanism. *Materials* **2021**, *14*, 7164. [CrossRef]
54. *HRN EN ISO 17892-4:2008*; Geotechnical Investigation and Testing—Laboratory Testing of Soil—Part 4: Determination of Particle Size Distribution. CEN: Brussels, Belgium; ISO: Geneva, Switzerland, 2008.
55. *HRN U.E9.024*; Highway Design and Construction. Construction of Pavement Base Courses Stabilized by Means of Portland Cement and Similar Chemicals. The Croatian Standards Institute: Zagreb, Croatia. (In Croatian)
56. Croatian Geological Survey. *Testing Report on the Mineral Composition of Sand by X-Ray Diffraction Method (XRD) No. 24/21*; Croatian Geological Survey: Zagreb, Croatia, 2021.
57. Guideline on the Use of Sand in Road Construction in the SADC Region AFCAP/GEN/028/C, InfraAfrica (Pty) Ltd., Botswana CSIR, South Africa TRL Ltd., UK Roughton International, UK CPP Botswana (Pty) Ltd., May 2013. Available online: https://assets.publishing.service.gov.uk/media/57a08a39e5274a27b20004bb/AFCAP-GEN028-C-Sand-in-Road-Construction-Final-Guideline.pdf (accessed on 3 December 2021).
58. Shalabi, F.I.; Mazher, J.; Khan, K.; Alsuliman, M.; Almustafa, I.; Mahmoud, W.; Alomran, N. Cement-Stabilized Waste Sand as Sustainable Construction Materials for Foundations and Highway Roads. *Materials* **2019**, *12*, 600. [CrossRef]
59. *HRN EN 14227-4:2013*; Hydraulically Bound Mixtures—Specifications—Part 4: Fly Ash for Hydraulically Bound Mixtures. CEN: Brussels, Belgium, 2013.
60. *HRN EN 933-10:2009*; Tests for Geometrical Properties of Aggregates—Part 10: Assessment of Fines—Grading of Filler Aggregates (Air Jet Sieving). CEN: Brussels, Belgium, 2009.
61. *HRN EN 450-1:2013*; Fly Ash for Concrete—Part 1: Definition, Specifications and Conformity Criteria. CEN: Brussels, Belgium, 2013.
62. Croatian Geological Survey. *Testing Report on the Mineral Composition of Ash Samples by X-Ray Diffraction Method (XRD) No. 65/19*; Croatian Geological Survey: Zagreb, Croatia, 2019.
63. *HRN EN 197-1:2012*; Part 1: Composition, Specifications and Conformity Criteria for Common Cements. CEN: Brussels, Belgium, 2012.
64. *HRN EN 13286-2:2010*; Unbound and Hydraulically Bound Mixtures—Part 2: Test Methods for Laboratory Reference Density and Water Content—Proctor Compaction. CEN: Brussels, Belgium, 2010.
65. *HRN EN 13286-47:2012*; Unbound and Hydraulically Bound Mixtures—Part 47: Test Method for the Determination of California Bearing Ratio, Immediate Bearing Index and Linear Swelling. CEN: Brussels, Belgium, 2012.
66. *HRN EN 13286-41:2003*; Unbound and Hydraulically Bound Mixtures—Part 41: Test Method for the Determination of the Compressive Strength of Hydraulically Bound Mixtures. CEN: Brussels, Belgium, 2003.
67. *HRN EN 13286-42:2003*; Unbound and Hydraulically Bound Mixtures—Part 42: Test Method for the Determination of the Indirect Tensile Strength of Hydraulically Bound Mixtures. CEN: Brussels, Belgium, 2003.
68. *HRS CEN/TS 13286-54*; Unbound and Hydraulically Bound Mixtures—Part 54: Test Method for the Determination of Frost Susceptibility. Resistance to Freezing and Thawing of Hydraulically Bound Mixtures. CEN: Brussels, Belgium, 2014.
69. *HRN EN 1367-1:2008*; Tests for Thermal and Weathering Properties of Aggregates—Part 1: Determination of Resistance to Freezing and Thawing. CEN: Brussels, Belgium, 2008.
70. *ASTM D5239-98*; Characterizing Fly Ash for Use in Soil Stabilization. ASTM: West Conshohocken, PA, USA, 1998.
71. *ASTM D6726-99a*; Standard Test Method for Using pH to Estimate the Soil-Lime Proportion Requirement for Soil Stabilization. ASTM: West Conshohocken, PA, USA, 1999.
72. Bell, F.G. Lime stabilization of clay minerals and soils. *Eng. Geol.* **1996**, *42*, 223–237. [CrossRef]
73. Enríquez, E.; Torres-Carrasco, M.; Cabrera, M.J.; Muñoz, D.; Fernández, J.F. Towards more sustainable building based on modified Portland cements through partial substitution by engineered feldspars. *Constr. Build. Mater.* **2021**, *269*, 121334. [CrossRef]

74. Kim, J.S.; Lee, J.Y.; Kim, Y.H.; Kim, D.; Kim, J.; Han, J.G. Evaluating the eco-compatibility of mortars with feldspar-based fine aggregate. *Case Stud. Constr. Mater.* **2022**, *16*, e00781. [CrossRef]
75. Pratt, A. Cementitious Compositions Containing Feldspar and Pozzolanic Particulate Material, and Method of Making Said Compositin. U.S. Patent Patent No.: US 8,066,813 B2, 29 November 2011.
76. Lahtinen, P. Fly Ash Mixtures as Flexible Structure Materials for Low-Volume Roads. Helsinki, Finnish Road Administration, Uusimaa Region. Finnra Reports 70. 2001. Available online: http://lib.tkk.fi/Diss/2001/isbn9512257076/isbn9512257076.pdf (accessed on 15 December 2021).
77. Institute of Civil Engineering of Croatia. Pavement Structure. In *General Technical Requirements for Road Work (GTR)*; IGH: Zagreb, Croatia, 2001; Volume 3. (In Croatian)
78. Parylak, K. Influence of Particle Structure on Properties of Fly Ash and Sand. *Mater. Sci.* **1992**, 1031–1041.
79. *ASTM C593-95*; Standard Specification for Fly Ash and Other Pozzolans for Use with Lime. ASTM: West Conshohocken, PA, USA, 2000.

Article

Assessment of the Possibility of Using Fly Ash from Biomass Combustion for Concrete

Jakub Jura and Malgorzata Ulewicz *

Faculty of Civil Engineering, Czestochowa University of Technology, Dabrowskiego 69 Street, PL 42-201 Czestochowa, Poland; jakub.jura@pcz.pl
* Correspondence: malgorzata.ulewicz@pcz.pl

Abstract: This article analyses the possibility of using fly ash from the combustion of wood–sunflower biomass in a fluidized bed boiler as an additive to concrete. The research shows that fly ash applied in an amount of 10–30% can be added as a sand substitute for the production of concrete, without reducing quality (compression strength and low-temperature resistance) compared to control concrete. The 28-day compressive strength of concrete with fly ash increases with the amount of ash added (up to 30%), giving a strength 28% higher than the control concrete sample. The addition of fly ash reduces the extent to which the compression strength of concrete is lowered after low-temperature resistance tests by 22–82%. The addition of fly ash in the range of 10–30% causes a slight increase in the water absorption of concrete. Concretes containing the addition of fly ash from biomass combustion do not have a negative environmental impact with respect to the leaching of heavy metal ions into the environment.

Keywords: concrete; mechanical properties; low-temperature resistance; fly ash; biomass

Citation: Jura, J.; Ulewicz, M. Assessment of the Possibility of Using Fly Ash from Biomass Combustion for Concrete. *Materials* **2021**, *14*, 6708. https://doi.org/10.3390/ma14216708

Academic Editor: F. Pacheco Torgal

Received: 29 September 2021
Accepted: 3 November 2021
Published: 7 November 2021

Publisher's Note: MDPI stays neutral with regard to jurisdictional claims in published maps and institutional affiliations.

Copyright: © 2021 by the authors. Licensee MDPI, Basel, Switzerland. This article is an open access article distributed under the terms and conditions of the Creative Commons Attribution (CC BY) license (https://creativecommons.org/licenses/by/4.0/).

1. Introduction

The development of industry causes a systematic increase in the amount of generated waste. Some of the wastes produced, such as fly ashes and slags from the combustion of conventional fuels (hard and brown coal), are used to produce composite materials with a cement matrix, such as cement mortars or concretes [1–10]. In recent years, there have also been a number of reports in the literature on the possibility of using waste from construction ceramics [11–13], sanitary and household ceramics [14–18], glass cullet [19–22] and polymer materials [23–25] to produce cement mortars and concretes.

There have also been reports of the possibility of using fly ashes from the co-combustion of hard coal and biomass in conventional or fluidized bed boilers for this purpose. Mortars and concretes with the addition of such ash usually achieve similar or lower strength values after 28 days of maturation (75% of the control samples [26], 98–84% [27], 72–93% [28], 98–46% [29]), and after a longer period (90–180 days) they increase their compressive strength, ultimately achieving a strength similar to [26,29,30] or higher than the control samples (2–20% higher than control samples [27], 5–12% [29]. The results obtained by the authors of these studies confirm that the ashes produced in co-combustion processes have a higher reactivity and can be a useful raw material in the production of cement matrix materials [29]. Currently, the physical and chemical properties of the ashes generated during combustion process are being tested, e.g., forest residues, the pulp and paper industry, sugar cane or corn cobs, and attempts are being made to develop methods for their management in various sectors of the economy [31–36].

There are few reports in the literature on the laboratory use of ashes from biomass combustion, including the production of composite materials with a cement matrix [37]. Most of the studies available in the literature concern the properties of ash and the possible use of fly ash from the combustion of sugar cane bagasse, most often used in the amount of 5–30% of the cement mass [38–41]. Reports show that the addition of such ash may both positively and negatively affect the mechanical and physical properties of materials with a cement matrix. The compressive strength of materials with such additives decreased, depending on the type of biomass used and the amount of fly ash added. Compressive strength was lower than the control samples (5–25% [42], 18% [43], 55% [44], 25% [45]) or higher than the control samples (3–14% [42], 30% [43], 5% [44], 1–17% [45], 17% [46], 13% [47]). The best results in terms of compressive strength were achieved by samples containing ashes from wood in the amount of 5% [42], 10% [44] and 20% [45], and in the case of sugar cane bagasse at 5–10% [38–41,43,47,48]), while the worst results were for samples containing ashes from wood in greater proportions (15% [42], 20% [43], 25% [45]) and for ash from the combustion of sugar cane bagasseused in proportions of 20–25% [38–41,43,47,48].

Mortars containing up to 30% ash usually showed higher resistance to freezing and thawing than the control samples (reduction of the drop in compressive strength up to 95% [46], down to 50% [47]). Currently, fluidized ashes generated during biomass combustion in fluidized bed boilers (classified as waste with the code 10 01 82), due to the different physicochemical properties determined by the diversity of incinerated biomass, are not usable and are therefore deposited in landfills, which places a burden on the natural environment. For these reasons, it is advisable to develop methods of managing these types of ashes. Therefore, this study attempts to evaluate the possibility of using waste fly ash from the combustion of biomass itself in a fluidized bed boiler with a circulating bed for the production of composite building materials with a cement matrix. Such a method of using this waste would significantly reduce the usage of natural resources and limit the negative environmental impact of waste ash.

2. Materials and Methods

2.1. Materials

Fly ash from the combustion process in a boiler with a circulating fluidized bed of biomass consisting of 80% waste wood and 20% sunflower was used for the research. Fly ash came from a power plant (GDF SUEZ, Połaniec, Poland) in the Świętokrzyskie Voivodeship and was tested in accordance with PN EN 450 1:2012. Ash roasting losses amounted to 2.9%. According to the PN-EN 451-2:2017-06 standard, the fineness of the samples as a residue on the 0.045 mm sieve is 7.9%, and the fly ash density is 2.35 g/cm^3. The elemental composition of the fly ash made with the use of the XRF X-ray spectrometer (Thermo Fisher Scientific, Waltham, MA, USA) is presented in Table 1.

Table 1. Percentage of oxides and elements in fly ash.

Oxide/Element	Content %	Oxide/Element	Content %
SiO_2	50.20	Na_2O	0.44
CaO	11.82	MnO	0.28
K_2O	7.99	TiO_2	0.30
Al_2O_3	12.29	SO_3	4.91
MgO	3.34	Cl	1.63
Fe_xO_y	3.50	Other	3.30

Fly ash consists mainly of silicon oxide (50.2%), aluminium oxide (12%), calcium oxide (11.82%) and almost 8% potassium oxide. The remaining compounds constitute less than 4%, which traces include Zn, Ba, CR, Sr, Zr, Cu, Rb, Ni and Pb, among other materials. The microstructure of the fly ash used for the tests (determined with an LEO Electron Microscopy Ltd. apparatus, Cambridge, UK) is presented in Figure 1, and reveals that the fly ash grains have an inhomogeneous structure with sharp edges, characteristic of crushed stone aggregate.

The TGA-DTA thermal analysis was also conducted for the fly ash used in the research (Figure 2).

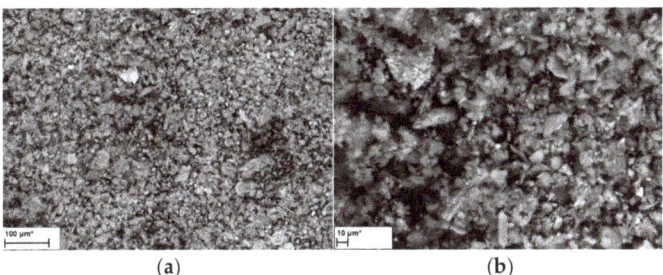

Figure 1. Fly ash microstructure in magnification: (**a**) 400×; (**b**) 1000×.

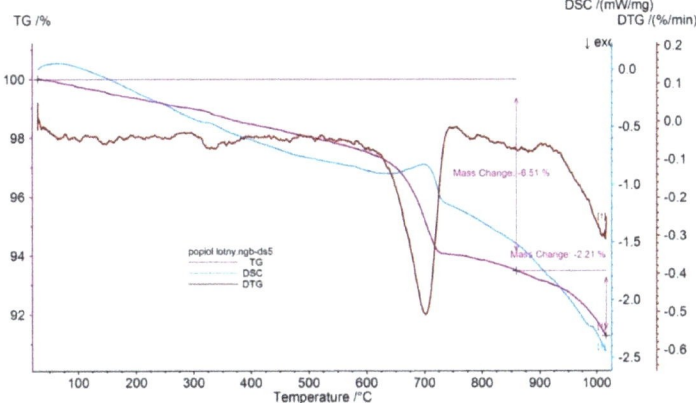

Figure 2. TGA-DTA thermogram for fly ash.

The test was carried out in a Jupiter STA 449 F5 (Netzsch, Selb, Germany) in the range from 30 to 1000 °C with a temperature increase rate of 10 °C/min in air with a gas flow rate of 100 cm^3/min. Up to a temperature of approximately 100 °C, the fly ash sample lost approximately 0.3% weight, which is related to the evaporation of water. At a temperature of about 350 °C, a peak was noticeable, which may have resulted from the transformation of Fe compounds. The peak was seen at a temperature of about 580–600 °C and corresponded to the conversion of α-quartz to β-quartz and the reaction between unreacted particles and activator trapped in the pores. Decomposition of aluminum hydroxide (Al(OH)$_3$) also took place in the same temperature range. At temperatures of 700 °C, endothermic or exothermic reactions occurred, which corresponded to the decomposition of CaCO$_3$, Ti(OH)$_4$ and Mg(OH)$_2$. The exothermic effect at about 900 °C may be related to the crystallization of amorphous ash phases. Observed changes were similar to those presented in [49] for a fly ash-based geopolymer. At a temperature of up to 900 °C, the loss of the sample mass (up to 6.51%) was related to the stripping of volatile parts. At temperatures above 900 °C, coal residues were burnt, and the fly ash sample lost more than 1% of its weight. The total weight loss was 8.7% in this temperature range. The DTG analysis also showed that in the case of the tested ash, the mass loss during heating was steady, and the greatest amplitude deviating from the entire line of the graph was for the temperature range of 650–800 °C. DSC Differential Scanning Calorimetry also enabled us to check the course of the formation of thermal effects in the sample. In the case of fly ash, the amount of heat obtained from the sample decreased almost steadily.

Two types of cements by CEMEX with high early strength were used for the tests: CEM I 42.5 R Portland cement and CEM II/A-V 42.5 R Portland ash cement.

Two types of coarse aggregate, i.e., gravel and basalt, as well as one type of fine aggregate, which was arenaceous quartz, were used for concrete tests. The composition of the control concrete with a gravel–sand (B1-0) and sand–basalt (B2-0) mixture of aggregates and CEM I cement, as well as that of a gravel–sand mixture of aggregates and CEM II per 1 m^3 of concrete is presented in Table 2.

Table 2. Composition of B1-0, B2-0 and B3-0 control concretes.

Concrete	B1-0	B2-0	B3-0
Ingredient		Quantity	
Cement CEM I, kg	364.20	339.50	-
Cement CEM II, kg	-	-	364.20
Water, dm^3	191.40	200.00	191.40
Aggregate—sand, kg	648.20	757.50	648.20
Aggregate G1; 2–8 mm, kg	662.00	-	662.00
Aggregate G2 8–16 mm, kg	541.70	-	541.70
Aggregate B1 2–8 mm, kg	-	668.90	-
Aggregate B2 8–16 mm, kg	-	547.30	-
Aggregate $\sum$, kg	1851.90	1973.70	1851.90
Plasticizer, dm^3	1.82	2.72	1.82

Table 3 presents the compositions of the designed concrete mixes containing fly ash. Fly ash was added to concrete in the amounts of 10%, 20%, 30% and 40% of the cement mass (used as a replacement for part of the sand calculated volumetrically), and the series of concretes were marked as B1-1, B1-2, B1-3 and B1-4, respectively, for the sand–gravel aggregates. In the case of sand–basalt aggregates, fly ash was added in the amounts of 10%, 20% and 30% of the cement mass, using ash as a substitute for part of the sand, and the series of concretes were marked as B2-1, B2-2 and B2-3. On the basis of the designed B1-0 control concrete, the members of the series were also made using CEM II/A-V 42.5 R ash cement, fly ash being used as a substitute for a part of the sand in the amounts of 10%, 20% and 30%, and were marked as B3-1, B3-2, B3-3, respectively.

Table 3. The composition of concretes with the addition of ash B1, B2 and B3 series.

Concrete	B1-1	B1-2	B1-3	B1-4	B2-1	B2-2	B2-3	B3-1	B3-2	B3-3
Ingredient					Quantity					
Cement CEM I, kg	364.20	364.20	364.20	364.20	339.47	339.47	339.47	-	-	-
Cement CEM II, kg	-	-	-	-	-	-	-	364.20	364.20	364.20
Water, dm^3	191.40	191.40	191.40	191.40	200.00	200.00	200.00	191.40	191.40	191.40
Aggregate—sand, kg	607.10	566.00	525.00	484.00	719.20	680.90	642.70	607.10	566.00	525.00
Aggregate G1 2–8 mm, kg	662.10	662.10	662.10	662.10	-	-	-	662.10	662.10	662.10
Aggregate G2 8–16 mm, kg	541.70	541.70	541.70	541.70	-	-	-	541.70	541.70	541.70
Aggregate B1 2–8 mm, kg	-	-	-	-	668.90	668.90	668.90	-	-	-
Aggregate B2 8–16 mm, kg	-	-	-	-	547.30	547.30	547.30	-	-	-
Aggregate $\sum$, kg	1810.9	1769.8	1728.8	1687.8	1935.4	1897.1	1858.8	1810.9	1769.8	1728.8
Plasticizer, dm^3	1.82	1.82	1.82	1.82	2.72	2.72	2.72	1.82	1.82	1.82
Fly ash, kg	36.40	72.80	109.20	145.60	33.95	67.89	101.84	36.40	72.80	109.20

2.2. Methods

Samples for the concrete mix tests were taken in accordance with the PN EN 12350 1:2011 standard. The air content in the concrete mixture was tested using the manometer method in accordance with the PN-EN 12350-7 standard.

Compressive strength tests of concretes were conducted (ToniTechnik 2030, Berlin, Germany) on 12 cubic samples with a side of 150 mm in accordance with PN EN 12390-3.

A constant load speed of 1.0 MPa/s was chosen. The load was applied to the specimen and continuously increased until the highest load was obtained. After the test, the testing machine, based on the entered dimensions of the sample, gave the compressive strength in MPa.

Concrete water absorption was tested in accordance with PN B 06250:1988. The tests were performed on six cubic samples (each series) with sides of 150 mm. The samples were disassembled after 24 h and were kept in water for the next 7 days, and then the samples were kept in the air for 21 days. The tested concrete samples were placed in a vessel on a grate at a distance of 15 mm from the bottom of the vessel. After obtaining a constant mass, the samples were placed in a laboratory dryer and dried at 105 °C to a constant mass, i.e., until the next weighings showed differences below 0.2% of the mass of the samples. The water absorption by weight is defined as the ratio of the mass of water penetrating into the saturated material to its dry mass and is given as a percentage.

The low-temperature resistance of concrete was tested on the basis of the PN-B-06250:1988 standard using a Toropol K-010 chamber (Toropol, Warsaw, Poland). 12 samples with a side of 100 mm were taken from each series of concretes. Samples saturated with water were weighed and then subjected to frost resistance testing. Six samples were left in water at 18 ± 2 °C throughout the test, and the remaining six were wiped of water and weighed with an accuracy of 0.2%. These concrete samples were frozen in air at (-18 ± 2 °C) for 4 h and then thawed in water (at $+18 \pm 2$ °C), also for 4 h. After 150 cycles of freezing and thawing the samples were reweighed and subjected to a compressive strength test.

Penetration of concrete with water under pressure was tested on the basis of the PN EN 12390-8 standard with the RatioTec WU60M apparatus (RatioTec, Essen, Germany), and three cubic samples with sides of 150 mm for each series were used for the tests. A water pressure of 500 kPa was applied to the samples for 72 h. After this time, the samples were taken out of the device and split in half. After the fracture surface had dried to a visible range of water, the maximum depth of water penetration was measured with an accuracy of 1 mm.

As part of the tests, a leaching analysis was also conducted based on the PN-EN-12457-2:2006 standard. For the test, samples containing 95% of grains smaller than 4 mm were prepared. For this purpose, the concrete samples were crushed in a jaw crusher and passed through a sieve with a mesh size of 4 mm. From the sample prepared in this way, an analytical sample (S) weighing 0.090 ± 0.005 kg was taken and placed in a vessel (bottle), and then distilled water (L) with electrical conductivity of 0.45 mS/m in the amount of 0.9 dm3 was added to it. According to the standard, the ratio of the washing liquid (L) to the solid phase (S) should be 10 dm^3 to 1 kg. The closed vessel was placed in a mixing device (roller table) for 24 ± 0.5 h. The vessel was then allowed to sit for 15 ± 5 min to allow the suspension to settle. The effluent was filtered through a membrane filter with a pore diameter of 0.045 µm. The concentrations of selected metal ions were determined by induction plasma atomic emission spectrophotometry (with an ICP-AES spectrometer) and the effluent pH was determined. Two samples were tested from the selected series.

3. Results

3.1. Properties of Concrete Mixes

During the preparation of concrete mixes, samples were taken to test the consistency as well as the air content in the mixes. The fall of the cone for the B1-0 control concrete mix was 145 mm, which qualifies it as belonging to the S3 consistency class. The air content in this concrete mix was 3.4%. B1 concrete mixes containing fly ash used as part of the sand were denser than the B1-0 mix. The air content in the concrete mix decreased with the increase of the amount of added ash and amounted to 2.7% for the B1 mix containing 30% ash (Table 4). Mixes with the sand–basalt aggregate behaved similarly to the B1 mixes, too. The addition of fly ash to concrete mixes based on CEM II cement (B3 series) also caused the condensation of the concrete mix and a reduction of the air content.

Table 4. Consistency class and air content of B1, B2 and B3 concrete mixtures.

Series	Consistency Class B2-0		Air Content, %
Ingredient	mm	Class	
B1-0	145	S3	3.4
B1-1	95	S2/S3	3.4
B1-2	60	S2	2.9
B1-3	25	S1	2.7
B1-4	10	S1	2.5
B2-0	150	S3	4.0
B2-1	100	S3	3.7
B2-2	60	S2	2.9
B2-3	30	S1	2.4
B3-0	140	S3	3.7
B3-1	90	S2	3.2
B3-2	50	S2	2.6
B3-3	20	S1	2.4

3.2. The Impact of Ash Addition on the Compressive Strength of Concrete

The concrete samples shaped into cubes with side dimensions of 150 mm were tested for compressive strength after 7, 28 and 56 days. After 7 days, the control concrete (B1-0) was marked by a compressive strength of 43.4 MPa. B1-0 concrete was the control concrete for the B1 series samples. All the concretes in the series with the addition of fly ash from biomass combustion in a fluidized bed boiler in the amounts of 10%, 20% and 30% of the weight of cement used instead of sand showed higher compressive strength after 7 days compared to the corresponding control concretes without the addition of ash (Figure 3).

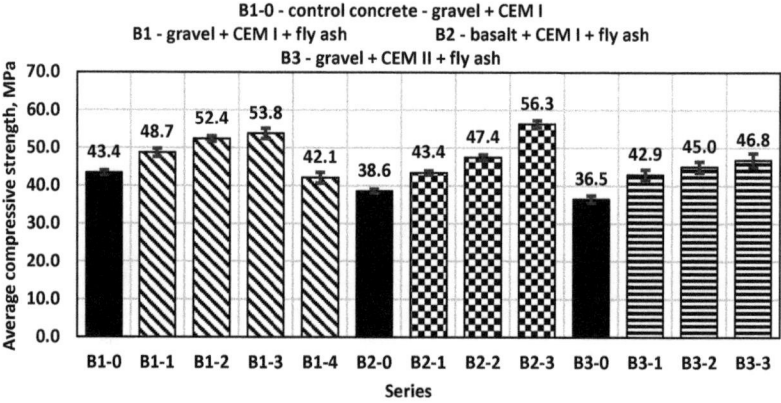

Figure 3. Average compressive strength after 7 days.

B1 series concretes with a sand–gravel mixture of aggregates containing up to 30% ash additive achieved a higher average compressive strength than the control concrete. The highest compressive strength was achieved by B1-3 concretes containing 30% fly ash and it was approximately 24% higher than the strength obtained with the sample from the control series (B1-0). On the other hand, the addition of ash in the amount of 40% caused a decrease in the compressive strength of the tested concretes by approximately 3% compared to the control concrete. The compressive strength results for B2 series concretes with a basalt–sand mixture of aggregates and the addition of fly ash obtained after 7 days confirmed the tendency observed for the B1 series concretes. The increase in strength for the B2-3 series was 45% compared to the B2-0 control concrete. For concrete samples of the B3 series, in which fly ash was used as an additive to concrete made on the basis of CEM II/A-V 42.5 R cement, the highest compressive strength was also obtained by samples containing 30% fly ash, achieving an average compressive strength 28% higher than the B3-0 control series. The addition of fly ash from biomass increased the compressive strength of concrete from 12% to over 45%. The smallest standard deviation for samples tested after 7 days was 0.55 for samples from series B2-0 and the highest was 1.79 for samples from series B3-3.

Compressive strength tests were also conducted for concrete samples after the standard time of 28 days of maturation. The control concrete (B1-0) with a sand–gravel mixture of aggregates obtained an average compressive strength of 49.7 MPa, while all concretes with the addition of fly ash in the amounts of 10%, 20%, 30% and 40% obtained higher compressive strength results than the control concrete (Figure 4).

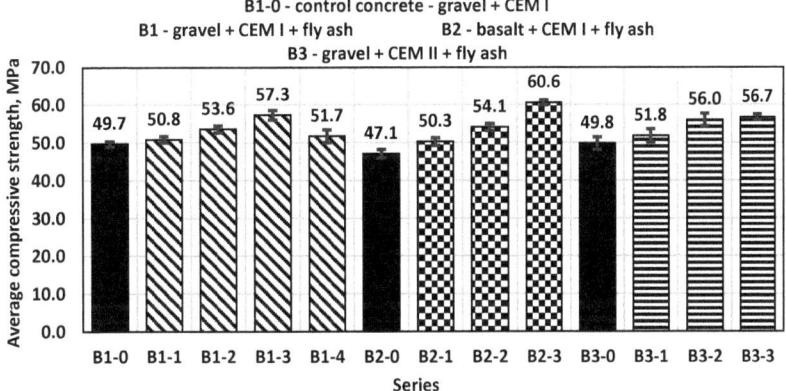

Figure 4. Average compressive strength after 28 days.

The highest value of compressive strength was achieved by B1-3 concrete series with the addition of 30% fly ash as a substitute for part of the sand. As shown in the figure, the ash content at the level of 30% of the cement mass is the limit, leading to an increase in the compressive strength of concrete; higher amounts of added ash caused a decrease in the value of this parameter. The obtained test results for the B2 and B3 concrete series showed a similar tendency to concrete samples from the B1 series and increased this parameter by approximately 28%. The smallest standard deviation for samples tested after 28 days was 0.65 for samples from series B2-3 and the highest was 1.72 for samples from series B3-1.

Compressive strength tests of concretes after 56 days of maturation were also conducted (Figure 5). At this stage, all concretes containing waste ash were marked by a higher compressive strength than the control concrete of the B1-0 series, and in the case of the B1 series of concrete, the increase in compressive strength was higher than in the case of the control concrete by 4–16% compared to the control concrete. The results obtained for concretes containing ash showed a similar tendency to concretes in the B2 and B3 series. Concretes with the addition of fly ash achieved higher compressive strengths than the control concrete, from 13 to 23%. Such a relationship may confirm that the ashes from the combustion of wood–sunflower biomass have a higher reactivity. The smallest standard deviation for samples tested after 56 days was 0.65 for samples from series B1-3 and the highest was 2.34 for samples from series B3-0.

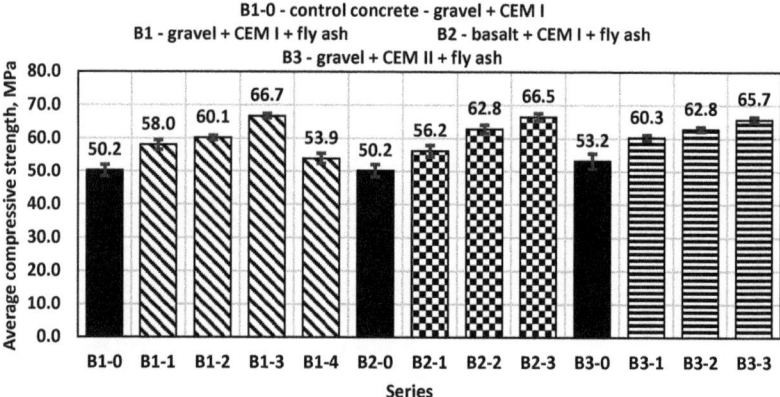

Figure 5. Average compressive strength after 56 days.

3.3. The Impact of Ash Addition on the Water Absorption of Concrete

Another property of the concretes tested was their water absorption, which was determined in accordance with the PN-B-06250 standard. Cubic samples with a side of 150 mm were prepared for the test. In the case of the B1-0 control concrete, the water absorption of the samples was 4.6%. The B1 series concretes, in which the fly ash from biomass combustion was added, showed a slightly higher water absorption, which increased along with the increase in the amount of added ash (Figure 6).

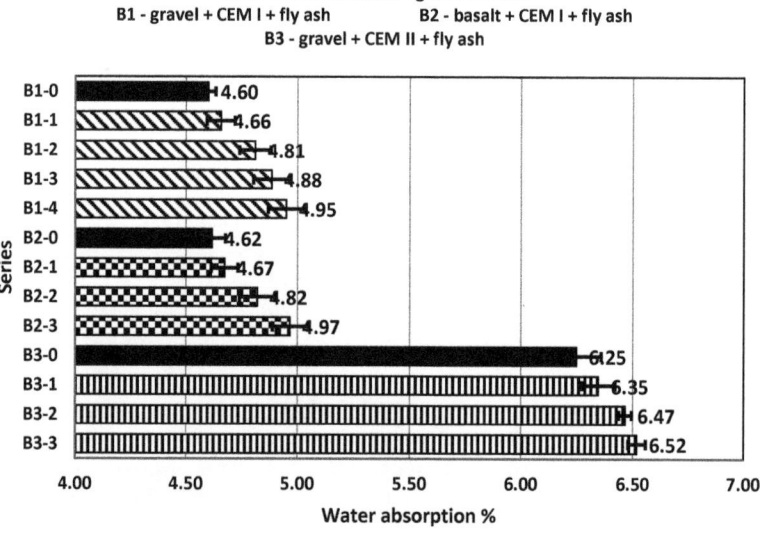

Figure 6. Water absorption of control concretes and concretes with the addition of fly ash.

The B2 series concretes also showed an increase in water absorption along with the increase of the added ash, however, the difference between the control concrete and the concrete containing 30% ash added was only 0.35%. The water absorption of B3 series concretes, in which CEM II ash cement was used, was higher than in the case of concretes made with CEM I cement. Additionally, in these series of concretes, an increase in the water absorption of the tested concretes was observed along with an increase in the amount of added ash. The water absorption of concrete should not exceed 5% in the case of concretes directly exposed to atmospheric conditions and 9% in the case of concretes sheltered from direct atmospheric conditions. According to these guidelines, the B1-0 control concrete and all B1 and B2 series concretes could be used outside, while the B3 series concretes could be used inside. The smallest standard deviation for samples was 0.03 for samples from series B1-0 and the highest was 0.11 for samples from series B3-0.

3.4. The Impact of Ash Addition on Concrete Low-Temperature Resistance

Low-temperature resistance tests of synthesized concretes were also conducted in accordance with the PN-B-06250:1988 standard. Closed pores play a key role in this, as they increase the resistance of concrete to freeze–thaw cycles. During freezing, ice is formed in open capillaries and the volume of the ice increases, the frozen water creating high hydrodynamic pressure and internal stresses. The decrease in compression strength in the control concretes after the freeze–thaw cycles was less than 19% (Figure 7).

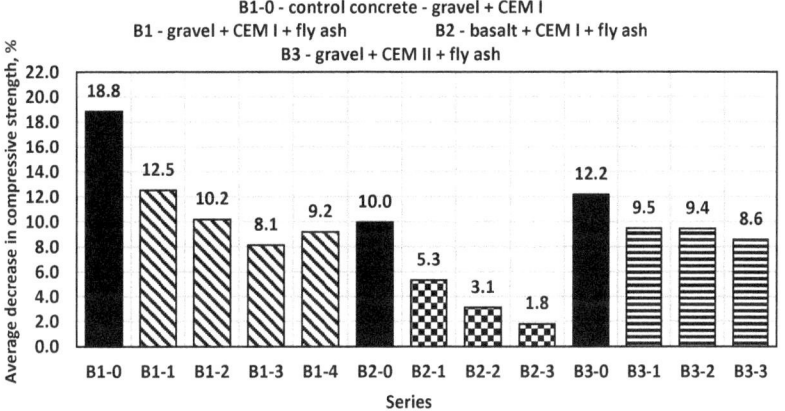

Figure 7. Decrease in the average compressive strength after frost resistance tests.

In the case of the B1 series of concretes containing the addition of 10%, 20%, 30% and 40% of fly ash, a decrease in compressive strength was obtained by 12.5%, 10%, 8% and 9%, respectively. An addition of 30% fly ash can be considered optimal, due to this proportion being associated the lowest strength drop. All B1 series concrete samples had no visible damage or cracks after visual inspection. The low-temperature resistance test did not cause a large loss of mass, which was in the range of 0.01–0.03%. The B2 series concretes, in which crushed basalt aggregate was used, showed even smaller drops in compression strength after low-temperature resistance tests than the B1 series concretes. The addition of 30% of ash reduced the decrease in compression strength to a value of only 1.8%. The B3 series concretes also had no visible damage or cracks, and in all samples the use of fly ash was associated with a lower decrease in compression strength after low-temperature resistance tests. The best result was obtained for the concrete containing 30% of fly ash (strength decrease approximately 8.5%). To sum up, on the basis of the obtained test results, it can be concluded that the use of ashes from the combustion of wood–sunflower biomass in a fluidized bed boiler as substitutes for a part of the sand content in concrete reduces the decrease in the compression strength of concrete after low-temperature resistance tests. In each series of concrete containing ash, the decrease in compression strength was lower than for the control concrete. This addition reduced the decrease in compression strength by up to 80% compared to the control samples.

3.5. Water Penetration of Concretes Containing Ash from Biomass Combustion

The B1-0 control concrete, influenced by water pressure, obtained an average water penetration depth of 42 mm (Figure 8). The B1 series concretes were marked by a greater depth of water penetration, which increased with the increase in the amount of added ash. In the case of B2 and B3 series concretes, the trend observed for the B1 series concretes is visible. The lowest water penetration depth was achieved by the B2-0 and B3-0 control concretes, and the highest by those containing 30% fly ash. It can be noticed that in the B3 series concretes with CEM II ash cement, the increase in water penetration depth with increasing ash content increased to a greater extent and reached higher values than in the case of concretes made on the basis of CEM I cement.

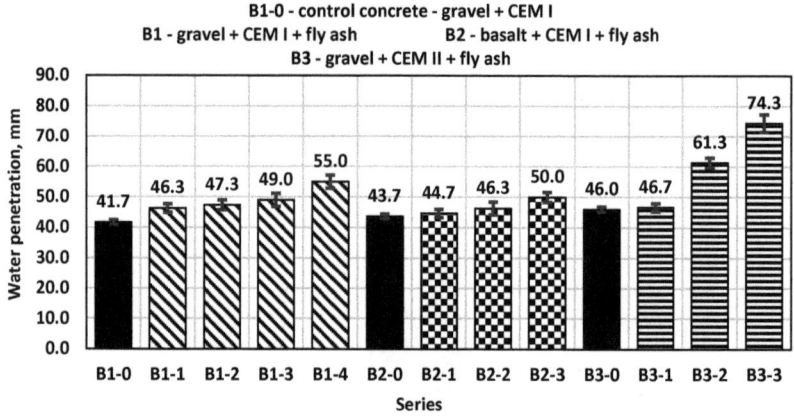

Figure 8. Water penetration of control concretes and concretes with the addition of fly ash.

3.6. Leaching Metal Ions from Concretes Containing Ash from Biomass Combustion

The test for the leaching of metal ions from concrete was conducted in accordance with the PN EN 12457 2:2006 standard. The leaching test was performed only for the control concrete (B1-0) and for concretes produced with the addition of fly ash in the amount of 30% of the cement mass (samples from the B1-3 series), assuming that the highest content in the eluate will occur for them. Two determinations were made for the tested concrete, maintaining the ratio of the volume of the liquid phase (L) to the mass of the solid phase (S) equal to 10:1. Calculations were made according to the standard assuming no moisture in the samples. After the process of leaching ions from the B1-0 control concrete, the pH of the solution was 10.10, while in the case of B1-3 concretes it was 9.67. The data presented in Table 5 show that the amount of leached toxic metal ions, such as Zn, Pb, Cu, Cr and Ba, from concretes containing fly ash (B1-3) is comparable to the amount of metal ions leached from the control concrete (B1-0). On the other hand, the content of toxic metal ions, such as Ni and Fe, from the concrete samples containing ash is slightly higher than from the control concrete samples. The obtained concentrations of Ni and Fe ions, as well as Zn, Pb, Cu, Cr and Ba metals, do not exceed the permissible values that must be met when discharging sewage into water or soil with substances particularly harmful to the environment, in accordance with the Regulation of the Minister of the Environment (Journal of Laws of 2014, item 1800). Thus, considering composite cracks and water penetration, concrete composites containing up to 30% fly ash do not pose a threat to the environment.

Table 5. Results of the analysis of ion leaching from concrete samples.

Element	Sample				Limit Values
	B1-0		B1-3		
	A	s	A	s	
	mg/kg		mg/kg		mg/dm^3
Zn	<0.005		<0.005		2
Cu	0.220	0.014	0.190	0.000	0.5
Ni	0.075	0.050	0.090	0.085	0.5
Ba	0.340	0.056	0.275	0.050	2
Pb	0.260	0.014	0.260	0.014	0.5
Cr	0.480	0.028	0.390	0.000	0.5
Fe	0.245	0.092	0.285	0.050	10
K	417.200	0.848	621.750	45.891	80
Na	184.900	29.699	186.550	8.273	800

A—released amount of component with L/S = 10, average based on two determinations; s—standard deviation based on two determinations.

3.7. Microstructure of Composites with a Cement Matrix Containing Ashes from Biomass Combustion

The tests were performed on broken surfaces of concrete samples. Figure 9 shows microscopic photos for the B1-0 concrete at 65× magnification and the X-ray energy dispersion analysis.

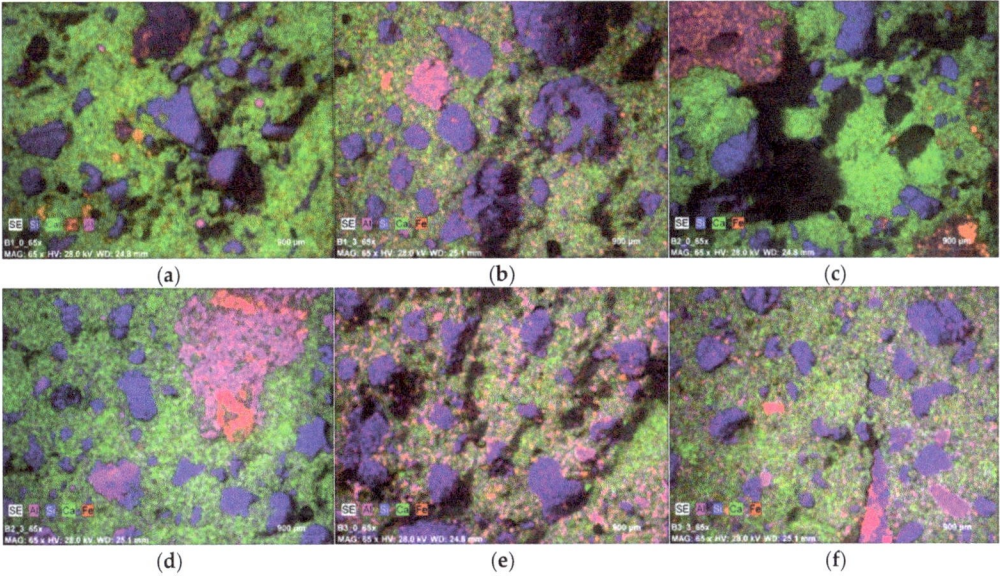

Figure 9. Maps of distribution of dominant elements in the studied area: (**a**) B1-0; (**b**) B1-3; (**c**) B2-0; (**d**) B2-3; (**e**) B3-0; (**f**) B3-3.

EDS analysis of the concrete surface performed in the area visible in the photo showed, apart from the presence of calcium (24.65%, green), a significant content of silicon (13.45%, blue), aluminium (2.31%; pink) and iron (2.72%, red). Sodium (1.56%) and potassium (0.39%) are present in lesser amounts. The remaining components (Mg, C, S, P) are present in a small amount (less than 0.7%). In the case of the B1-3 concrete samples containing 30% fly ash from biomass combustion, microscopic photos show a very similar structure to concrete without the addition of ash, which results from very fine granulation of the ash. The aggregate grains distributed evenly in the concrete are also clearly visible in these concretes. EDS analysis of the B1-3 concrete surface confirmed similar amounts of elements compared to the control concrete. In the B2-0 control concrete, in which sand was used as fine aggregate and a basalt mixture as coarse aggregate, microscopic photos showed basalt grains with irregular and sharp edges and the present fine aggregate with oval shapes; the largest share in the chemical composition is calcium (18.91%) and silicon (18.36%). In the case of the B2-3 concrete with 30% ash added, coarse basalt aggregate and quartz sand are also visible, and in terms of chemical composition, ash-free concrete (B2-0 series) and concrete with 30% fly ash added (B2-3 series) do not differ significantly from each other. The analysis of the B3 series surfaces showed that the microstructure of the B3-0 concrete, made on the basis of CEM II cement, has a very similar microstructure and appearance to the B1-0 concrete made on the basis of CEM I cement. The B3-3 concrete containing 30% fly ash under the microscope did not differ noticeably in structure from the B3-0 concrete. The EDS analysis of the B3-0 series concrete surfaces made on the basis of CEM II ash cement is also marked by a similar chemical composition. This concrete consists of approximately 23% calcium and 17% silicon, and similar results were obtained for samples with the addition of the B3-3 ash. The obtained results are not very precise guides, as only a few measuring points on the surface were anyalysed. In addition, the concrete structure is not homogenous throughout the sample volume. However, it can be concluded that non-destructive research methods using modern techniques for testing material structures are a way to effectively assess the impact of structure on the strength of concrete.

4. Conclusions

The use of fly ash from the wood–sunflower biomass combustion process in a circulating fluidized bed boiler for the production of building materials is a desirable solution in terms of sustainable construction.

1. The conducted tests of the properties and structure of composite materials with a cement matrix containing waste fluidized ashes from the combustion of wood–sunflower biomass (in the proportion of 80% wood waste and 20% sunflower waste) and the analysis of the obtained results enabled the conclusion that this ash, in the amount of 10% and up to 30%, can be added as a sand replacement for the production of composite materials with a cement matrix, such as concretes (made on the basis of CEM I or CEM II/A-V 42.5R cement), without reducing their quality (compression strength and low-temperature resistance) in relation to materials produced without the addition of waste ash.
2. The microstructure of concretes made on the basis of CEM I 42.5R cement containing fly ash from biomass combustion in the amount of 10–30% is marked by a microstructure very similar to the control concrete made without the addition of waste ash.
3. The addition of fly ash in the amount of 10–30% causes the concrete mix to condense and reduce its air content. The compression strength of concrete samples with sand–gravel and sand–basalt mixtures of aggregates made with the tested additive is higher than that obtained for the control samples.
4. The use of fly ash in the range of 10–30% causes a slight increase in the water absorption of concrete samples made with both sand–gravel and sand–basalt mixes of aggregates. The addition of waste ashes increases the depth of water penetration of the samples of all tested concrete series.
5. The addition of fly ash from biomass to concretes in amounts up to 30% reduces the decrease in compression strength of these concretes after low-temperature resistance tests and does not cause the formation of fragments, cracks and defects in the sample. In addition, concrete composites containing the addition of ash from biomass combustion, taking into account the crack of the composite and the penetration of water,

do not show a negative impact on the natural environment related to the leaching (release) of heavy metals into the environment.
6. The use of waste ash has a positive effect on the natural environment by reducing the demand for natural resources. The use of fly ash in the amount of 30% of the cement mass as a sand substitute, depending on the type of aggregate mixture, reduces sand usage by 115–120 kg/m^3, i.e., by about 15–20%.

Author Contributions: Investigation, software, methodology, writing original draft preparation, J.J.; conceptualization, methodology, analysis, writing review and editing, M.U. All authors have read and agreed to the published version of the manuscript.

Funding: This research received no external funding.

Institutional Review Board Statement: Not applicable.

Informed Consent Statement: Not applicable.

Data Availability Statement: The data presented in this study are available on request from the corresponding author.

Acknowledgments: The authors would like to thank GDF SUEZ Połaniec for making waste materials available for research.

Conflicts of Interest: The authors declare no conflict of interest. The funders had no role in the design of the study; in the collection, analyses, or interpretation of data; in the writing of the manuscript; or in the decision to publish the results.

References

1. Sekar, T.; Ganesan, N.; Nampoothire, N. Studies on strength characterization on utilization of waste materials as coarse aggregate in concrete. *Int. J. Eng. Sci. Technol.* **2011**, *3*, 5436–5440.
2. Argiz, C.; Menéndez, E.; Sanjuán, M.A. Effect of mixes made of coal bottom ash and fly ash on the mechanical strength and porosity of Portland cement. *Mater. Constr.* **2013**, *63*, 49–64.
3. Baran, T.; Drożdż, W. Evaluation of properties of domestic calcareous fly ash and its processing methods. *Roads Bridges Drog. Mosty* **2013**, *12*, 5–15.
4. Cheriaf, M.; Cavalcante, J.; Péra, J. Pozzolanic properties of pulverized coal combustion bottom ash. *Cem. Concr. Res.* **1999**, *29*, 1387–1391. [CrossRef]
5. Dziuk, D.; Giergiczny, Z.; Garbacik, A. Calcareous fly ash as a main constituent of common cements. *Roads Bridges Drog. Mosty* **2013**, *12*, 57–69.
6. Fan, W.J.; Wang, X.Y.; Park, K.B. Evaluation of the Chemical and Mechanical Properties of Hardening High-Calcium Fly Ash Blended Concrete. *Materials* **2015**, *8*, 5933–5952. [CrossRef] [PubMed]
7. Gazdiča, D.; Fridrichováa, M.; Kulíseka, K.; Vehovská, L. The potential use of the FBC ash for the preparation of blended cements. *Procedia Eng.* **2017**, *180*, 1298–1305. [CrossRef]
8. Kocak, Y.; Nas, S. The effect of using fly ash on the strength and hydration characteristics of blended cement. *Constr. Build. Mater.* **2014**, *73*, 25–32. [CrossRef]
9. Strigáč, J.; Števulová, N.; Mikušinec, J.; Sobolev, K. The fungistatic properties and potential application of by-product fly ash from fluidized bed combustion. *Constr. Build. Mater.* **2018**, *159*, 351–360. [CrossRef]
10. Yao, Z.T.; Ji, X.S.; Sarker, P.K.; Tang, J.H.; Ge, L.Q.; Xia, M.S.; Xi, Y.Q. A comprehensive review on the applications of coal fly ash. *Earth Sci. Rev.* **2015**, *141*, 105–121. [CrossRef]
11. Anderson, D.J.; Smith, S.T.; Au, F.T.K. Mechanical properties of concrete utilising waste cerami cascoarse aggregate. *Constr. Build. Mater.* **2016**, *117*, 20–28. [CrossRef]
12. Collivignarelli, M.C.; Cillari, G.; Ricciardi, P.; Miino, M.C.; Torretta, V.; Rada, E.C.; Abbà, A. The production of sustainable concrete with the use of alternative aggregates: A review. *Sustainability* **2020**, *12*, 7903. [CrossRef]
13. Subramani, T.; Suresh, B. Experimental investigation of using ceramic waste as a coarse aggregate making a light weight concreto. *Int. J. Appl. Innov. Eng. Manag.* **2015**, *4*, 153–162.
14. Halicka, A.; Ogrodnik, P.; Zegardlo, B. Using ceramic sanitary ware waste as concrete aggregate. *Constr. Build. Mater.* **2013**, *48*, 295–305. [CrossRef]
15. Medina, C.; Frías, M.; Sánchez de Rojas, M.I. Microstructure and properties of recycled concretes using ceramic sanitary ware industry waste as coarse aggregate. *Constr. Build. Mater.* **2012**, *31*, 112–118. [CrossRef]
16. Medina, C.; Frías, M.; Sánchez de Rojas, M.I.; Thomas, C.; Polanco, J.A. Gas permeability in concrete containing recycled ceramic sanitary ware aggregate. *Constr. Build. Mater.* **2012**, *37*, 597–605. [CrossRef]

17. Ray, S.; Haque, M.; Soumic, S.A.; Mita, A.F.; Rahman, M.D.M.; Tanmoy, B.B. Use of ceramic wastes as aggregates in concrete production: A review. *J. Build. Eng.* **2021**, *43*, 102567. [CrossRef]
18. Ulewicz, M.; Halbiniak, J. Application of waste from utilitarian ceramics for production of cement mortar and concrete. *Physicochem. Probl. Miner. Process.* **2016**, *52*, 1002–1010.
19. Adhikary, S.K.; Ashish, D.K.; Rudžionis, Z. Expanded glass as light-weight aggregate in concrete—A review. *J. Clean. Prod.* **2021**, *313*, 127848. [CrossRef]
20. Omoding, N.; Cunningham, L.S.; Lane-Serff, G.F. Effect of using recycled waste glass coarse aggregates on the hydrodynamic abrasion resistance of concrete. *Constr. Build. Mater.* **2021**, *268*, 121177. [CrossRef]
21. Walczak, P.; Małolepszy, J.; Reben, M.; Rzepa, K. Mechanical properties of concrete mortar based on mixture of CRT glass cullet and fluidized fly ash. *Procedia Eng.* **2015**, *108*, 453–458. [CrossRef]
22. Walczak, P.; Małolepszy, J.; Reben, M.; Szymański, P.; Rzepa, K. Utilization of Waste Glass in Autoclaved Aerated Concrete. *Procedia Eng.* **2015**, *122*, 302–309. [CrossRef]
23. Kishore, K.; Gupta, N. Application of domestic & industrial waste materials in concrete: A review. *Mater. Today Proc.* **2020**, *26*, 2926–2931. [CrossRef]
24. Pietrzak, A.; Ulewicz, M. The Effect of the Addition of Polypropylene Fibres on Improvement on Concrete Quality. *MATEC Web Conf.* **2018**, *183*, 02011. Available online: http://doi.org/10.1051/matecconf/201818302011 (accessed on 7 November 2021).
25. Ulewicz, M.; Pietrzak, A. Properties and Structure of Concretes Doped with Production Waste of Thermoplastic Elastomers from the Production of Car Floor Mats. *Materials* **2021**, *14*, 872. [CrossRef]
26. Johnson, A.; Catalan, L.J.J.; Kinradeb, S.D. Characterization and evaluation of fly-ash from co-combustion of lignite and wood pellets for use as cement admixture. *Fuel* **2010**, *89*, 3042–3050. [CrossRef]
27. Saraber, A.J. Fly Ash from Coal and Biomass for Use in Concrete Origin, Properties and Performance. Ph.D. Thesis, Delft University of Technology, Delft, The Netherlands, 2017. [CrossRef]
28. Tkaczewska, E.; Mróz, R.; Łój, R. Coal–biomass fly ashes for cement production of CEM II/A-V 42,5R. *Constr. Build. Mater.* **2012**, *28*, 633–639. [CrossRef]
29. Tkaczewska, E.; Małolepszy, J. Hydration of coal–biomass fly ash cement. *Constr. Build. Mater.* **2009**, *23*, 2694–2700. [CrossRef]
30. Wang, S.; Miller, A.; Llamazos, E.; Fonseca, F.; Baxter, L. Biomass fly ash in concrete: Mixture proportioning and mechanical properties. *Fuel* **2008**, *87*, 365–371. [CrossRef]
31. Azam, M.; Jahromy, S.; Raza, W.; Wesenauer, F.; Schwendtner, K.; Winter, F. Comparison of the Characteristics of Fly Ash Generated from Bio and Municipal Waste: Fluidized Bed Incinerators. *Materials* **2019**, *12*, 2664. [CrossRef]
32. Barišić, I.; Netinger Grubeša, I.; Dokšanović, T.; Marković, B. Feasibility of Agricultural Biomass Fly Ash Usage for Soil Stabilisation of Road Works. *Materials* **2019**, *12*, 1375. [CrossRef]
33. Girón, R.P.; Suárez-Ruiz, I.; Ruiz, B.; Fuente, E.; Gil, R.R. Fly Ash from the Combustion of Forest Biomass (Eucalyptus globulus Bark): Composition and Physicochemical Properties. *Energy Fuels* **2012**, *26*, 1540–1556. [CrossRef]
34. Steenari, B.M.; Schelander, S.; Lindqvist, O. Chemical and leaching characteristics of ash from combustion of coal, peat and wood in a 12 MW CFB—A comparative study. *Fuel* **1999**, *78*, 249–258. [CrossRef]
35. Mazurkiewicz, A.; Mokrzycki, M. Fly ash from energy production—A waste, by product and raw material. *Miner. Resour. Manag.* **2015**, *31*, 139–150.
36. Steenari, B.M.; Lindqvist, O. Stabilisation of biofuel ashes for recycling to forest soil. *Biomass Bioenergy* **1997**, *13*, 39–50. [CrossRef]
37. Memon, S.A.; Khan, M.K. Ash blended cement composites: Eco-friendly and sustainable option for utilization of corncob ash. *J. Clean. Prod.* **2018**, *175*, 442–455. [CrossRef]
38. Khalil, J.M.; Aslam, M.; Ahmad, S. Utilization of sugarcane bagasse ash as cement replacement for the production of sustainable concrete—A review. *Constr. Build. Mater.* **2021**, *270*, 121371. [CrossRef]
39. Priya, K.L.; Ragupathy, R. Effect of sugarcane bagasse ash on strength properties of concrete. *Int. J. Res. Eng. Technol.* **2016**, *5*, 2321–7308.
40. Rukzon, S.; Chindaprasirt, P. Utilization of bagasse ash in high-strength concrete. *Mater. Des.* **2012**, *34*, 45–50. [CrossRef]
41. Jagadesh, P.; Ramachandramurthy, A.; Murugesan, R. Evaluation of mechanical properties of Sugar Cane Bagasse Ash concrete. *Constr. Build. Mater.* **2018**, *176*, 608–617. [CrossRef]
42. Carevic, I.; Baricevic, A.; Štirmer, N.; Šantek Bajto, J. Correlation between physical and chemical properties of wood biomass ash and cement composites performances. *Constr. Build. Mater.* **2020**, *256*, 119450. [CrossRef]
43. Ganesana, K.; Rajagopala, K.; Thangavelb, K. Evaluation of bagasse ash as supplementary cementitious material. *Cem. Concr. Compos.* **2007**, *29*, 515–524. [CrossRef]
44. Hamid, Z.; Rafiq, S. An experimental study on behavior of wood ash in concrete as partial replacement of cement. *Mater. Today Proc.* **2021**, *46*, 3426–3429. [CrossRef]
45. Rajamma, R.; Ball, R.J.; Tarelho, L.A.C.; Allen, G.C.; Labrincha, J.A.; Ferreira, V.M. Characterization and use of biomass fly ash in cement-based materials. *J. Hazard. Mater.* **2009**, *172*, 1049–1060. [CrossRef] [PubMed]
46. Jura, J. Influence of Type of Biomass Burned on the Properties of Cement Mortar Containing Fly Ash. *Constr. Optim. Energy Potential* **2020**, *9*, 77–82.
47. Nagrockiene, D.; Daugela, A. Investigation into the properties of concrete modified with biomass combustion fly ash. *Constr. Build. Mater.* **2018**, *174*, 369–375. [CrossRef]

48. Rerkpiboon, A.; Tangchirapat, W.; Jaturapitakkul, C. Strength, chloride resistance, and expansion of concretes containing ground bagasse ash. *Constr. Build. Mater.* **2015**, *101*, 983–989. [CrossRef]
49. Burduhos Nergis, D.D.; Abdullah, M.M.A.B.; Sandu, A.V.; Vizureanu, P. XRD and TG-DTA Study of New Alkali Activated Materials Based on Fly Ash with Sand and Glass Powder. *Materials* **2020**, *13*, 343. [CrossRef]

Article

An Experimental and Empirical Study on the Use of Waste Marble Powder in Construction Material

Muhammad Sufian [1,*], Safi Ullah [1], Krzysztof Adam Ostrowski [2,*], Ayaz Ahmad [2,3,*], Asad Zia [4], Klaudia Śliwa-Wieczorek [2], Muhammad Siddiq [1] and Arsam Ahmad Awan [5]

1. School of Civil Engineering, Southeast University, Nanjing 210096, China; safi@seu.edu.cn (S.U.); siddiq@seu.edu.cn (M.S.)
2. Faculty of Civil Engineering, Cracow University of Technology, 24 Warszawska Str., 31-155 Cracow, Poland; klaudia.sliwa-wieczorek@pk.edu.pl
3. Department of Civil Engineering, Abbottabad Campus, COMSATS University Islamabad, Islamabad 22060, Pakistan
4. School of Civil Engineering, Zhengzhou University, Zhengzhou 450001, China; asadzia005@gs.zzu.edu.cn
5. MM Pakistan (Pvt) Limited, Lahore 54000, Pakistan; arsam.ahmad@mmpakistan.com
* Correspondence: drsufian@seu.edu.cn (M.S.); krzysztof.ostrowski.1@pk.edu.pl (K.A.O.); ayazahmad@cuiatd.edu.pk (A.A.)

Citation: Sufian, M.; Ullah, S.; Ostrowski, K.A.; Ahmad, A.; Zia, A.; Śliwa-Wieczorek, K.; Siddiq, M.; Awan, A.A. An Experimental and Empirical Study on the Use of Waste Marble Powder in Construction Material. *Materials* **2021**, *14*, 3829. https://doi.org/10.3390/ma14143829

Academic Editor: Malgorzata Ulewicz

Received: 12 May 2021
Accepted: 6 July 2021
Published: 8 July 2021

Publisher's Note: MDPI stays neutral with regard to jurisdictional claims in published maps and institutional affiliations.

Copyright: © 2021 by the authors. Licensee MDPI, Basel, Switzerland. This article is an open access article distributed under the terms and conditions of the Creative Commons Attribution (CC BY) license (https://creativecommons.org/licenses/by/4.0/).

Abstract: Marble is currently a commonly used material in the building industry, and environmental degradation is an inevitable consequence of its use. Marble waste occurs during the exploitation of deposits using shooting technologies. The obtained elements most mainly often have an irregular geometry and small dimensions, which excludes their use in the stone industry. There is no systematic way of disposing of these massive mounds of waste, which results in the occurrence of landfills and environmental pollution. To mitigate this problem, an effort was made to incorporate waste marble powder into clay bricks. Different percentage proportions of marble powder were considered as a partial substitute for clay, i.e., 5–30%. A total of 105 samples were prepared in order to assess the performance of the prepared marble clay bricks, i.e., their water absorption, bulk density, apparent porosity, salt resistance, and compressive strength. The obtained bricks were 1.3–19.9% lighter than conventional bricks. The bricks with the addition of 5–20% of marble powder had an adequate compressive strength with regards to the values required by international standards. Their compressive strength and bulk density decreased, while their water absorption capacity and porosity improved with an increased content of marble powder. The obtained empirical equations showed good agreement with the experimental results. The use of waste marble powder in the construction industry not only lowers project costs, but also reduces the likelihood of soil erosion and water contamination. This can be seen to be a crucial factor for economic growth in agricultural production.

Keywords: marble waste; bricks; clay; compressive strength; marble powder; eco-friendly materials

1. Introduction

Marble is a crystalline metamorphic rock that can be formed into different shapes and sizes for flooring, monumental and decorative purposes. According to a survey, the demand for marble stone worldwide hit 816 million m^3 in 2016. In 2019, the global demand for marble was estimated at USD 55,420 million, and it is projected to hit USD 68,790 million by the end of 2026. This is an annual increase of 3.1% [1]. Pakistan's marble and granite deposits are projected to be equal to 300 billion tons. In addition, the country's gross monthly marble output is approximately 1 million tons, with 2000 processing units and 1225 quarries in service. As a consequence, in a country like Pakistan, where the marble industry is huge, pollution is a major concern. Pakistan has a range of marble processing plants, where marble waste is generated on a regular basis. This marble waste, in the form of slurry and mud, is discarded in open spaces (without adequate disposal) and has harmful environmental consequences [2,3]. The powdered form of marble stone decreases soil fertility by increasing alkalinity, endangering plants, and wreaking havoc on the environment. There is also a major loss in fauna and flora, as the slurry of marble accumulation impacts foliage and plant leaves, possibly drying out already grown trees and bushes [4–6]. The valuable powdered form is discarded by the marble industry, which has an impact on financial development and even results in environmental pollution [7]. Similarly, the high production of clay bricks in Pakistan is also causing soil degradation. As a consequence, the use of marble wastes as an alternate material in the manufacturing of bricks may be both inexpensive and beneficial for the atmosphere.

1.1. The Use of Various Types of Waste Materials in Clay Bricks

Many studies have used various wastes as additives in the manufacturing of clay bricks in order to test their various properties. The use of waste materials for construction purposes can be helpful in reducing the risk of shortages in natural resources and for improving the environment [8–10]. The addition of different solid wastes for the production of fired clay bricks were tested. As a result, bricks with improved properties, such as thermal conductivity, density, porosity, and water absorption, were obtained. According to Kadir A.A. and Mohajerani A., brick is a material with a relatively high mechanical strength, which can be recycled and reused in the production of other construction materials [11]. Muñoz Velasco P. et al. carried out a study regarding the use of various sorts of wastes in fired clay bricks and found the use of waste in clay bricks was eco-friendly and, in some cases, improved their properties [12]. Some studies related to the use of various types of wastes and low-cost materials are presented below.

1.2. The Use of Rice Husk Ash

Fernando P.R. compared the physical and chemical properties of rice husk ash bricks with traditional bricks and found that the addition of 5% of rice husk ash improves the compressive strength and water absorption properties of the bricks [13]. Rao B.J. used different wastes, such as fly ash, rice husk, sawdust, and bagasse in order to make clay bricks. Experimental tests were carried out to check the physical and mechanical properties of the resultant bricks. It was shown that the addition of waste materials to the brick mixture affects the consistency of freshly formed bricks [14].

1.3. The Use of Rice Fly Ash, Silica Fume, Wood Dust, Slags and Dry Grass

Kadir A.A. and Sarani N.A. assessed numerous waste materials, such as fly ash, rubber, limestone, wood dust, and sludge. They noticed that such waste had a beneficial effect on the manufactured lightweight bricks, and also enhanced the thermal conductivity of fired clay bricks [15]. Zhang L. presented a review of the research work concerning the use of waste materials for the production of bricks. Many waste materials, including fly ash and slag, as well as different methods, were studied with regards to the production of bricks [16]. Baspinar M.S. et al. found that silica fume in fired clay bricks improves the strength and efflorescence properties of the bricks. In addition, the potency of silica fume depends on the temperature of the fire [17]. Abbas S. et al. prepared bricks using clay and up to 25% of fly ash. It was found that the compressive strength of the bricks with the addition of fly ash was lower than the bricks without fly ash. In contrast to the clay bricks, the fly ash bricks were lighter and had less efflorescence [18]. Phonphuak N. prepared clay bricks with different percentages of dry grass and tested them in order to find out about the different properties of the resultant product. As a result, it was revealed that the bulk density and compressive strength of the bricks decreased with an increase in the content of dry grass [19].

1.4. The Use of Wood Ash, Marble Powder, and Other Types of Waste Powders

Oorkalan A. et al. used different proportions of waste products of wood ash, ceramic powder, and marble powder in clay bricks. Compressive strength and water absorption tests were conducted. It was concluded that the raw materials were useful and increased the brick's compressive strength and durability [20]. Ngayakamo B.H. et al. prepared fired clay bricks by using granite and eggshell powder in order to reduce environmental pollution. The obtained findings showed that the fired clay bricks made of 20% granite and 10% eggshell powder had the best performance [21].

1.5. The Use of Glass Wastes

Incorporating glass waste into fired clay bricks is an excellent way to develop eco-friendly bricks and other enhanced materials for the construction industry. Xin Y. et al. manufactured fired clay bricks with a partial addition of glass waste. They discovered that as the glass content rose and the particle size decreased, the compressive strength of the fired clay bricks improved significantly [22]. Mobili A. et al. used glass-reinforced plastic dust as a partial replacement for clay to produce fired clay bricks. The addition of GRP dust improved the water absorption capacity and decreased the effect of the firing of clay bricks [23]. Kazmi S. M. S. et al. used waste glass sludge to manufacture clay bricks and concluded that the bricks made of waste glass sludge achieved a higher compressive strength and flexural strength than the reference clay bricks. Moreover, adding waste glass sludge made the bricks lighter and increased their tolerance to efflorescence, sulphate attack, and freeze-thaw cycles [24].

1.6. The Use of Natural Fibers

It has been proven that some natural fibers can improve the strength properties of composite materials and also be used to produce cheap bricks from fired clay [25]. Kadir A.A. et al. made low-cost clay bricks with different amounts of coconut fiber in order to assess their physical and chemical properties. It was shown that the use of coconut fibers as an additive to the brick mixture allows this waste to be effectively used, and at the same time, bricks with good quality to be obtained [26]. Kadir A. A. et al. investigated the physical and mechanical properties of palm kernel shells to be used as a substitute for clay in fired clay bricks. Palm kernel husk was considered to be a waste that could be used in the production of bricks. This is because, with its addition, a product of acceptable quality was obtained [27].

1.7. The Use of Marble Dust in Clay Bricks

An experimental study was conducted on marble waste (in different proportions) as an additive to bricks. It was proven that the marble waste had a good impact on the chemical, physical, and mechanical strength of the resultant bricks [4,28,29]. Considering the mechanical properties of bricks containing wastes, it has been shown that the addition of marble powder to a mixture has no prominent impact on their load-bearing capacity [30]. Kathiresan M. et al. used marble sludge powder as a partial replacement for clay in order to prepare modified clay bricks with enhanced durability and strength properties. It was discovered that the use of marble sludge powder in brick manufacturing results in safe and environmentally sustainable recycled materials [31]. It is worth mentioning that Seghir N.T. et al. used waste marble powder as a partial replacement for cement in mortar in order to assess the properties of the resultant product, i.e., compressive strength, density, and apparent porosity. The obtained results showed that the addition of waste marble powder reduced the density and compressive strength of the elements, while at the same time improving their porosity [32]. Ramachandran G. et al. used granite and marble wastes in fly ash bricks to determine their strength properties and stated that the use of such wastes as binding materials is helpful in minimizing the risk of pollution [33]. Rehman W. et al. stated that marble waste could be used to prepare concrete bricks. Furthermore, it was shown that the compressive strength of bricks depends on the mutual mass proportions of marble dust, sand, and cement [34].

1.8. The Use of Marble Powder in Cement-Based Materials

Many researchers have used marble waste and brick waste in different types of concrete and cement mortar. It has been determined that concrete made with the addition of waste marble powder can be ecological. This is consistent with sustainable development strategies, which are now a global trend [35,36]. According to Shah M.U. et al., ultrafine brick waste has a high degree of pozzolanic ability and can be used as an addition to cement. In paper [37], it was shown that the mechanical properties of cement paste containing 5% and 10% of waste burnt brick powder were higher than for the reference mixture containing no waste. According to [38], the use of marble powder improved carbonation and water resistance, decreased shrinkage strain, and reduced the content of cement. Sadek M. D. et al. used waste marble and granite powders as mineral additives in self-compacting concrete. According to the researchers, waste powders should be used as mineral additives in self-compacting concrete to improve its performance [39]. Rodrigues R. et al. investigated the mechanical properties of concrete with different marble sludge-to-cement substitution ratios. Concrete containing marble sludge had lower mechanical properties than the reference samples [40]. The effect of marble powder as a partial replacement for cement on the mechanical properties and toughness of high-performance concrete was examined by Talah A. et al. It was found, when compared to the reference concrete, that marble powder is a good additive for the manufacturing of concrete with improved mechanical properties [41,42]. Elmaghraby M. S. and Ismail A. I. M. explored the chemical, mechanical, and mineralogical properties of the waste kaolinitic sand used to prepare concrete. According to their test results, some mixtures can be used in industrial furnaces at 1500 °C in order to produce concrete [43]. In summary, it is worth emphasizing that the use of waste marble powder for the production of many types of concrete would be beneficial for the natural environment due to the positive aspect of waste disposal. Thus, it is possible to obtain good concrete properties in terms of workability, strength, permeability, and microstructural performance [44].

2. Research Significance

The Earth's ecosystems are being destroyed, and there is a continuing increase in water pollution, bad air quality, and ground contamination. Pollution is not only a health issue, but also a major obstacle for sustainability. To minimize this challenge, researchers and engineers need to be more focused on the effective use of waste materials in the construction industry. The use of waste materials is one of the most critical steps of sustainability, because it helps to minimize the impact of environmental degradation, save renewable resources, lower the overall cost of building projects, and bring economic value to waste materials. It is more and more clear that advances in the production of waste derived materials benefit society. Marble and granite factories generate a large volume of wastes, including sludge and other residues, which pose a serious threat to the environment by polluting soil and water. In this paper, industrial waste in the form of marble powder was used as the base material for producing bricks. These bricks are inexpensive, offer good compressive strength, and are lightweight. The current research project aims to identify the waste-related commodity arrangements and treatment scenarios that are suitable for the production of marble powder-based slurry for sustainable bricks. This study assesses the properties of the final product after incorporating waste marble powder. As a result of this study, the use of waste marble powder in the construction industry might lead to a sustainable and environmentally friendly material with better properties. The key points of this study were to examine the effects of waste marble powder on the different properties of sustainable marble clay bricks, i.e., water absorption ability, bulk density, apparent porosity, efflorescence, and compressive strength. When optimizing the composition of waste marble bricks, it is possible to minimize the risk of environmental contamination by reducing the amount of marble waste. The use of waste marble powder would be advantageous due to the use of wastes, as well as for obtaining good strength and low-cost bricks beneficial for sustainable construction.

3. Materials and Experimental Methodology

The materials used in this research were marble powder, which was collected from a local company in Peshawar, Pakistan, and clay from a kiln in the district of Peshawar, where the bricks were prepared. The type of soil in this region is clay loam containing 30.9% sand, 37.8% silt and 31.3% clay [45]. The research methodology of this project included the preparation of raw materials, the dosing of a mixture of marble powder and clay waste, the formation of bricks, the drying process, the firing and cooling process, and laboratory tests. To manufacture marble clay bricks, many sites were visited, with one of the finest kilns in the region being selected. The place where the soil was collected was first cleaned and excavated to a depth of five feet. The required quantity of earth was then taken, cleaned from stones, and converted into powder form using a small earth crushing roller available in the kiln. The pulverized clay was sprinkled with water and kept in the conditions of an open laboratory space for 24 h before it was used, as shown in Figure 1a. The clay was prepared at the site and then manually mixed with different amounts of marble powder, see Figure 1b. The tempering process was carried out by adding the appropriate amount of water to the clay, which was then mixed thoroughly to make a homogenous mixture. The pressing of the mixture was done with the help of peoples' feet. The mixing time was about 15 min. A temperature of 25 ± 1 °C and humidity of 48 ± 2% was recorded during this work. The marble powder and clay proportions are given in Table 1. The steel molds were filled with tempered clay and pressed hard in order to appropriately fill the corners of the mold. An extra clay on the top surface of the molds was removed with the help of a wooden plank. The molds were then lifted up and properly shaped raw bricks were made on the ground, as shown in Figure 2a. Before the molding process, the ground was levelled, and sand was sprinkled over it. The molded bricks were dried using the hack drying method, i.e., the wet bricks were arranged in rows on their edges on slightly raised ground (hacks) in such a way that an appropriate space was kept between the rows for air and heat circulation. It was ensured that there was no sudden drying caused by direct exposure to the sun and wind, and a portable cover was provided to protect the bricks from rain. The naturally dried bricks were then taken for firing to a kiln called the 'bull trench kiln', the temperature of which was 1000–1100 °C. The burnt bricks were taken from the kiln and kept in an open place for cooling. A total of 105 brick samples were prepared; 15 samples for each marble proportion. All the samples were then safely deposited in a laboratory. The standard brick size available in Pakistan is shown in Figure 2b.

Figure 1. View of: (**a**) clay sample, and (**b**) mixing of clay and marble dust.

Table 1. Proportions of marble powder in the clay bricks.

Brick Groups	Number of Bricks	% of Marble by Weight	% of Clay by Weight	Marble Weight (kg)	Clay Weight (kg)
A	15	0	100	0	50
B1	15	5	95	2.5	47.5
B2	15	10	90	5	45
B3	15	15	85	7.5	42.5
B4	15	20	80	10	40
B5	15	25	75	12.5	37.5
B6	15	30	70	15	35

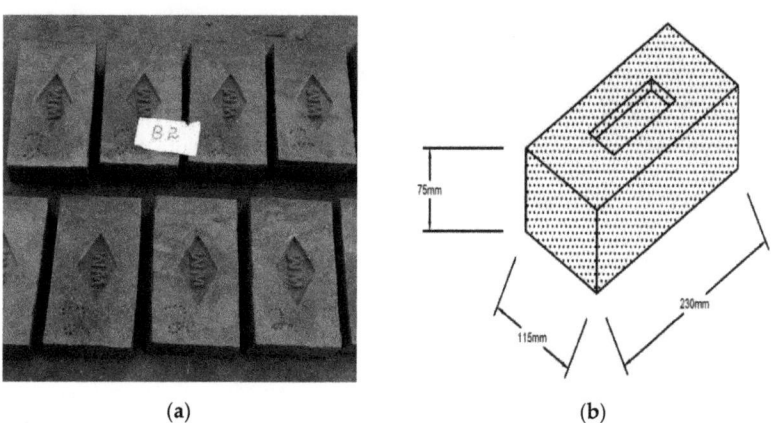

Figure 2. View of: (**a**) prepared marble clay bricks, and (**b**) standard brick size.

3.1. Energy Dispersive X-ray (EDX) Test of Marble Powder

To check the elements of marble powder, the energy dispersive X-ray test (EDX) was conducted. The contact area from which the X-rays were released was within the spectrum of 1 µm^3 for SEM-EDX. X-ray photons were examined spectroscopically in order to obtain basic information about the sample. Various images of regions were taken throughout the course of the EDX test, and the associated peaks for each element are displayed in Figure 3. The greatest density at the extreme peak at 2000 (eV) represents the oxides with the atomic value of 64.76%, while the second highest peak at about 3000 (eV) indicates the carbon element content with a value of 18.33%. Thirdly, the minor peaks seen at various energy levels, i.e., in the initial area until 1800 (eV), 6000 (eV) and 10,000 (eV), suggest a calcium content of about 16.71%. The lowest peaks imply a magnesium content of about 0.20%. The peak intensity of oxides was greater due to their preferred crystal orientation, while the other components' peak intensity was lower due to their randomized crystal orientations. The values of elements measured in weight % and atomic % are given in Table 2. The first letter—C, O, Ca and Mg—denotes elements identified during the EDX examination, while the second letter, K, denotes the shell of the element.

Table 2. Energy dispersive X-ray test report.

Element	Weight %	Atomic %
CK	11.40	18.33
OK	53.66	64.76
MgK	0.25	0.20
CaK	34.69	16.71
Total	100%	99.80%

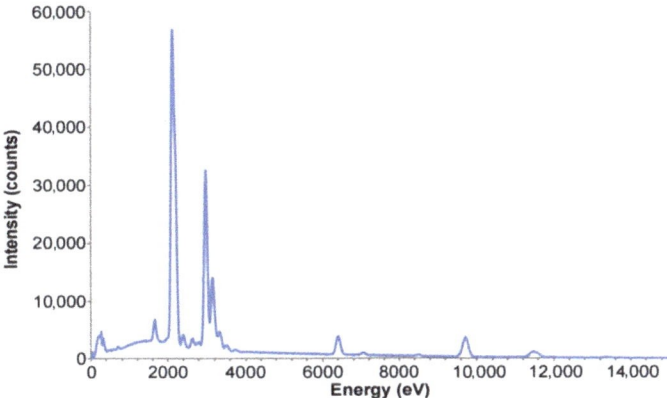

Figure 3. EDX test result.

3.2. Water Absorption Test

The water absorption tests were carried out on each of the groups of the analyzed brick samples. Three specimens of bricks from each mix group were tested. The ASTM C67-07 standard, 2007 [46] was followed. All the specimens of bricks were checked for loose debris and chips and were then wiped clean before the test. The specimens were dried in a ventilated oven at a temperature of 105 °C to 115 °C (Figure 4a). The specimens were cooled at room temperature and their weight was noted. The dried specimens were fully submerged in water from waterworks at a room temperature of 25 ± 1 °C for 24 h (Figure 4b). The specimens were then detached and wiped with a damp cloth and weighed again.

(**a**) (**b**)

Figure 4. Water absorption test: (**a**) drying the samples in the oven, and (**b**) soaking the samples in water.

3.3. Efflorescence Test

The efflorescence test was performed on three brick samples taken from each mix proportion group, according to the specifications of the ASTM C67-07 standard, 2007 [46]. Thin flat-bottomed trays with appropriately distilled water were used to saturate the brick specimens. The brick specimens were vertically submerged in water up to 25 mm, as presented in Figure 5. The entire procedure took place in a well-ventilated room with a temperature of 25 ± 1 °C and a humidity of 48 ± 2%. This process continued until all the water in the trays had been absorbed by the brick specimens and the surplus water had evaporated. When the water had been absorbed and the brick appeared to be dry, a similar quantity of water was again placed in the trays and allowed to dry as before. The bricks were inspected for efflorescence after the second evaporation. The purpose of this test was to check the white crystalline salty deposits on the surface of the bricks. The tolerance of the bricks to the salt attack is determined based on the amount of salt efflorescence on their surfaces. The goal of this test was to see whether the surface of the marble powder bricks had a white crystalline salty coating. The creation of salt deposits on the surface of bricks is caused by the presence of water (moisture) in either the bricks or the surrounding environment, which dissolves the salts contained within the bricks and transports them to the bricks' surface due to capillary action. When saltwater reaches the surfaces of the bricks, the water evaporates due to air, in turn leaving a fine white powdered salt layer. This indicates that the occurrence of efflorescence is dependent on the circumstances of exposure, the type of bricks (dense or porous), and the nature of the components employed in the production of the bricks.

Figure 5. Efflorescence test.

3.4. Bulk Density Test

The bulk density test was conducted to find the tapped bulk density or bulk volume of the marble clay bricks according to the ASTM C134-95 standard [47]. The apparatus used in this test included a flat metal rule with a square at one end for measuring the brick samples, an oven for drying purposes, and a balance to weigh the brick samples. The three dimensions, i.e., length, width and thickness of each brick unit were measured carefully and noted. After the measurements, the brick samples were put inside the oven to dry at a controlled temperature of 110 °C for 2 h. The samples were then cooled inside the room and weighed. The air temperature was recorded to be 26 ± 1 °C. This procedure was repeated two times, and then the bulk density of the marble clay bricks was calculated by dividing the dry weight of the samples (in kg) by the samples' volume (in m^3).

3.5. Porosity Test

This test was performed according to the ASTM C67-07 [46] and ASTM C20-00 standards [48]. The brick specimens were dried at 110 °C in an oven and weighed after being cooled at room temperature. The specimens were then placed in distilled water and boiled for 2 h. Afterwards, they were cooled and weighed while still immersed in the water. After obtaining the suspended weight, the bricks were removed from the water immediately, blotted lightly with a moistened towel, and weighed. This procedure was done at an air temperature of 26 ± 1 °C.

3.6. Compressive Strength Test

To determine the compressive strength of the marble clay bricks, a compression test was performed according to the ASTM C67/C67M standard [48] by using a universal testing machine, as illustrated in Figure 6. Prior to the compression test, the unevenness of the bricks' surfaces was removed in order to make the surfaces smooth and parallel. All the specimens were submerged in water at room temperature for 24 h. The specimens were then removed, and any surplus moisture was drained out. The frog and all the voids in the bed face were filled with plaster of Paris (1:2) and left to dry for the next 48 h. At the time of testing, the dimensions of each specimen were measured using a scale. The specimens were placed with their flat faces in a horizontal position, with the mortar filled face facing upwards, carefully positioned between the plates of the testing machine. The load was applied axially at a uniform rate of 14 N/mm^2 per minute until failure occurred. A total of 3 specimens were tested for each proportion group, and the average of the obtained results was taken as the final compressive strength. The compression test was performed in a laboratory at a temperature of 26 ± 1 °C and a humidity of 48 ± 2%.

Figure 6. Compressive strength test.

4. Test Results and Discussion

4.1. Water Absorption Test Results

The results of the water absorption test are given in Figure 7a. Due to the fact that all the specimens were fully submerged in water from waterworks for the same amount of time (24 h), only water absorption percentages are shown on the graph's Y-axis in the graphs. From the test results, it was noted that an increased content of marble powder raised the water absorption capacity of the bricks. The water absorption capacity of the marble powder bricks was respectively 4.13%, 6.73%, 11.13%, 13.75%, 15.46%, and 18.14% higher than the reference bricks (with a 0% content of the marble powder). The permissible maximum water absorption of a brick, according to ASTM specifications, is 20% of its total weight. However, it should be remembered that the maximum water absorption value depends on national requirements and may differ depending on the location. It should be noted that this value depends on the open porosity of bricks. Moreover, in recent years, the requirements for the maximum water absorption of bricks have become more stringent. For example, on the basis of PN-B 12011: 1997 [49] (recent requirements in Poland for the tenth class of bricks), PN-EN 771-1 [50] (actual requirements in Europe), and requirements for second-class bricks in Pakistan [51], the maximum water absorption is 24%, 22% and 25%, respectively. The waste marble clay bricks exceeded the value according to ASTM in all the tested group samples (i.e., with the addition of 5% to 30% of marble waste). An empirical equation was established between the water absorption capacity and the marble powder content, in order to compare the correlation coefficient value with the experimental results, as presented in Figure 7b. The experimentally calculated values in the current study were used to develop an empirical relation between the water absorption and marble waste percentage used in the bricks. Using the derived empirical equation, the expected water absorption for each percentage of marble waste was determined. Furthermore, the water absorption values for the samples with the addition of marble waste, which were obtained from the experimental and empirical equations, are compared in Figure 7b. The correlation coefficient (R^2) value is equal to 98%, confirming a good compliance of the obtained empirical results with the experimental data.

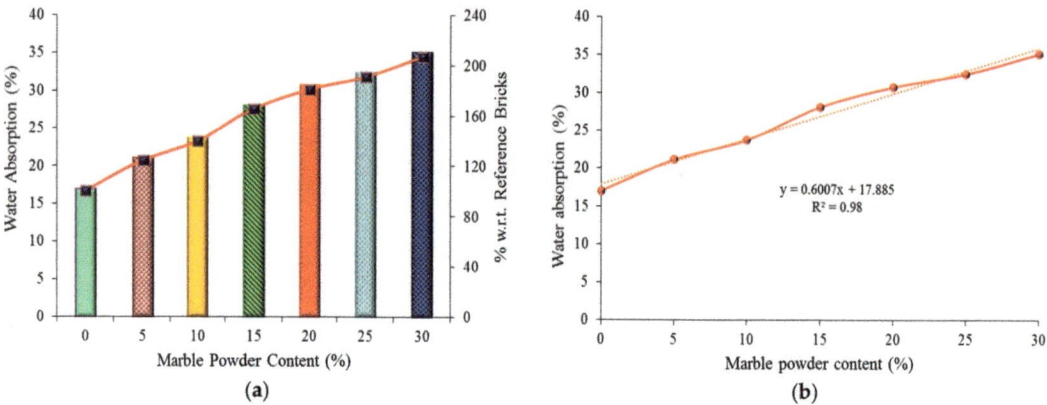

Figure 7. Graphical representation of the water absorption of the bricks: (**a**) experimental results, and (**b**) empirical results.

4.2. Efflorescence Test Results

From the conducted detailed observations and results, a slight efflorescence was seen, i.e., about 10–15% of the exposed areas of the B5 and B6 bricks were covered with a thin deposit of salt, as shown in Figure 8. The rest of the bricks had no perceptible deposits of efflorescence on their surfaces. It was shown that the addition of marble waste to bricks did not significantly affect the amount of salt efflorescence.

Figure 8. The effect of efflorescence.

4.3. Bulk Density Test Results

The results of the bulk density of the clay bricks with the addition of waste marble powder are shown in Figure 9a. The bulk density of the waste marble clay bricks was lower by 4.5%, 8.7%, 9.9%, 10.74%, 12.1%, and 21.5%, respectively when compared with the reference bricks. The weight of waste marble powder bricks was measured, and it was found that the marble powder bricks are 1.3% to 19.9% lighter than the reference bricks that contain 0% of marble powder. The empirical equation is used for evaluating the bulk density with regards to the marble powder content used in the bricks (Figure 9b). The correlation coefficient (R^2) value is equal to 90%, confirming a good compliance of the obtained empirical results with the experimental data.

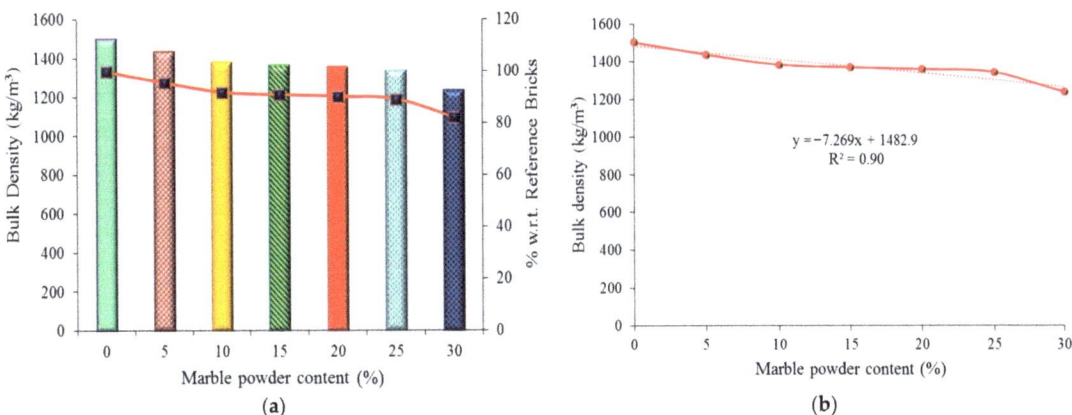

Figure 9. Graphical representation of bulk density: (**a**) experimental results, (**b**) empirical results.

4.4. Porosity Test Results

The obtained porosity results (in %) are presented in Figure 10a. The porosity ranged from 18.5% to 33.3%. The porosity of the marble powder bricks was higher when compared to the reference bricks. The porosity range increased with increasing marble powder content in the bricks. This is due to carbon dioxide (CO_2) being released during the calcination process of the calcium carbonate ($CaCO_3$). Figure 10b shows the empirical results of the porosity of the marble powder bricks. The correlation coefficient (R^2) is equal to 94%, confirming a good compliance of the obtained empirical results with the experimental data.

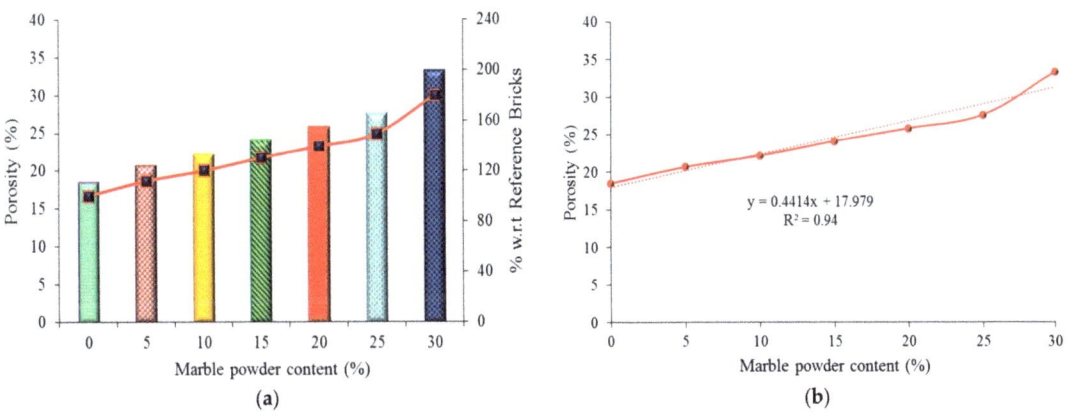

Figure 10. Graphical representation of porosity: (**a**) experimental results, (**b**) empirical results.

4.5. Compressive Strength Test Results

The results of the compression test indicate that the compressive strength decreased with an increase in the marble powder content in the bricks. The highest compressive strength of 20.4 MPa was obtained by reference sample "A", which contained a 0% marble powder content. The specimens of group "B1", with a 5% marble powder content, achieved a compressive strength of 12.2 MPa. In turn, groups B2, B3, B4, B5, and B6, with a 10%, 15%, 20%, 25%, and 30% marble powder content, achieved a compressive strength of 9.6 MPa, 7.8 MPa, 6.5 MPa, 5.9 MPa, and 4.5 MPa, respectively. The compressive strength results are given in Figure 11a. The compressive strength of the waste marble clay bricks was 40.2%, 52.9%, 61.8%, 68.1%, 71%, and 77.9% lower than the compressive strength of the reference bricks. The reason for this could be the increase in the porosity of the bricks, which is caused by an increase in the amount of marble powder. This in turn results in a decrease of their compressive strength. Vertical and diagonal cracks were formed in the case of applying the maximum compressive load. To evaluate the compressive strength results of the marble powder bricks, an empirical equation describing the compressive strength in relation to the marble powder content was developed. It is shown in Figure 11b. The correlation coefficient (R^2) is equal to 95%, which confirms the good compliance of the obtained empirical results with the experimental data.

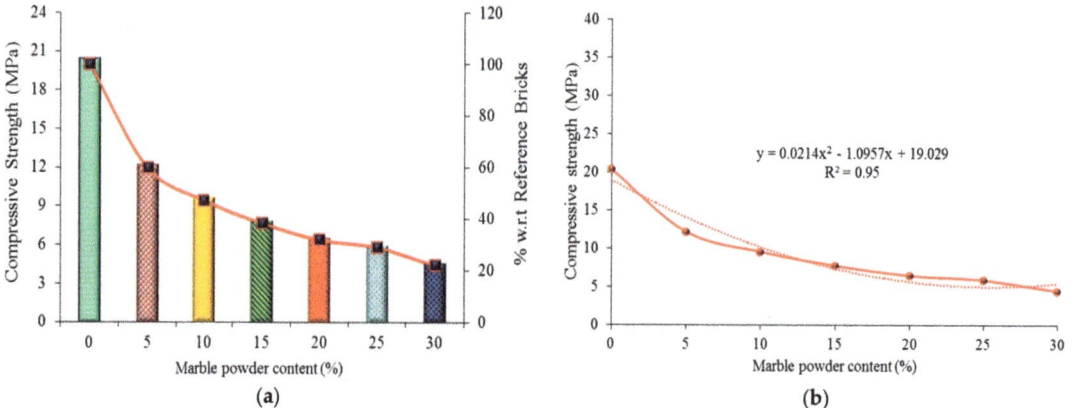

Figure 11. Graphical representation of compressive strength: (**a**) experimental results, and (**b**) empirical results.

5. Comparison of the Experimental Results and Empirical Results

To evaluate the characteristics of the marble powder bricks, i.e., water absorption, bulk density, porosity, and compressive strength, the experimental results and empirical results were compared, as shown in Figure 12. Figure 12a shows a graphical comparison of the experimental results and empirical results of the water absorption properties of the bricks. The statistical error/difference ranged from −2.86% to 2.37%. The comparison of the experimental results and empirical results of bulk density is given in Figure 12b. The error ranges from −4.35% to 5.27%. The porosity results of the marble powder bricks, which were obtained experimentally and empirically, are shown in Figure 12c. The error ranges from −6.24% to 5.12%. There is an increasing trend in the obtained results regarding the bricks that have a content of marble powder of up to a 25%. The compressive strength results calculated from the empirical equation were very good. The difference between them and the empirical results presented in Figure 12d ranges from −15% to 20%. Overall, the results obtained from the empirical equations showed very good compatibility with the experimental results, which is confirmed by the level of the maximum error.

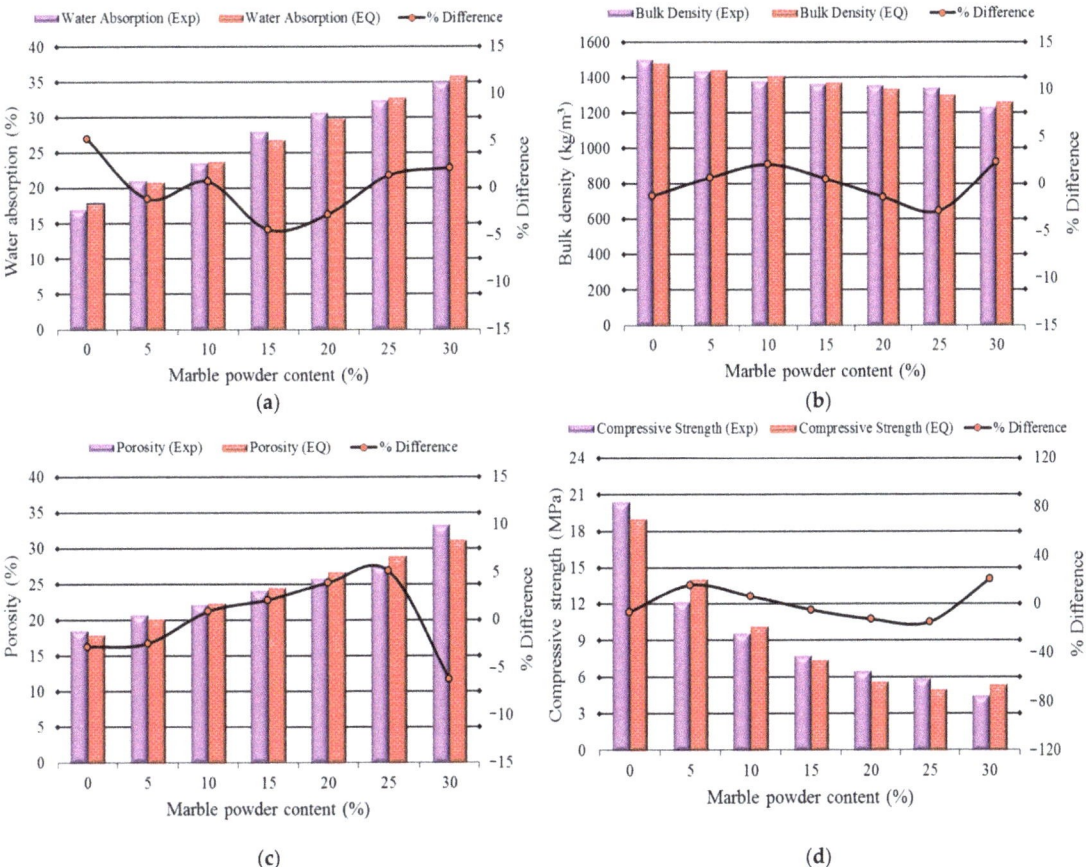

Figure 12. A comparison of the experimental results and those obtained from the laboratory tests for: (**a**) water absorption; (**b**) bulk density; (**c**) porosity (**d**) compressive strength.

6. Application of Waste Marble Powder in Eco-Friendly Bricks
6.1. Practical Use of the Designed Bricks in the Construction Industry

Marble clay sustainable bricks can be used both in the exteriors and interiors of buildings. In RCC frame structures, marble powder bricks may be used in partition walls due to the fact that such walls do not need to withstand too excessive loads. Depending on the climatic conditions, in some countries, bricks containing waste marble powder can be used for innovative home flooring and for street surface elements. These bricks are easy to recycle, which means they reduce the cost of making new construction materials for new buildings and leave minimal debris behind, in turn reducing the chance of potential pollution. Such bricks can be used in the construction of pavements for pedestrians. According to the results of this work, adding waste marble powder to bricks reduces their compressive strength, which enables them to be used in emergency situations where high strength is not necessary, e.g., in infrastructure facilities such as shelters and temporary hospitals, rooms for construction site workers, or elements of small architecture. For such purposes, a marble clay sustainable brick with a marble powder content of 5% to 20% is suggested for use as a sustainable building material.

6.2. Cost Comparison

The price of the raw materials, such as marble powder and clay soil, was analyzed. The cost of transportation is not included in this cost analysis, since it differs regarding the place of production. It was found that the addition of waste marble powder in the manufacturing of bricks is cost effective. The use of waste marble powder in bricks not only reduces the cost of their production, it also contributes to the obtaining of lighter elements, as well as improving their structural performance. The price of the raw materials was given by the different dealers of constructional materials: the marble powder was about USD 0.06 per kilogram, and the clay was about USD 0.09 per kilogram. Therefore, the marble powder was one third cheaper when compared to the clay that was used for the manufacturing of the bricks. Using waste marble powder as a partial substitute for clay reduces the content of the clay, and thus reduces the cost of the bricks. The price of the bricks was reduced by 1.6%, 3.2%, 4.7%, 6.3%, 7.9% and 9.5% with the addition of marble powder of 5%, 10%, 15%, 20%, 25% and 30%, respectively. Table 3 presents a comparison of prices between marble powder and clay per ton quantity (in US dollars).

Table 3. Cost comparison of marble powder and clay.

Brick Sample	Marble Powder by Weight %	Clay by Weight %	Marble Powder Price per Ton of Mixture $	Clay Price per Ton of Mixture (USD)	Total Cost of Mix per Ton (USD)
A	0	100	0.0	92.90	92.90
B1	5	95	3.18	88.26	91.44
B2	10	90	6.36	83.61	89.97
B3	15	85	9.54	78.97	88.51
B4	20	80	12.72	74.32	87.04
B5	25	75	15.9	69.70	85.60
B6	30	70	19.08	65.03	84.11

Note: 1 ton of clay = USD 92.9, 1 ton of marble powder = USD 63.6.

7. Conclusions

Waste materials from marble stone are a major source of concern for stone factories and municipality managements. As a result, the use of waste marble powder as an additive for the production of bricks can improve the natural environment due to the recycling of this waste. For this purpose, the addition of marble powder in different proportions for the manufacturing of bricks was analyzed. The key findings of this study are as follows:

1. The water absorption capacity of the bricks increased with an increase in the content of marble powder.
2. In general, the addition of marble waste to bricks does not significantly affect the amount of salt efflorescence. However, a minor effect of efflorescence was observed when 25% and 30% of marble powder was used in the bricks. The effect of efflorescence was related to the porosity of the bricks, i.e., if the bricks are porous, the effect of efflorescence will be more visible.
3. The bulk density of the bricks declined with a rise in the amount of marble powder, causing the porosity to increase and the weight of the bricks to decrease.
4. The amount of marble powder in clay bricks influences their compressive strength, i.e., as the percentage of marble powder in clay bricks increases, the compressive strength decreases due to increased porosity.
5. The marble powder bricks were lighter in weight than the reference clay bricks. The weight of the bricks decreased as the marble powder content increased. This drop in the weight of the brick samples can result in a large savings for the construction industry, i.e., a higher quantity of bricks and lower structural load.
6. The conducted empirical analysis showed a great compatibility with the laboratory test results regarding the following properties of bricks, i.e., water absorption, bulk density, porosity and compressive strength.

7. Marble clay bricks can be used in emergency situations where high strength is not needed, such as refugee camps, emergency hospitals during flooding and earthquakes, street flooring, and pavements. In such situations, marble clay sustainable bricks with a 5% to 20% marble powder content as are suggested for use as a sustainable construction material.
8. Due to high porosity, the water absorption capacity of marble bricks increased. Therefore, because of this property, it is recommended that marble bricks be used for construction in such areas and countries where the moisture rate in air is relatively low or the bricks are protected against moisture.
9. The use of waste marble powder as a partial replacement for clay in the manufacturing of bricks is cost efficient, as the cost of bricks decreases with the inclusion of marble powder, which has a direct influence on project costs. This study suggests using 5% to 20% waste marble powder as a clay substitute, which reduces the cost of bricks by 1.6% to 6.3%, respectively.

Author Contributions: Conceptualization, M.S. (Muhammad Sufian) and A.Z.; methodology, M.S. (Muhammad Sufian), S.U. and A.A.; software, M.S. (Muhammad Sufian) and A.A.; validation, A.A., M.S. (Muhammad Siddiq); formal analysis, A.Z. and M.S. (Muhammad Siddiq); investigation, M.S. (Muhammad Sufian), K.A.O. and S.U.; resources, M.S. (Muhammad Sufian), K.Ś.-W., A.A.A. and A.A.; data curation, M.S. (Muhammad Sufian); writing—original draft preparation, M.S. (Muhammad Sufian), A.A., K.A.O.; writing—review and editing, A.Z., K.A.O., A.A.A., M.S. (Muhammad Siddiq) and S.U.; visualization, K.Ś.-W., A.A. and S.U.; supervision, K.A.O.; project administration, M.S. (Muhammad Sufian); funding acquisition, K.A.O. All authors have read and agreed to the published version of the manuscript.

Funding: The article processing charge (APC) was funded by the Faculty of Civil Engineering Cracow University of Technology.

Institutional Review Board Statement: Not applicable.

Informed Consent Statement: Not applicable.

Data Availability Statement: The data presented in this article are available within the article and can be requested from the corresponding authors.

Acknowledgments: The authors acknowledge the assistance and support provided by all the persons and organizations that helped throughout this research work.

Conflicts of Interest: The authors declare no conflict of interest.

References

1. More, A. Marble Market Share and Size 2021 Global Growth, New Updates, Trends, Industry Expansion, Demand by Regions Opportunities. Available online: https://www.marketwatch.com/press-release/marble-market-size-2021-with-global-market-top-players-industry-share-trend-industry-news-business-growth-and-statistics-with-research-methodology-by-forecast-to-2024-2021-06-26 (accessed on 22 April 2021).
2. Dodge, H. Decorative Stones for Architecture In The Roman Empire. *Oxf. J. Archaeol.* **1988**, *7*, 65–80. [CrossRef]
3. TDAP A Report on Marble & Granite. 2010. Available online: https://www.scribd.com/document/154135371/Tdap-Report-on-Marble-and-Granite (accessed on 2 May 2021).
4. Sütçü, M.; Alptekin, H.; Erdogmus, E.; Er, Y.; Gencel, O. Characteristics of fired clay bricks with waste marble powder addition as building materials. *Constr. Build. Mater.* **2015**, *82*, 1–8. [CrossRef]
5. Demirel, B. The effect of the using waste marble dust as fine sand on the mechanical properties of the concrete. *Phys. Sci. Int. J.* **2010**, *5*, 1372–1380.
6. Mann, H.S.; Brar, G.S.; Mann, K.S.; Mudahar, G.S. Experimental Investigation of Clay Fly Ash Bricks for Gamma-Ray Shielding. *Nucl. Eng. Technol.* **2016**, *48*, 1230–1236. [CrossRef]
7. ERPS Kushwah. Disposal system of Marble Slurry for Clean and Green Environment. *Int. J. Eng. Sci. Res. Technol. Sci.* **2014**, *3*, 500–503.
8. Shakir, A.A.; Naganathan, S.; Mustapha, K.N. Development of Bricks from Waste Material: A Review Paper. *Aust. J. Basic Appl. Sci.* **2013**, *7*, 812–818.
9. Ma, B.; Wang, J.; Tan, H.; Li, X.; Cai, L.; Zhou, Y.; Chu, Z. Utilization of waste marble powder in cement-based materials by incorporating nano silica. *Constr. Build. Mater.* **2019**, *211*, 139–149. [CrossRef]

10. Zia, A.; Khan, A.A. Effectiveness of Bagasse Ash for Performance Improvement of Asphalt Concrete Pavements. *SN Appl. Sci.* **2021**, *3*, 502. [CrossRef]
11. Kadir, A.A.; Mohajerani, A. Bricks: An Excellent Building Material for Recycling Wastes—A Review. In Proceedings of the IASTED International Conference Environmental Management and Engineering, Banff, AB, Canada, 6–8 July 2011; pp. 108–115. [CrossRef]
12. Velasco, P.M.; Ortiz, M.P.M.; Giró, M.M. Fired clay bricks manufactured by adding wastes as sustainable construction material—A review. *Constr. Build. Mater.* **2014**, *63*, 97–107. [CrossRef]
13. Fernando, P.R. Experimental Investigation of the Effect of Fired Clay Brick on Partial Replacement of Rice Husk Ash (RHA) with Brick Clay. *Adv. Recycl. Waste Manag.* **2017**, *2*, 2–5.
14. Rao, B.J. Effect of different Waste Materials Addition on the Properties of Clay Bricks. *Int. J. Res. Appl. Sci. Eng. Technol.* **2019**, *7*, 2552–2557. [CrossRef]
15. Kadir, A.A.; Sarani, N.A. An Overview of Wastes Recycling in Fired Clay Bricks. *Int. J. Integr. Eng.* **2012**, *4*, 53–69.
16. Zhang, L. Production of bricks from waste materials—A review. *Constr. Build. Mater.* **2013**, *47*, 643–655. [CrossRef]
17. Baspinar, M.S.; Demir, I.; Orhan, M. Utilization potential of silica fume in fired clay bricks. *Waste Manag. Res.* **2010**, *28*, 149–157. [CrossRef]
18. Abbas, S.; Saleem, M.A.; Kazmi, S.M.; Munir, M.J. Production of sustainable clay bricks using waste fly ash: Mechanical and durability properties. *J. Build. Eng.* **2017**, *14*, 7–14. [CrossRef]
19. Phonphuak, N. Application of Dry Grass for Clay Brick Manufacturing. *Key Eng. Mater.* **2017**, *757*, 35–39. [CrossRef]
20. Oorkalan, A.S.; Gopinath, V.; Abhilash, M.; Manikandan, P.U.H. Experimental investigations of bricks using ceramic powder, marble dust and wood ash. *Int. Res. J. Eng. Technol.* **2020**, *7*, 2914–2920.
21. Ngayakamo, B.H.; Bello, A.; Onwualu, A.P. Development of eco-friendly fired clay bricks incorporated with granite and eggshell wastes. *Environ. Chall.* **2020**, *1*, 100006. [CrossRef]
22. Xin, Y.; Mohajerani, A.; Kurmus, H.; Smith, J.V. Possible recycling of waste glass in sustainable fired clay bricks: A review. *Int. J.* **2021**, *20*, 57–64.
23. Mobili, A.; Giosuè, C.; Tittarelli, F. Valorisation of GRP Dust Waste in Fired Clay Bricks. *Adv. Civ. Eng.* **2018**, *2018*, 1–9. [CrossRef]
24. Kazmi, S.M.S.; Abbas, S.; Nehdi, M.L.; Saleem, M.A.; Munir, M.J. Feasibility of Using Waste Glass Sludge in Production of Ecofriendly Clay Bricks. *J. Mater. Civ. Eng.* **2017**, *29*, 04017056. [CrossRef]
25. Zia, A.; Ali, M. Behaviour of fiber reinforced concrete in controlling the rate of cracking in canal-lining. *Constr. Build. Mater.* **2017**, *155*, 726–739. [CrossRef]
26. Kadir, A.A.; Zulkifly, S.N.M.; Abdullah, M.M.A.B.; Sarani, N.A. The Utilization of Coconut Fibre into Fired Clay Brick. *Key Eng. Mater.* **2016**, *673*, 213–222. [CrossRef]
27. Abdul Kadir, A.; Sarani, N.A.; Abd Kadir, S.A. Incorporation of Palm Kernel Shell into Fired Clay Brick. In Proceedings of the MATEC Web of Conferences, Bandung, Indonesia, 28–29 November 2017.
28. Bilgin, N.; Yeprem, H.; Arslan, S.; Bilgin, A.; Günay, E.; Marşoglu, M. Use of waste marble powder in brick industry. *Constr. Build. Mater.* **2012**, *29*, 449–457. [CrossRef]
29. Memon, N.A.; Abro, F.R.; Bhutto, M.A.; Sumadi, S.R. Marble powder as stabilizer in natural clayey soils. *J. Sci. Int.* **2015**, *27*, 4105–4110.
30. Kannadason, R.; Ravindra, J. Effect of Partial Replacement of Marble Powder and Rice Husk Ash in Brick Material. *Int. J. Emerg. Technol. Eng. Res.* **2017**, *5*, 187–190.
31. Kathiresan, M.; Gunasekar, M.; Sonia, M.T. Experimental study on manufacturing bricks by using marble sludge powder for acid resistance test. *SSRG Int. J. Civ. Eng.* **2017**, 120–124.
32. Seghir, N.T.; Mellas, M.; Sadowski, Ł.; Krolicka, A.; Żak, A.; Ostrowski, K. The Utilization of Waste Marble Dust as a Cement Replacement in Air-Cured Mortar. *Sustainability* **2019**, *11*, 2215. [CrossRef]
33. Aarthi, G.M.; Arthi, S.; Ishwarya, S.K.S. Experimental Study on Fly Ash Bricks with Hollow Balls. *Int. Res. J. Eng. Technol.* **2020**, *7*, 798–802.
34. Rehman, W.; Riaz, M.; Ishaq, M.; Faisal, M. Utilization of Marble Waste Slurry in the Preparation of Bricks. *J. Pakistan Inst. Chem. Eng.* **2014**, *42*, 47–54.
35. Ashish, D.K. Concrete made with waste marble powder and supplementary cementitious material for sustainable development. *J. Clean. Prod.* **2019**, *211*, 716–729. [CrossRef]
36. Singh, M.; Choudhary, K.; Srivastava, A.; Sangwan, K.S.; Bhunia, D. A study on environmental and economic impacts of using waste marble powder in concrete. *J. Build. Eng.* **2017**, *13*, 87–95. [CrossRef]
37. Shah, M.; Usman, M.; Hanif, M.; Naseem, I.; Farooq, S. Utilization of Solid Waste from Brick Industry and Hydrated Lime in Self-Compacting Cement Pastes. *Materials* **2021**, *14*, 1109. [CrossRef] [PubMed]
38. Li, L.; Huang, Z.; Tan, Y.; Kwan, A.; Liu, F. Use of marble dust as paste replacement for recycling waste and improving durability and dimensional stability of mortar. *Constr. Build. Mater.* **2018**, *166*, 423–432. [CrossRef]
39. Sadek, D.M.; El-Attar, M.M.; Ali, H.A. Reusing of marble and granite powders in self-compacting concrete for sustainable development. *J. Clean. Prod.* **2016**, *121*, 19–32. [CrossRef]
40. Rodrigues, R.; de Brito, J.; Sardinha, M. Mechanical properties of structural concrete containing very fine aggregates from marble cutting sludge. *Constr. Build. Mater.* **2015**, *77*, 349–356. [CrossRef]

41. Talah, A.; Kharchi, F.; Chaid, R. Influence of Marble Powder on High Performance Concrete Behavior. *Procedia Eng.* **2015**, *114*, 685–690. [CrossRef]
42. Aliabdo, A.A.; Elmoaty, A.E.M.A.; Auda, E.M. Re-use of waste marble dust in the production of cement and concrete. *Constr. Build. Mater.* **2014**, *50*, 28–41. [CrossRef]
43. Elmaghraby, M.; Ismail, A. Utilization of Some Egyptian Waste Kaolinitic Sand as Grog for Bricks and Concrete. *Silicon* **2015**, *8*, 299–307. [CrossRef]
44. Nandhini, K.; Karthikeyan, J.; Nandhini, K.; Karthikeyan, J. Influence of Industrial and Agricultural by-Products as Cementitious Blends in Self-Compacting Concrete—A Review. *Silicon* **2021**, 1–22. [CrossRef]
45. Rahim, H.U. Comparative Analysis of Soil Physio-Chemical Properties of Two Different Districts Peshawar and Swabi, KP, Pakistan. *Int. J. Environ. Sci. Nat. Resour.* **2017**, *7*, 7. [CrossRef]
46. American Society for Testing and Materials. ASTM Committee C-15 on Manufactured Masonry Units. In *Test Methods for Sampling and Testing Brick and Structural Clay*; C15 Committee; ASTM Int.: West Conshohocken, PA, USA, 2008; pp. 1–12. [CrossRef]
47. ASTM C134. *Test Methods for Size, Dimensional Measurements, and Bulk Density of Refractory Brick and Insulating Firebrick*; C08 Committee; ASTM Int.: West Conshohocken, PA, USA, 2016; Volume 95, pp. 1–4. [CrossRef]
48. ASTM C20-00. *Standard Test Methods for Apparent Porosity, Water Absorption, Apparent Specific Gravity, and Bulk Density of Burned Refractory Brick and Shapes by Boiling Water*; ASTM Int.: West Conshohocken, PA, USA, 2015; pp. 1–3. [CrossRef]
49. PN-B-12011:1997. Wyroby Budowlane Ceramiczne—Cegły Kratówki. Available online: https://sklep.pkn.pl/pn-b-12011-1997p.html (accessed on 20 June 2021).
50. PN-EN 771-1. *Requirements for Masonry Components—Part 1: Ceramic Masonry Components*; iTeh Standards: Etobicoke, ON, Canada, 2011.
51. Available online: https://pndajk.gov.pk/uploadfiles/downloads/brick%20works%2011.pdf (accessed on 20 June 2021).

Article

Properties of Concretes Incorporating Recycling Waste and Corrosion Susceptibility of Reinforcing Steel Bars

Zinoviy Blikharskyy [1], Khrystyna Sobol [2], Taras Markiv [2] and Jacek Selejdak [1,*]

[1] Faculty of Civil Engineering, Czestochowa University of Technology, 69 Str. Dabrowskiego, 42-201 Czestochowa, Poland; zinoviy.blikharskyy@pcz.pl

[2] Lviv Polytechnic National University, Bandera Str. 12, 79000 Lviv, Ukraine; khrystynasobol@ukr.net (K.S.); taras.y.markiv@lpnu.ua (T.M.)

* Correspondence: jacek.selejdak@pcz.pl; Tel.: +48-34-350-924

Abstract: In this paper, properties of concretes incorporating recycling waste and corrosion susceptibility of reinforcing steel bars were studied. It was established that fineness of ground granulated blast furnace slag (GGBFS) and fly ash (FA) and their simultaneous combination have an influence on the kinetics of strength development of Portland cements and concretes. The compressive strength of concrete containing 10% by mass of GGBFS and 10% by mass of FA even exceeds the compressive strength of control concrete by 6.5% and concrete containing 20% by mass of GGBFS by 8.8% after 56 days of hardening. The formation of the extra amount of ettringite, calcium hydrosilicates as well as hydroaluminosilicates causes tightening of a cement matrix of concrete, reducing its water absorption, and improving its resistance to freezing and thawing damage.

Keywords: Portland cement; ground granulated blast furnace slag; fly ash; pozzolanic reaction; corrosion; rebar

1. Introduction

Environmental problems are becoming more and more serious from year to year. Despite the remarkable development of scientific and technological progress, some industrial processes result in the significant production of waste, which cannot be landfilled completely and present a big problem for both the producer and the environment [1]. The waste accumulation is observed every year all over the world. Most of these industrial wastes are by-products such as granulated blast furnace slag and fly ash. The building material industry is one of the areas where such wastes can be utilized in large quantities.

The cementing materials and products on their basis have played a very important role in the development of construction for many years. The Portland cements and concretes are the most used materials in the construction industry [2,3]. The technology of Portland cement production, which remains the main component of concrete on the basis of a mineral binder, causes significant greenhouse gases emission, which results in global warming and climate change [4,5]. The cement production process generates almost 7% of the global greenhouse gas emissions [6–10]. Using waste as supplementary cementitious materials (SCMs) instead of clinker in the technology of cement production and cement in concrete will contribute to the partial solution of the above-mentioned problems [11–13]. Ground granulated blast-furnace slag and fly ash are among the types of waste that are utilized successfully in the construction industry and are attributed to the supplementary cementitious materials.

GGBFS is rather valuable waste and can be recycled in environment processes, agriculture, and the construction industry. Ground granulated blast furnace slag has begun to be used widely since the discovery of its latent hydraulic properties. Many researchers have been conducting lots of experiments to determine the properties of cements and concretes containing GGBFS and fly ash [14–21]. According to Ivashchyshyn et al. [15] and

Kurdowski [22], properties of cements and concretes containing blast-furnace slag depend on the chemical composition, fineness of slag, and the amount of vitreous structure in slag. Kumar et al. [23] and Zhu et al. [21] also concluded that the efficiency of slag depends on its particle size. Pal et al. [17] indicated that the fineness of slag and slag/cement ratio affect the strength of concrete incorporating GGBFS. Osborne [24] confirmed many benefits of the effective use of GGBFS in concretes such as reduced heat evolution, lower permeability, higher strength at later ages, and increased corrosion resistance. It was also pointed out that the use of very high levels of Portland cement replacement with slag can influence the corrosion susceptibility of reinforcing steel. Menéndez et al. [25] proved the efficiency of ternary blended cement incorporating limestone filler (LF) and blast-furnace slag. LF in such systems causes an increase of hydration at early stages while slag contributes to hydration at medium and later stages. Domenico et al. [26] studied the behavior of reinforced concrete (RC) beams containing electric arc furnace (EAF) slag under the load. EAF slag was used as a coarse aggregate. It was concluded that ultimate flexural and shear capacity of the above-mentioned beams was higher than the corresponding traditional RC beams, due to the improved properties of EAF concrete.

FA is a by-product of electricity generating plants. It is used both in the technology of cement and concrete production [20,27]. It has positive influence on the properties of fresh and hardened concretes such as workability, strength, drying shrinkage, thermal properties, and abrasion resistance [28,29]. Ghais et al. [30] state that the addition of 10% by mass of fly ash increases the concrete strength while the higher amount reduces it. The above-mentioned research was focused on the utilization both ground and unground (unactivated mechanically) fly ash. The usage of unactivated fly ash makes it possible to avoid the extra operations such as its grinding, which reduces the consumption of electricity and, as a result, natural energy resources, which are still used for electricity production, and greenhouse gas emissions. Unactivated fly ash allows for the particle size distribution to be optimized in cements and concretes because it plays a role as a mineral addition with pozzolanic properties and as a microfiller [29].

Sanytsky et al. [20] and Li et al. [31] concluded that the combined use of GGBFS and FA in both cement and concrete is very effective. On one hand, some researchers have reported that the binding system containing Portland cement, GGBFS, and FA provided higher strength at all ages than the ones incorporating GGBFS and FA separately [31,32]. On the other hand, Jeong et al. [33] did not find any differences in the cement system incorporating GGBFS and FA.

Fernández et al. [34] also pointed out that the synergistic effect between GBFS and CFA depends on the chemical composition of Portland cement in the cementitious system, which consists of these constituents. Thus, chemical compositions and the fineness of Portland cement, ground granulated blast-furnace slag, and fly ash can have an influence on the properties of the cementitious system.

The use of GGBFS and FA may also have an influence on the initial pH of the cementitious system and the protective passive layer, which is formed on the reinforcing steel bars due to the alkaline environment in concrete [35]. In the last few decades, corrosion of rebars in reinforced concrete structures is one of the main factors, which causes their extensive premature deterioration and has a serious effect on the serviceability and safety [36,37]. However, there have not been many publications in the literature related to the effect of the amount of SCMs on the corrosion susceptibility of reinforcing steel bars in concretes, in spite of the tendency of the significant reduction of clinker factor (CF) in cements to follow their contemporary tendencies. It can have an influence on the stability of steel reinforcement bars in such concretes. Moreover, it is very difficult to obtain a high strength class of cements with low CF and it can result in the increase in cement consumption in a concrete, cause higher creep and shrinkage deformation and, as a result, influence the durability. The development of cement with a lower substitution level of clinker and a higher compressive strength provides a reduction of its consumption in concretes and an improvement in the mechanical and durability properties of concretes as well as the stability of steel reinforcement bars in reinforced concretes.

According to Yeau and Kim [38], the corrosion probability of reinforcing steel bars in Type V cement concrete was higher than in Type I cement concrete. It was concluded that the corroded surface area depends on the amount of GGBFS replaced and the resistance to steel corrosion is better in Type I cement concrete with higher amount of GGBFS. Topçu and Boğa [39] studied the corrosion performance of steel reinforcing bars in concrete incorporating 25 and 50% by mass of GGBFS. This was done in C1 (uncontrolled relative humidity and temperatures) and C2 (in water with 20 ± 2 °C temperature) curing conditions and at ages of 28 as well as 90 days. The researchers concluded that the deterioration occurrence times extend in concrete containing 25% by mass of GGBFS for both curing conditions. The half-cell potentials of concretes incorporating 25% by mass of GGBFS, which were exposed to C2 curing conditions, have not passed through the area of the possible corrosion unlike other series that have passed through this area. According to Song and Saraswathy [40], the reduction in the pH of concrete containing GGBFS had no significant effect on the corrosion resistance of embedded reinforcement. The researchers also observed that the steel reinforcement bar embedded in cement that incorporated unactivated fly ash suffered severe corrosion [41]. However, concrete incorporating activated fly ash improved corrosion-resistance properties. Polder [42] summarized that replacement of the clinker with 50–70% by mass of GGBFS and 20–30% by mass of fly ash resulted in a higher chloride penetration resistance and electrical resistivity, decreasing the risk of corrosion in environments contaminated with chloride. Composite cements containing 25% by mass of both slag and fly ash behave similarly. Thus, the literature review shows that in some cases, contradictory results in experimental studies have been obtained.

This is why the aim of this research was to establish the effect of fineness of GGBFS and unactivated fly ash on the strength of Portland cements and properties of concretes as well as the level of Portland cement replacement without the reduction of compressive strength in comparison with the control cement and concrete without SCMs and to study the behavior of reinforcing steel bars in concretes incorporating such recycling waste.

2. Materials and Methods

Portland cement CEM I 42.5 (PRJSC Dyckerhoff Cement Ukraine, Zdolbuniv, Ukraine), ground granulated blast-furnace slag (ArcelorMittal Kryvyi Rih, Kryvyi Rih, Ukraine), and unactivated fly ash (Class F) derived from the Burshtyn power plant (Zakhidenergo, Halych district, 6 kilometres south-east from Burshtyn, Ivano-Frankivsk region, Ukraine) were used in this study as supplementary cementitious materials. The specific surfaces of GGBFS and unactivated FA were 310, 380, 510 and 310, 510 m^2/kg, respectively. The commercially available lignosulfonate based plasticizer (PC "Zastava", Chervonograd, Ukraine) with a specific gravity of 1.18 and solid content of 35% was used in the research. The properties of Portland cement are presented in Table 1.

Table 1. Physical and mechanical properties of Portland cement.

Specific Surface [m^2/kg]	Residue on Sieve 008 [%]	Water Demand [%]	Setting Time [min]		Compressive Strength [MPa]	
			Initial	Final	2 days	28 days
390	2.8	29.0	150	240	29.5	53.5

The Portland cements of the second type CEM II, especially CEM II/A-S, is very popular in the construction market. The content of additions can range between 6–20% by mass in this cement according to EN 197-1. This is why the maximum amount of Portland cement (20% by mass) was replaced with GGBFS in the first stage of the research.

The properties of Portland cements and aggregates were determined according to Ukrainian standards [43–47]. The results of the tests of the aggregates are reported in Table 2.

Table 2. Properties of aggregates.

Aggregate Type	Density [g/cm³]	Bulk Density [kg/m³]	Voidage [%]	Dust and Clay Particles [%]	Water Absorption [%]	Fineness Modulus
Fine (quartz sand)	2.65	1438	45.7	0.4	-	1.85
Coarse (granite gravel, 5–20 mm)	2.68	1370	48.9	0.3	0.6	-

The specific surface of Portland cement, GGBFS and FA was determined using the Tovarov device (Engineering firm "Integral", Podolsk, Russia). The Vicat apparatus (PC "Biomedservice", Poltava, Ukraine) was used to determine the setting time. The specific surfaces of GGBFS and fly ashes and residues in sieve 008 (the mesh opening size was 0.08 mm) are shown in Table 3.

Table 3. Composition of Portland cements and the fineness of SCMs.

Mixture Identification	Portland Cement [% by Mass]	GGBFS [% by Mass]	FA [% by Mass]	Specific Surface, [m²/kg] GGBFS	Specific Surface, [m²/kg] FA	Residue on Sieve 008, [%] GGBFS	Residue on Sieve 008, [%] FA
C0	100	0	0	-	-	-	-
CS310	80	20	0	310	-	6.0	-
CS380	80	20	0	380	-	5.4	-
CS500	80	20	0	500	-	4.8	-
CF5(310)	95	0	5	-	310	-	14.4
CF5(510)	95	0	5	-	510	-	1.3
CF10(310)	90	0	10	-	310	-	14.4
CF10(510)	90	0	10	-	510	-	1.3
CF15(310)	85	0	15	-	310	-	14.4
CF15(510)	85	0	15	-	510	-	1.3
CS15(380)F5(310)	80	15	5	380	310	5.4	14.4
CS10(380)F10(510)	80	10	10	380	510	5.4	1.3

The chemical composition of Portland cement, ground granulated blast-furnace slag and fly ash, which was determined by x-ray spectrometer ARL 9800 XP (Thermo Fisher Scientific, Waltham, MA, USA), is shown in Figure 1.

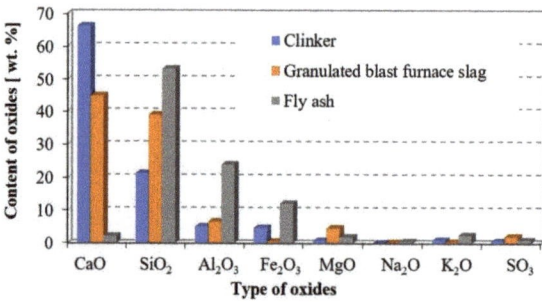

Figure 1. Chemical composition of clinker, granulated blast-furnace slag, fly ash.

Fly ash in comparison with Portland cement and GGBFS is characterized by a higher amount of silica, alumina, and iron oxides, which provide its pozzolanic activity.

The morphology and chemical composition of the main mineral components of the studied materials were determined using scanning electron microscope SEM FEI Quanta 250 FEG, equipped with EDS (FEI Company, Hillsboro, OR, USA).

The mortar (cement:sand = 1:3, W/C = 0.39) was prepared, and three prisms 40 × 40 × 160 mm³ were molded according to [45] to determine the compressive strength of Portland cement (Table 1) and each Portland cement composition (Table 3). The test specimens in the mold had been left for at least 24 ± 1 h at a temperature of (20 ± 1) °C and relative humidity ≥ 90%, protected against shock, vibration, and dehydration. After removal from the mold, the test specimens were cured until the test in water at a temperature of (20 ± 1) °C. The results of the mortar tests are the mean value of six specimens.

Concrete mixes were designed according to [48]. Compressive strength of concrete was determined according to [49]. The 100 mm cubes were molded for each mixture proportion (Table 4). The test specimens in the mold have been left for at least 24 h at a temperature of (20 ± 5) °C, protected against dehydration, shock, and vibration. After removal from the mold, the test specimens were cured until test at a temperature of (20 ± 3) °C and relative humidity of (95 ± 5) %. The reported results are the mean value of three specimens. The standard deviation of obtained results was 0.6–1.5 MPa.

Table 4. Mixture proportions.

Mixture Identification	Portland Cement [kg/m^3]	GGBFS [kg/m^3]	GGBFS [%]	FA [kg/m^3]	FA (%)	Plasticizer [% of Cement]	W/C
C	350	0	-	-	-	0.9	0.59
CS20(380)	280	70	20	-	-	0.9	0.59
CS15(380)F5(310)	280	52.5	15	17.5	5	0.9	0.55
CS10(380)F10(510)	280	35	10	35	10	0.9	0.53

Note: the following are valid for all mixtures: fine aggregate = 690 kg/m^3 and coarse aggregate (5–20 mm) = 1150 kg/m^3.

The water absorption and the freeze–thaw resistance of concretes were determined according to [50] and [51], respectively. The water absorption test was carried out in 28 days of concrete hardening. Three specimens of each concrete composition were dried in an oven to constant mass, then placed in a desiccator to cool and were finally weighed. Then, specimens were immersed in water at the temperature of 18–22 °C and stored there until full saturation.

Twelve specimens (six basic and six control specimens) of each concrete composition were molded and cured 28 days according to the Ukrainian standard (at the temperature of 20 ± 3 °C and the relative humidity of 95 ± 5 °C) to determine the freeze–thaw resistance of concretes. Then, the accelerated method was used according to [51] to determine the freeze–thaw resistance of concretes. Specimens cured 28 days at the temperature of 20 ± 3 °C and the relative humidity of 95 ± 5 °C were immersed in a 5% solution of sodium chloride to the 1/3 of their height and stored 24 h, then to 2/3 of their height and stored 24 h, and then the specimens were immersed in the solution completely and stored for 48 h. Then, all specimens were removed from the solution and the compressive strength of control specimens was determined before the freeze–thaw test of the basic specimens. The basic specimens were placed in a freezing chamber (HS280/75) and were frozen gradually up to −50 °C. The basic specimens were subjected to five freeze-thaw cycles. According to [52], five cycles of freeze–thaw in such a condition correspond to 100 freeze–thaw cycles in water at −20 °C [52]. After five freeze-thaw cycles, the compressive strength of the basic specimens was determined. Then, the compressive strength reduction was calculated.

The effect of Portland cement replacement with 10% by mass of GGBFS and 10% by mass of FA on the corrosion susceptibility of reinforcing steel in concrete was studied by the potentiostatic electro-chemical test according to [53]. Concrete specimens were prepared according to the recommendations in Annex A of the above-mentioned standard. The maximum size of coarse aggregate was 10 mm. Two mixtures were prepared: a reference control fine-grained concrete without SCMs (RCC) and a reference test one (RTC), which contained 10% by mass of GGBFS and 10% by mass of FA. The following mix-proportion was used: C = 350 kg/m^3, S = 800 kg/m^3, G (5–10 mm) = 1150 kg/m^3, W = 175 kg/m^3, plasticizer (0,9% by mass of a cement). The consistency class of fine-grained concrete mix was S1. Twenty percent by mass of Portland cement was replaced with 10% by mass of GGBFS and 10% by mass of FA in a reference test fine-grained concrete. Nine similar specimens (prisms of 40 × 40 × 160 mm^3) of each mixture were molded. A smooth steel bar (working electrode) with the diameter of 6 mm and length of 120 mm was placed in the center of the mold and secured to prevent movement during filling and compaction. Before making the specimens, steel bars were cleaned using emery paper and then degreased with acetone. All specimens were cured 28 days at the temperature of 20 ± 3 °C and relative humidity of 95 ± 5 °C. Three specimens were tested after 28 days of hardening in the above-mentioned conditions. The other six specimens of each mix-proportion were kept in conditions of daily alternating wetting and drying (3 h of total immersion in water and 21 h in the conditions, where the test was conducted). Three specimens were tested after three and six months. Before measurement, specimens were saturated with drinking water by boiling them for 3 h. Before the beginning of the test, concrete was chipped from one end of the beam specimen, exposing the reinforcing bar by 20 mm ± 10 mm. Then the working, reference, and counter electrodes were connected to the polarization circuit. Measurements of the magnitude of the current in microamperes were carried out in concrete specimens in (60 ± 5) min after turning on the potentiostat. The output of the potentiostat was raised, and the current recorded when passing the range from the steady-state potential to plus 1000 mV for 60 min. For each test specimen, the surface area of the working electrode, which was in direct contact with the concrete, was calculated. Using the calculated area of the rebar, the current density was determined.

3. Experimental Results and Discussions

The strength of building materials is an important characteristic, which defines their quality. This study focused mainly on the influence of SCMs on the compressive strength of Portland cements and properties of concretes as well as the corrosion susceptibility of reinforcing steel bars. It is a well-known fact that there are two schemes that show how GGBFS can be added to Portland cement. It can be mixed with constituents of cement by intergrinding or after previous separate grinding to the appropriate specific surface. The second way was used as it is more effective. It allows for the achievement of a more complete realization of potential hydraulic properties of GGBFS, because the size of the slag particles remained rather coarse after intergrinding of clinker and GGBFS due to the poor grindability of the GGBFS. Therefore, the Portland cement and GGBFS, ground to the appropriate specific surface previously, were mixed in the proportions that are shown in Table 3.

As can be seen from Figure 2, the fineness of GGBFS had a significant influence on the compressive strength of a cement. The compressive strength of mortar (CS310) on the basis of Portland cement containing 20% by mass of GGBFS with a specific surface of 310 m^2/kg was 34, 21 and 16% lower than the control mortar (C) after 2, 7, and 28 days of hardening, respectively. The increase of fineness of GGBFS caused gradual growth of strength at both early and later ages in comparison with CS310. The compressive strength of mortar CS380 was reduced by 6% after 28 days of hardening, while CS500 almost reached the compressive strength of the control mortar. However, GGBFS with a specific surface of 380 m^2/kg was used for further research, because the process of slag grinding up to 500 m^2/kg would be a rather expensive and simultaneously time consuming process. Moreover, GGBRS is becoming more and more expensive on the market year by year, and scientists are looking for SCMs to replace it.

Widespread by-products such as fly ash, which needs to be utilized, were used to improve the properties of cements. The specific surface of fly ash can be different and range widely at the thermal power plant. The unactivated fly ash with specific surfaces of 310 m^2/kg and 510 m^2/kg was used in the research. The Portland cement CEM I 42.5 was replaced with 5% (CF5(310), CF5(510)), 10% (CF10(310), CF10(510)) and 15% (CF15(310), CF15(510)) by mass of FA. The obtained results are presented in Figure 3.

A reduction in compressive strength was observed in mortars CF10(310) and CF15(310). Nicoara et al. [13] observed the same tendency. It was also found by Sanytsky et al. [20] that the replacement level of Portland cement with fly ash was increased due to its mechanical activation up to 40% by mass, but the strength reduction took place at all ages in comparison with activated Portland cement. It should be noted that the compressive strength of mortar CF5(310) decreased at the early age (two and seven days), but the growth of the kinetics of strength development was observed later, and it even slightly exceeded the compressive strength of the control mortar C at 28 days. The substitution levels of Portland cement for fly ash can be different. This depends on the specific surface of fly ash, because the size of the particles is very important. The reaction takes place on the surface and as a result, finer materials react faster [54,55]. The fly ash with a specific surface of 310 m^2/kg was rather coarse and 5% by mass of Portland cement could only be replaced without the compressive strength reduction. At the same time, the usage of fly ash with a higher specific surface (510 m^2/kg) made it possible to increase the replacement level of Portland cement without the reduction of strength after 28 days. The results in Figure 3 indicate that the tendency remained the same when the fly ash with the specific surface of 510 m^2/kg was used, but the replacement level could be increased to 10% by mass.

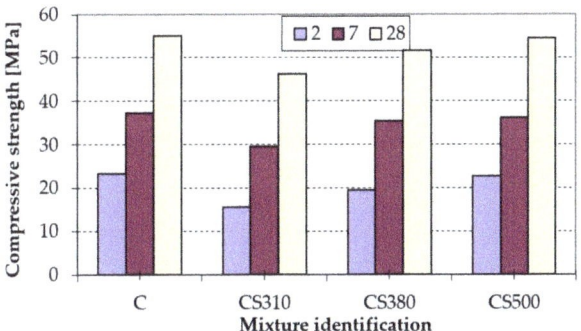

Figure 2. Compressive strength of Portland cements.

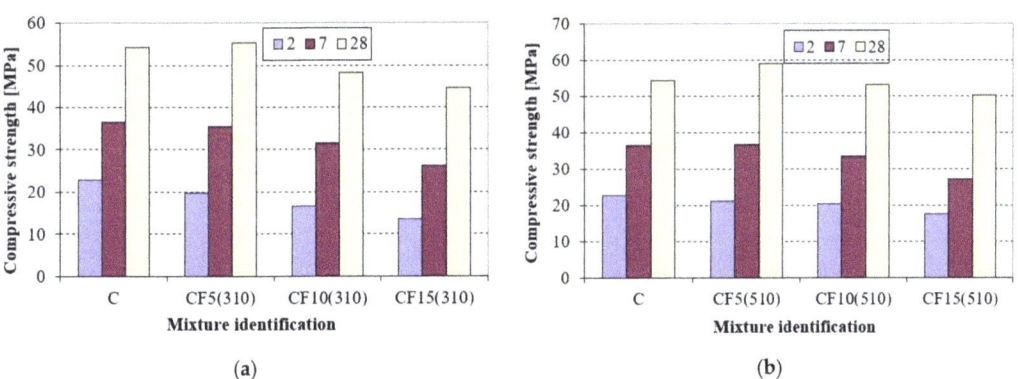

Figure 3. Compressive strength of Portland cements incorporating fly ash with specific surfaces of 310 m2/kg (a) and 510 m^2/kg (b).

The literature review and analysis of the obtained results show the necessity of combining GGBFS and FA. Compressive strength of mortars containing 10 and 15% by mass of GGBFS as well as 5 and 10% by mass of FA with specific surfaces of 310 (CS15(380)F5(310)) and 510 m^2/kg (CS10(380)F10(510)) was studied. The mortar incorporating 20% by mass of GGBFS (CS20(380)) was used for comparison. The above-mentioned mortars containing the same amount of SCMs (20% by mass) was studied and compared with control mortar C. The results in Figure 4 indicate that mechanical strength of mortar containing 20% by mass of GGBFS is lower in comparison with control concrete.

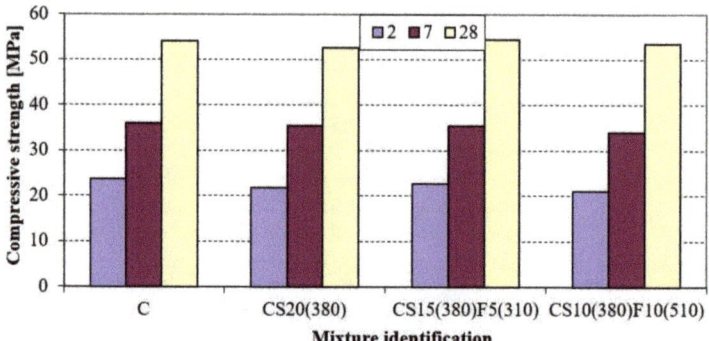

Figure 4. Compressive strength of Portland cements.

However, the kinetics of the strength gain of mortars in which Portland cement was replaced with both GGBFS and fly ash was lower for up to two days, but due to the synergetic effect of the combination of these two by-products, accelerated, leveled CS10(380)F10(510), and even exceeded CS15(380)F5(310) in the compressive strength of control mortar at later ages (28 days). The existence of this effect was also confirmed by many researchers [15,20,34,55].

Scanning electron microscopy was used to study the peculiarities of microstructure formation of cement paste without SCMs and paste incorporating 15% by mass of GGBFS and 5% by mass of fly ash. As shown in Figure 5a, many hydration products are formed. The plate-like portlandite is traditionally observed in the pure cement paste without SCMs as a result of the reactions between main minerals of clinker such as alite and belite and water, which was confirmed by the EDS spectra (Figure 5b). The structure of cement paste incorporating GGBFS and fly ash became denser due to the formation of an additional amount of fiber-like crystals of hydrosilicates, ettringite needles, and hydroaluminosilicates in the non-clinker part, which was confirmed by the results of microprobe analysis (Figure 5e), because GGBFS and FA are the source of active Al_2O_3 and SiO_2. The above-mentioned hydration products appear in the vacant space of cement matrix and also on the fly ash surface (Figure 5c). Unactivated fly ash also optimizes the particle size distribution in the binding system and provides the prolonged time of cement hydration, which results in obtaining a more refined and compact structure over time. The colmatation of pores and the growth of the cement paste strength take place as a result (Figure 5d).

GGBFS and FA are also used successfully in the technology of concrete production as cement replacements. In this work, a concrete mix design was carried out. Concrete mixture proportions are presented in Table 4. The slump of the mixtures was maintained in the range of 190–200 mm.

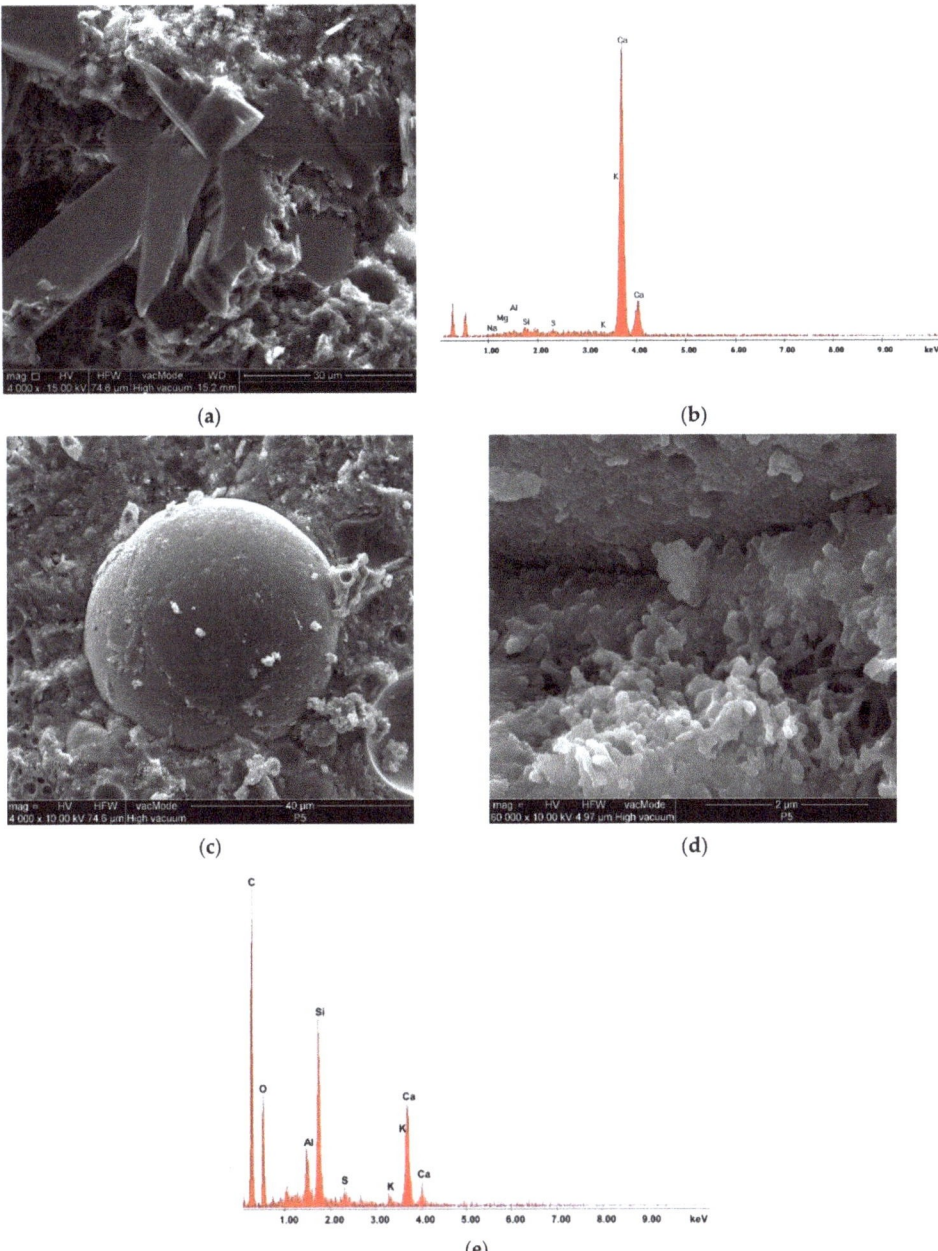

Figure 5. Microstructure and EDS spectra of Portland cement paste: (**a**,**b**) without SCMs; (**c**–**e**) containing 15% by mass of GGBFS and 5% by mass of FA.

As shown in Table 4, the replacement of 20% by mass of Portland cement with GGBFS (310) resulted in the reduction of compressive strength by 34.9%. The targeted workability was obtained at the same W/C ratio. Concrete incorporating 15% by mass of GGBFS and 5% by mass of fly ash (310) had better workability because the spherical particles of

FA produced the ball bearing effect. It allows the designed slump at lower W/C ratio to be obtained, which was equal to 0.55. When the content of FA was increased from 5% (F5(310)) to 10% (F10(510)) by mass and the amount of GGBFS (380) was decreased from 15% to 10% by mass, the W/C ration of the concrete decreased from 0.59 to 0.55, causing a slight increase in compressive strength after two days of hardening in comparison with concrete containing 20% by mass of GGBFS (380). It should be noted that the percentages of compressive strength reduction of concretes containing GGBFS and FA reduced after 14 days, then leveled and even exceeded the strength of C and CS20(380) after 56 days of hardening. Ghais et al. [30] also confirmed that the addition of 10% by mass of fly ash increased the concrete strength. Tan et al. [32] concluded that the compressive strength of concrete incorporating 10% by mass of finely ground fly ash and 10% by mass of finely ground granulated blast furnace slag increased at all ages. Better particle packing takes place and a denser structure is formed, resulting in an improvement in the long-term compressive strength and the highest strength (50.6 MPa) of concrete CS10(380)F10(510) at 56 days of hardening (Figure 6).

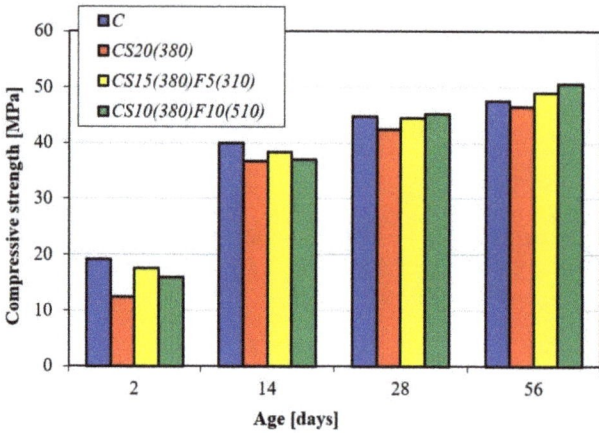

Figure 6. Compressive strength of concretes.

The water absorption confirmed the positive effect of the rational selection of SCMs and their combination including the individual properties of all constituents to obtain a synergetic effect. Figure 7 shows that concretes containing GGBFS and FA had the lowest water absorption due to the tighter microstructure.

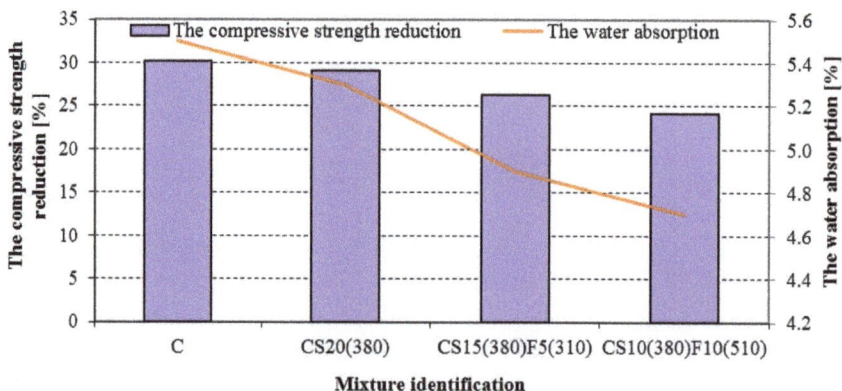

Figure 7. The water absorption and compressive strength reduction of concretes after 100 freezing/thawing cycles.

The freeze–thaw resistance of concretes after 100 freezing/thawing cycles was also investigated because it defines the durability of concretes. If concrete in a water-saturated state is exposed to the alternate freezing and thawing, the water freezes and expands by about 9%. The absence of enough space to accommodate this extra volume results in a disruptive pressure, which will be created. If the internal stresses exceed the tensile strength of concrete, they can eventually cause its cracking, scaling, and crumbling [56,57]. The hydraulic and osmotic pressure will cause the further increase of the disruption. If the water freezes in the completely water filled pores, the unfrozen water has to be transported to empty spaces through the capillary and gel pores of the cement paste, creating a hydraulic pressure, which depends on the distance between the pores and the fineness of the capillary pores. A longer distance and finer pores result in the pressure increase, which will be further enhanced by osmotic pressure causing the external and internal damage of concrete [58].

The compressive strength reduction of designed concretes is reported in Figure 7. As can be seen, the compressive strength reduction of concretes decreased gradually from 30.2 for control concrete to 24.1% for concrete incorporating 10% by mass of GGBFS (380) and 10% by mass of FA(510) after 100 freezing/thawing cycles. Łukowski et al. [59] observed some worsening of the frost resistance of concrete incorporating slag, but it is not clear from their study what the specific surface of GGBS was. Lindvall et al. [60] reported that concrete with pulverized fuel ash (PFA) has similar frost resistance to CEM I concrete if the maximum content of PFA is 25% by mass and $(w/c)_{eq} \leq 0.45$. The higher resistance of concrete CS10(380)F10(510) to freezing and thawing damage is obtained due to the improvement of the pore structure. The higher amount of secondary CSH gel obtained due to the pozzolanic reaction results in a decrease in the capillary porosity of concrete incorporating GGBFS and FA, which confirms the efficiency of the simultaneous combination of these SCMs.

The replacement of cement with SCMs in concretes also results in the reduction of the clinker and as a result, calcium hydroxide, which is produced during Portland cement hydration and is responsible for creating a protective passive layer on the rebars due to the alkaline environment. This is why the corrosion susceptibility of reinforcing steel bars in such concretes was studied. As can be seen from Figure 8, the deference between the current density of the working electrode surface was very small up to 0.5 V. If the output of the potentiostat is raised to values that are higher than 0.5 V, the corrosion current density increased more for concrete containing 10% by mass of GGBFS (380) and 10% by mass of FA (510) after 28 days of hardening in comparison with the reference control fine-grained concrete. The current density was 40.9 and 63.5 $\mu A/cm^2$ at 1 V for RCC and RTC, respectively. The content of calcium hydroxide in RCC was lower because 20% by mass of Portland cement was replaced with SCMs and the pozzolanic reaction also consumed some amount of $Ca(OH)_2$, reducing its content in hardened concrete, and likely a thinner protective passive layer formed on the rebars. Over time, the current density deference of compared concretes was reduced to 7.1 $\mu A/cm^2$ after 90 days and even almost levels after 180 days at a potential of 1 V due to the formation of the tighter structure of concretes incorporating GGBFS and FA. Polder [42] also observed the decrease in the risk of the steel reinforcing bar corrosion in such a binding system, even in chloride contaminated environments.

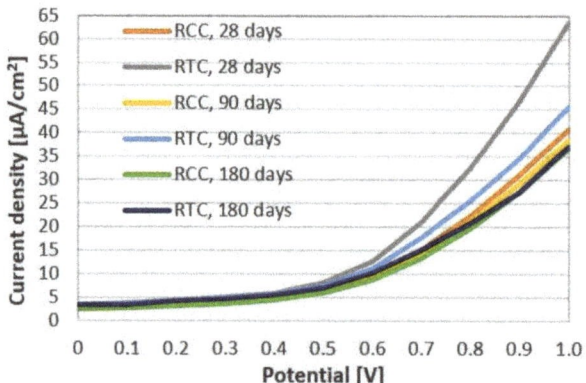

Figure 8. The current density of the working electrode surface in concretes at 28, 90, and 180 days of hardening.

4. Conclusions

The experimental work was done to evaluate the influence of a granulated blast-furnace slag and fly ash on the properties of Portland cements and concretes as well as the corrosion susceptibility of the reinforcing steel bars. The above-mentioned recycling waste was used for a partial replacement of binding materials in this article. The following conclusions can be drawn from the study:

- The increase of fineness of GGBFS from 310 to 500 m^2/kg resulted in the growth of Portland cement strength by 47 and 18% at 2 and 28 days of hardening, respectively.
- The kinetics of strength development of Portland cements and concretes incorporating GGBFS and FA is lower at the early age of structure formation, but accelerates at later ages due to the pozzolanic reaction that takes place.
- The synergetic effect of the compressive strength development was revealed in Portland cement and concrete containing granulated blast-furnace slag and fly ash.
- The replacement level of clinker in cements and Portland cement in concretes with GGBFS and FA depends on their specific surface.
- The obtained results show the efficiency of unactivated fly ash usage for the cement and concrete production and the necessity of permanent control of fly ash fineness to determine the replacement level of the clinker in a cement and a cement in a concrete.
- Simultaneous use of supplementary cementitious materials of different nature activity such as GGBFS (latent hydraulic) and FA (pozzolanic) allows for the improvement in the microstructure of concretes due to the formation of an extra amount of ettringite and tobermorite-like low-basic calcium hydrosilicates as well as hydroaluminosilicates in the non-clinker part of cement paste. Unactivated fly ash also allows the particle size distribution to be optimized in cements and concretes to archive the prolong time of cement hydration, which results in obtaining a more refined and compact structure over time due to its role both as a mineral addition with pozzolanic properties and as a microfiller. As a result, the durability and the stability of rebars in such concretes increase, which conforms to the strategy of sustainable development in the construction.
- Incorporation of cement replacement materials such as GGBFS and FA in concretes results in the reduction of water absorption from 5.5% to 4.7% and, as a result, in the increase of resistance to freezing and thawing damage, improving its durability. The decrease in compressive strength reduction was observed from 30.2 for concrete C to 24.1% for concrete CS10(380)F10(510) after 100 freeze–thaw cycles.
- Concretes containing 20% by mass of SCMs will have a better corrosion resistance in comparison with the concrete on the basis of Portland cement CEM I. Steel rein-

forcement protection does not worsen in concretes incorporating 10% by mass of GGBFS and 10% by mass of FA due to the low Portland cement substitution rate and their tighter microstructure. The obtained results concerning the corrosion susceptibility of reinforcing steel bars in concrete incorporating GGBFS and FA will be taken into account in the future developments of this study in the real reinforced concrete elements.

Author Contributions: Conceptualization, Z.B. and K.S.; Methodology, T.M. and J.S.; Software, T.M. and J.S.; Validation, Z.B., K.S., and T.M.; Formal analysis, K.S. and J.S.; Investigation, K.S. and T.M.; Resources, Z.B. and J.S.; Data curation, K.S. and T.M.; Writing—original draft preparation, K.S. and T.M.; Writing—review and editing, J.S. and Z.B.; Visualization, Z.B. and J.S.; Supervision, J.S. and K.S.; Project administration, T.M. and Z.B. All authors have read and agreed to the published version of the manuscript.

Funding: This research received no external funding.

Institutional Review Board Statement: Not applicable.

Informed Consent Statement: Not applicable.

Data Availability Statement: The data presented in this study are available on request from the corresponding author.

Conflicts of Interest: The authors declare no conflict of interest.

References

1. Scharff, H. Landfill reduction experience in The Netherlands. *Waste Manag.* **2014**, *34*, 2218–2224. [CrossRef] [PubMed]
2. Gambhir, M.L. *Concrete Technology. Theory and Practice*, 4th ed.; Tata McGraw Hill Education Private Limited: New Delhi, India, 2009; pp. 1–3.
3. Samad, S.; Shah, A. Role of binary cement including Supplementary Cementitious Material (SCM), in production of environmentally sustainable concrete: A critical review. *Int. J. Sustain. Built Environ.* **2017**, *6*, 663–674. [CrossRef]
4. Mehta, P.K. High-performance, high-volume fly ash concrete for sustainable development. In *Proceedings of the International Workshop on Sustainable Development and Concrete Technology, Beijing, China, 20–21 May 2004*; Wang, K., Ed.; Ames: Center for Transportation Research and Education, Iowa State University: Beijing, China, 2004; pp. 3–14.
5. Saafan, M.A.; Etman, Z.A.; El Lakany, D.M. Microstructure and durability of ground granulated blast furnace slag cement Mortars. *Iran. J. Sci. Technol. Trans. Civ. Eng.* **2020**. [CrossRef]
6. Malhotra, V.M. Role of supplementary cementing materials in reducing greenhouse gas emissions. In *Concrete Technology for a Sustainable Development in the 21st Century*; Gjorv, O.E., Sakai, K., Eds.; E&FN Spon: London, UK, 2000; pp. 226–235.
7. Malhotra, V.M. Introduction: Sustainable development and concrete technology. *Concr. Int.* **2002**, *24*, 22.
8. Huntzinger, D.N.; Eatmon, T.D. A life-cycle assessment of Portland cement manufacturing: Comparing the traditional process with alternative technologies. *J. Clean. Prod.* **2009**, *17*, 668–675. [CrossRef]
9. Worrell, E.; Price, L.; Martin, N.; Hendriks, C.; Meida, L.O. Carbon dioxide emissions from the global cement industry. *Annu. Rev. Energy Environ.* **2001**, *26*, 303–329. [CrossRef]
10. Miller, S.A.; Horvath, A.; Monteiro, P.J. Readily implementable techniques can cut annual CO_2 emissions from the production of concrete by over 20%. *Environ. Res. Lett.* **2016**, *11*, 074029. [CrossRef]
11. Aïtcin, P.-C. 4-Supplementary cementitious materials and blended cement. In *Science and Technology of Concrete Admixtures*; Aïtcin, P.-C., Flatt, R.J., Eds.; Woodhead Publishing: Cambridge, UK, 2016; pp. 53–73. [CrossRef]
12. Arrigoni, A.; Panesar, D.K.; Duhamel, M.; Opher, T.; Saxe, S.; Posen, I.D.; MacLean, H.L. Life cycle greenhouse gas emissions of concrete containing supplementary cementitious materials: Cut-off vs. substitution. *J. Clean. Prod.* **2020**, *263*, 121465. [CrossRef]
13. Nicoara, A.I.; Stoica, A.E.; Vrabec, M.; Šmuc Rogan, N.; Sturm, S.; Ow-Yang, C.; Gulgun, M.A.; Bundur, Z.B.; Ciuca, I.; Vasile, B.S. End-of-life materials used as supplementary cementitious materials in the concrete industry. *Materials* **2020**, *13*, 1954. [CrossRef]
14. Taylor, R.; Richardson, I.G.; Brydson, R.M.D. Composition and microstructure of 20-year-old ordinary Portland cement-ground granulated blast-furnace slag blends containing 0 to 100% slag. *Cem. Concr. Res.* **2010**, *40*, 971–983. [CrossRef]
15. Ivashchyshyn, H.; Sanytsky, M.; Kropyvnytska, T.; Rusyn, B. Study of low-emission multi-component cements with a high content of supplementary cementitious materials. *East. Eur. J. Enterp. Technol.* **2019**, *4*, 39–47. [CrossRef]
16. Sanytsky, M.; Kropyvnytska, T.; Ivashchyshyn, H.; Rykhlitska, O. Eco-efficient blended cements with high volume supplementary cementitious materials. *Bud. Archit.* **2019**, *18*, 5–14. [CrossRef]
17. Pal, S.C.; Mukherjee, A.; Pathak, S.R. Investigation of hydraulic activity of ground granulated blast furnace slag in concrete. *Cem. Concr. Res.* **2003**, *33*, 1481–1486. [CrossRef]

18. Krivenko, P.; Petropavlovskyi, O.; Kovalchuk, O.; Rudenko, I.; Konstantynovskyi, O. Enhancement of alkali-activated slag cement concretes crack resistance for mitigation of steel reinforcement corrosion. In *The International Conference on Sustainable Futures: Environmental, Technological, Social and Economic Matters (ICSF 2020), E3S Web of Conferences, Kryvyi Rih, Ukraine, 20–22 May 2020*; Semerikov, S., Chukharev, S., Sakhno, S., Striuk, A., Osadchyi, V., Solovieva, V., Vakaliuk, T., Nechypurenko, P., Bondarenko, O., Danylchuk, H., Eds.; EDP Science: Paris, France, 2020; Volume 166, p. 06001. [CrossRef]
19. Krivenko, P.; Petropavlovskyi, O.; Rudenko, I.; Konstantynovskyi, O. The influence of complex additive on strength and proper deformations of alkali-activated slag cements. *Mater. Sci. Forum* **2019**, *968*, 13–19. [CrossRef]
20. Sanytsky, M.; Sobol, K.; Markiv, T. Модифіковані Композиційні Цементи*(Modified Composite Cements)*, 1st ed.; Lviv Polytechnic National University Publishing House: Lviv, Ukraine, 2010; pp. 41–46. (In Ukrainian)
21. Zhu, G.; Zhong, Q.; Chen, G.; Li, D. Effect of particle size of blast furnace slag on properties of portland cement. *Procedia Eng.* **2012**, *27*, 231–236. [CrossRef]
22. Kurdowski, W. *Cement and Concrete Chemistry*; Springer: Dordrecht, The Netherlands, 2014. [CrossRef]
23. Kumar, S.; Kumar, R.; Bandopadhyay, A.; Alex, A.C.; Kumar, B.R.; Das, S.K.; Mehrotra, S.P. Mechanical activation of granulated furnace slag and its effect on the properties and structure of Portland slag cement. *Cem. Concr. Compos.* **2008**, *30*, 679–685. [CrossRef]
24. Osborne, G.J. Durability of Portland blast-furnace slag cement concrete. *Cem. Concr. Compos.* **1999**, *21*, 11–21. [CrossRef]
25. Menéndez, G.; Bonavetti, V.; Irassar, E.F. Strength development of ternary blended cement with limestone filler and blast-furnace slag. *Cem. Concr. Compos.* **2003**, *25*, 61–67. [CrossRef]
26. Domenico, D.D.; Faleschini, F.; Pellegrino, C.; Ricciardi, G. Structural behavior of RC beams containing EAF slag as recycled aggregate: Numerical versus experimental results. *Constr. Build. Mater.* **2018**, *171*, 321–337. [CrossRef]
27. Stechyshyn, M.; Sanytsky, M.; Poznyak, O. Durability properties of high volume fly ash self-compacting fiber reinforced concretes. *East. Eur. J. Enterp. Technol.* **2015**, *3*, 49–53. [CrossRef]
28. Corinadelsi, V.; Moriconi, G. Influence of mineral additions on the performance of 100% recycled aggregate concrete. *Constr. Build. Mater.* **2009**, *23*, 2869–2876. [CrossRef]
29. Poon, C.S.; Lam, L.; Wong, Y.L. A study on high strength concrete prepared with large volumes of low calcium fly ash. *Cem. Concr. Res.* **2000**, *30*, 447–455. [CrossRef]
30. Ghais, A.; Ahmed, D.; Siddig, E.; Elsadig, I.; Albager, S. Performance of concrete with fly ash and kaolin inclusion. *Int. J. Geosci.* **2014**, *5*, 1445–1450. [CrossRef]
31. Li, G.; Zhao, X. Properties of concrete incorporating fly ash and ground granulated blast-furnace slag. *Cem. Concr. Comp.* **2003**, *25*, 293–299. [CrossRef]
32. Tan, K.; Pu, X. Strengthening effects of finely ground fly ash, granulated blast furnace slag, and their combination. *Cem. Concr. Res.* **1998**, *28*, 1819–1825. [CrossRef]
33. Jeong, Y.; Park, H.; Jun, Y.; Jeong, J.-H.; Oh, J.E. Microstructural verification of the strength performance of ternary blended cement systems with high volumes of fly ash and GGBFS. *Constr. Build. Mater.* **2015**, *95*, 96–107. [CrossRef]
34. Fernández, Á.; García Calvo, J.L.; Alonso, M.C. Ordinary Portland Cement composition for the optimization of the synergies of supplementary cementitious materials of ternary binders in hydration processes. *Cem. Concr. Compos.* **2018**, *89*, 238–250. [CrossRef]
35. Poursaee, A. *Corrosion of Steel in Concrete Structures*; Woodhead Publishing: Cambridge, UK, 2016. [CrossRef]
36. Blikharskyy, Z.; Selejdak, J.; Blikharskyy, Y.; Khmil, R. Corrosion of reinforce bars in RC constructions. *Syst. Saf. Hum. Tech. Facil. Environ.* **2019**, *1*, 277–283. [CrossRef]
37. Blikharskyy, Y.; Selejdak, J.; Kopiika, N. Non-uniform corrosion of steel rebar and its influence on reinforced concrete elements' reliability. *Prod. Eng. Arch.* **2020**, *26*, 67–72. [CrossRef]
38. Yeau, K.; Kim, E. An experimental study on corrosion resistance of concrete with ground granulate blast-furnace slag. *Cem. Concr. Res.* **2005**, *35*, 1391–1399. [CrossRef]
39. Topçu, I.; Boğa, A. Effect of ground granulate blast-furnace slag on corrosion performance of steel embedded in concrete. *Mater. Des.* **2010**, *31*, 3358–3365. [CrossRef]
40. Song, H.-W.; Saraswathy, V. Studies on the corrosion resistance of reinforced steel in concrete with ground granulated blast-furnace slag–An overview. *J. Hazard Mater.* **2006**, *B138*, 226–233. [CrossRef] [PubMed]
41. Saraswathy, V.; Song, H.W. Effectiveness of fly ash activation on the corrosion performance of steel embedded in concrete. *Mag. Concr. Res.* **2007**, *59*, 651–661. [CrossRef]
42. Polder, R. Effects of slag and fly ash on reinforcement corrosion in concrete in chloride Environment-research from the Netherlands. *Heron 57* **2012**, *3*, 197–210.
43. DSTU B V.2.7-188:2009. In *Building Materials. Cements. Methods of Determination of Fineness*; Ukrarkhbudinform: Kyiv, Ukraine, 2010.
44. DSTU B V.2.7-185:2009. In *Building Materials. Cements. Methods of Determination of Normal Thickness, Setting Time and Soundness*; Ukrarkhbudinform: Kyiv, Ukraine, 2010.
45. DSTU B V.2.7-187:2009. In *Building Materials. Cements. Methods of Determination of Bending and Compression Strength*; Ukrarkhbudinform: Kyiv, Ukraine, 2010.

46. DSTU B V.2.7-71-98. In *Building Materials. Mountainous Rock Road Metal and Gravel, Industrial Waste Products for Construction Works. Methods of Physical and Mechanical Tests*; Ukrarkhbudinform: Kyiv, Ukraine, 1998.
47. DSTU B V.2.7-232:2010. In *Building Materials. Sand for Construction Work Testing Methods*; Ukrarkhbudinform: Kyiv, Ukraine, 2010.
48. DSTU-N B V.2.7-299:2013. In *Guidelines for Appointments of the Heavy Concrete*; Ukrarkhbudinform: Kyiv, Ukraine, 2014.
49. DSTU B V.2.7-214:2009. In *Building Materials. Concrete. Methods of Determining the Strength of Control Samples*; Ukrarkhbudinform: Kyiv, Ukraine, 2010.
50. DSTU B V.2.7-170:2008. In *Building Materials. Concretes. Methods of Determination of Middle Density, Moisture Content, Water Absorptions Porosity and Watertightness*; Ukrarkhbudinform: Kyiv, Ukraine, 2009.
51. DSTU B V.2.7-49-96. In *Building Materials. Concretes. Rapid Methods for Determination of Frost Resistance by Repeated Alternated Freezing and Thawing*; Ukrarkhbudinform: Kyiv, Ukraine, 1996.
52. DSTU B V.2.7-47-96. In *Building Materials. Concretes. Methods for Determination of Frost Resistance. General Requirements*; Ukrarkhbudinform: Kyiv, Ukraine, 1996.
53. DSTU B V.2.7-171:2008. In *Building Materials. Admixtures for Concretes and Building Mortars. General Specifications*; Ukrarkhbudinform: Kyiv, Ukraine, 2008.
54. Lothenbach, B.; Scrivener, K.; Hooton, R.D. Supplementary cementitious materials. *Cem. Concr. Res.* **2011**, *41*, 1244–1256. [CrossRef]
55. Rivera, R.A.; Sanjuán, M.Á.; Martín, D.A. Granulated blast-furnace slag and coal fly ash ternary portland cements optimization. *Sustainability* **2020**, *12*, 5783. [CrossRef]
56. Sun, Z.; Scherer, G.W. Effect of air voids on salt scaling and internal freezing. *Cem. Concr. Res.* **2010**, *40*, 260–270. [CrossRef]
57. Coussy, O.; Monteiro, P.J.M. Poroelastic model for concrete exposed to freezing temperatures. *Cem. Concr. Res.* **2008**, *38*, 40–48. [CrossRef]
58. Ramachandran, V. *Concrete Admixtures Handbook*; Noyes Publications: New Jersey, NJ, USA, 1995; p. 1111.
59. Łukowski, P.; Salih, A.; Sokołowska, J. Frost resistance of concretes containing ground granulated blast-furnace slag. *MATEC Web Conf.* **2018**, *163*, 05001. [CrossRef]
60. Lindvall, A.; Espin, O.; Löfgren, I. Properties of concretes with pulverized fly ash and ground granulated blast furnace slag. In Proceedings of the Nordic Mini Seminar: Durability Aspects of Fly Ash and Slag in Concrete, Oslo, Norway, 15–16 February 2012; pp. 15–16. Available online: https://www.researchgate.net/publication/257939233_PROPERTIES_OF_CONCRETES_WITH_PULVERIZED_FLY_ASH_AND_GROUND_GRANULATED_BLAST_FURNACE_SLAG (accessed on 18 April 2021).

Review

Management of Solid Waste Containing Fluoride—A Review

Małgorzata Olejarczyk [1,2], Iwona Rykowska [1] and Włodzimierz Urbaniak [1,*]

1. Faculty of Chemistry, Adam Mickiewicz University, ul. Uniwersytetu Poznańskiego 8, 61-614 Poznań, Poland; malgorzata.olejarczyk@amu.edu.pl (M.O.); obstiwo@amu.edu.pl (I.R.)
2. Construction Company "Waciński" Witold Waciński, ul. Długa 15, 83-307 Kiełpino, Poland
* Correspondence: wlodzimierz.urbaniak@amu.edu.pl

Abstract: Technological and economic development have influenced the amount of post-production waste. Post-industrial waste, generated in the most considerable amount, includes, among others, waste related to the mining, metallurgical, and energy industries. Various non-hazardous or hazardous wastes can be used to produce new construction materials after the "solidification/stabilization" processes. They can be used as admixtures or raw materials. However, the production of construction materials from various non-hazardous or hazardous waste materials is still very limited. In our opinion, special attention should be paid to waste containing fluoride, and the reuse of solid waste containing fluoride is a high priority today. Fluoride is one of the few trace elements that has received much attention due to its harmful effects on the environment and human and animal health. In addition to natural sources, industry, which discharges wastewater containing F− ions into surface waters, also increases fluoride concentration in waters and pollutes the environment. Therefore, developing effective and robust technologies to remove fluoride excess from the aquatic environment is becoming extremely important. This review aims to cover a wide variety of procedures that have been used to remove fluoride from drinking water and industrial wastewater. In addition, the ability to absorb fluoride, among others, by industrial by-products, agricultural waste, and biomass materials were reviewed.

Keywords: solidification/stabilisation; fluoride removal; defluorination techniques; adsorption; industrial waste

Citation: Olejarczyk, M.; Rykowska, I.; Urbaniak, W. Management of Solid Waste Containing Fluoride—A Review. *Materials* **2022**, *15*, 3461. https://doi.org/10.3390/ma15103461

Academic Editor: Igor Cretescu

Received: 18 March 2022
Accepted: 9 May 2022
Published: 11 May 2022

Publisher's Note: MDPI stays neutral with regard to jurisdictional claims in published maps and institutional affiliations.

Copyright: © 2022 by the authors. Licensee MDPI, Basel, Switzerland. This article is an open access article distributed under the terms and conditions of the Creative Commons Attribution (CC BY) license (https://creativecommons.org/licenses/by/4.0/).

1. Introduction

According to the circular economy principles, issues related to the correct and effective management of production waste are currently among the fundamental problems [1–3].

The following article comprehensively presents various materials used to neutralize fluorine ions, including waste materials. Moreover, a new material proposed by the authors was presented here, which is made of two industrial waste materials that form a sorbent for fluorine adsorption, and after use, it can be used in building material.

Environmental pollution due to the mismanagement of solid waste is a global problem. Many publications on specific waste streams have been published in the scientific literature to quantify their environmental impact [4–10]. N. Ferronato [10], in a review work, assessed the global problems associated with various waste materials, pointing out how they affect the environment, what their relation is to human health, and how they influence sustainable development. The results shown by the authors provide a reference point for scientists and stakeholders to quantify comprehensive effects and plan integrated solid waste collection and treatment systems to make it easier to achieve sustainable development at the global level [10].

The efficient management and further utilization of waste materials becomes a significant problem for the industry and is growing as the amount of waste materials is increasing, and management costs are rising for both the industry and local administrations [11–19].

Therefore, recycling and reusing industrial waste and by-products are of great importance [20–23]. In fact, in reducing environmental problems and increasing economic benefits, there is a great need for technologies to transform waste materials into products of commercial value [24–29].

For example, some waste materials are converted in the scope of solidification and stabilization (S/S) processes. The solidification aids in changing the physical state of waste, from a liquid to a solid material, by encapsulation, thus decreasing the level of migration to the environment. The stabilization, by applications of some chemical reactions, migrates dangerous materials to less soluble or less toxic forms [30].

There are several S/S processes and methods proposed, tested, and implemented in practice [31–34]. These solutions, however, still need a lot of research to increase their effectiveness/performance and long-term effects [31]. The massive usage of S/S products, e.g., as construction materials, is still blocked by the potential risk of migrating the contaminants to the environment, including the toxic materials. New research, however, points out low leachability factors, which indicates that the S/S waste (contaminant source) can be regarded as an environmentally sustainable material with potential beneficial uses in construction [35].

Therefore, research and development is needed for the wide production and utilization of construction materials from various nonhazardous or hazardous waste materials [36].

In our opinion, special attention should be paid to waste containing fluoride, and the reuse of solid waste containing fluoride is a high priority today.

As one of the most extended elements on earth [37], fluorine (F) is widely used in the chemical industry, which in turn has produced large amounts of fluorine-containing hazardous waste. Fluoride, which is the most electro-negative element in the halogen family, is considered to be one of the main environmental pollutants due to its low biodegradability, high reactivity, and popularity [38]. One of the sources of introducing fluoride into the environment is the industry, which discharges sewage containing F− ions to surface waters and contributes to an increase in the concentration of fluoride in waters and environmental pollution [39].

Fluoride is one of the few trace elements that has received much attention due to its harmful effects on the environment and human and animal health [40–43]. Sabine [44] Guth et al. reviewed the available literature to critically assess the risks to human health from fluoride exposure, with a focus on developmental toxicity. Several factors, such as pH, alkalinity, chemical composition of aquifers, hardness, etc., determine the presence and concentration of fluoride in water resources [45–51].

Table 1 summarizes the review publications that have been published over the past decade on fluoride removal from both drinking and industrial wastewater, followed by the impact of fluoride waste to the environment and human health, and finally defluorination techniques. It presents the state of the art in the field of fluorinated waste management in one place. Moreover, it summarizes most of the techniques already proposed. The effectiveness of various materials for fluoride removal has been reviewed, taking into account key factors such as pH, initial fluorine concentration, surface area, particle size, and temperature, as well as the occurrence of counterions influencing the process of defluorination [39,52–62].

Natural and anthropogenic processes contribute to the release of fluorine compounds into the environment, causing the fluoride concentration in the soil to be much higher than the limit values, which is further followed by health and environmental problems in many regions of the world.

Table 1. A summary of review publications that have been published over the past decade on the removal of fluoride from drinking water and industrial wastewater.

Authors	Title	Aim
Habuda-Stanić M. et al., 2014 [52]	Review on Adsorption of Fluoride from Aqueous Solution	A list of various adsorbents (oxides and hydroxides, biosorbents, geomaterials, carbonaceous materials, and industrial by-products) and their modifications is discussed. This survey showed that various adsorbents, especially binary and trimetal oxides and hydroxides, have good potential for fluoride removal from aquatic environments.
Waghmare S.S. et al., 2015 [53]	Fluoride removal by industrial, agricultural and biomass wastes as adsorbents: a review	Reviews the fluoride uptake capacities of industrial by-products, agricultural wastes, and biomass materials from plants, grass, etc., and their modified forms as adsorbents in batch and column performance.
Tomar V. et al., 2013 [54]	A critical study on efficiency of different materials for fluoride removal from aqueous media	An extensive list of adsorbents for fluoride removal is compiled. In particular, nanomaterial-based adsorbents might be promising adsorbents for environmental and purification purposes.
Kumar P.S., 2019 [39]	Treatment of fluoride-contaminated water: a review	Reviews the origin of fluoride, the analysis of fluoride derivatives, and the technologies to remove fluoride from water, using different adsorbent types.
Nagendra Rao C.R. 2003 [58]	Fluoride and environment—a review	Current information on fluoride presence in the environment and its effects on human health, as well as basic methods of defluoridation.
Schlesinger W.H. et al., 2020 [59]	Global Biogeochemical Cycle of Fluorine	Synthesis of what is currently known about the natural and anthropogenic fluxes of fluorine.
He J. et al., 2020 [60]	Review of fluoride removal from water environment by adsorption	The recent developments in fluoride removal from the water environment by adsorption methods. Based on the review, four technical strategies of adsorption method, including nano-surface effect, structural memory effect, anti-competitive adsorption, and ionic sieve effect, were proposed.
Bhatnagar A. et al., 2011 [61]	Fluoride removal from water by adsorption—a review	An extensive list of various adsorbents from literature has been compiled, and their adsorption capacities under various conditions (pH, initial fluoride concentration, temperature, contact time, adsorbent surface charge, etc.) for fluoride removal are presented.
Bodzek M. et al., 2018 [39]	Fluorine in the Water Environment-Hazards and Removal Methods, Engineering and Protection of Environment	Detailed information on recent researchers' efforts in the field of fluoride removal during potable water production. The contaminant elimination methods have been broadly divided in three sections, i.e., coagulation/precipitation, adsorption, and membrane techniques. Both precipitation with the use of calcium salts or coagulation with aluminium sulphate and ferric salts followed by sedimentation are used for fluorine removal. In electrocoagulation, a coagulant is generated in situ by means of oxidation of anode usually made of aluminium or iron.
Wang L. et al. 2019 [62]	A Review on Comprehensive Utilization of Red Mud and Prospect Analysis	Comprehensive utilization methods for reducing red mud (RM) environmental pollution and divides the comprehensive utilization of RM into three aspects: the effective extraction of valuable components, resource transformation, and environmental application.

2. Anthropogenic Sources of Contamination with Fluorine Compounds

In many countries around the world, high levels of fluoride are the result of discharges of sewage polluted with fluoride [52].

Such wastewater is usually produced by industry: superphosphate fertilizers [63–65]; glass and ceramics production processes [66,67]; aluminium and zinc smelters [68–70];

steel production; uranium enrichment plants; coal-fired power plants; beryllium extraction plants; oil refineries [61,69–72]; the photovoltaic solar cell industry [61,73–78]; the production of high-tech silicon-based semiconductors [61,75–78]; and municipal waste incineration plants through HF emissions caused by the incineration of fluorinated plastics, fluorinated textiles, or CaF_2 in sludge [79]. Fluorine is also used in electroplating. In addition, it is used as a melting point depressant in metallurgical furnaces in the smelting process. Water from mines can be a significant source of fluoride.

Chlorofluorocarbons (CFS) have been used extensively as gas in deodorants and coolants in refrigerators. However, due to their destructive effect on the ozone layer, some of these compounds are withdrawn from use. Fluoride also migrates to the environment due to the use of pesticides (e.g., cyhalothrin, fenfluthrin, and tefluthrin) [21]. It is also liberated into the environment in the brick production process [76].

It is estimated that about 30% of pharmaceuticals (including antibiotics, antidepressants, drugs against asthma, and atresia) are based on fluoride. The next big emitters of fluoride are cooling gases used in air conditioning, ventilation, and cooling devices contain fluorine in their composition [80,81]. Fluor is released into the atmosphere by burning hard coal, brown coal, and fuel oil. Then, industrial dust containing soluble fluorides and gaseous compounds (including HF) is emitted [82]. Wastewater from these industries has a higher $F-$ concentration than natural waters, starting from ten thousand mg/L, and in the case of phosphate production, fluoride concentrations in wastewater can reach up to 3000 mg/L [83].

The combustion of biomass releases fluoride into the atmosphere, which is the main stream of this atmospheric pollutant, which has not been characterized before. The emission of fine particles (PM 2.5) of water-soluble fluorine ($F-$) from the biomass combustion was assessed at the Fourth Fire Laboratory in Missoula Experiment (FLAME-IV) using X-ray energy dispersive scanning electron microscopy (SEM-EDX) and ion chromatography with conductivity detection. Based on recent assessments of global biomass combustion, they estimated that biomass combustion releases 76 Gg $F-$ per year into the atmosphere, with an upper and lower limit of 40–150 Gg $F-$ per year. The estimated $F-$ flux from biomass combustion is comparable to fluoride emission from coal combustion and other anthropogenic sources. These data show that biomass combustion is the primary source of fluoride released into the atmosphere in the form of fine particles that can be transported over long distances [37].

As the aforementioned fluoride-originated pollutants raise several health problems, the World Health Organization (WHO) determined the acceptable level of fluoride content in drinking water at the level of 1.5 mg/L [45]. However, the concentration of fluorides in industrial wastewater mostly exceeds these WHO guidelines, reaching even thousands of milligrams per litre [40,84,85]. Thus, fluoride pollution in the aquatic environment, caused by natural and artificial activities, has been a significant problem worldwide. Searching for new, effective ways to remove of fluoride-originated waste from water seems to be very important [60].

3. Selected Types of Reagents for Fluoride Removal

Several conventional techniques may be pointed here, such as adsorption [61,67–73,86–91], chemical precipitation [86,92], coagulation and precipitation methods [72,93–99], ion exchange [100–111], and electrocoagulation [69,77,86,112–123], as well as more advanced membrane processes [83,124–131], reverse osmosis [132–134], and electrochemical treatment [69,115–123]. In general, such compounds as CaCl2 and CaO are added to precipitate fluoride in wastewater.

Each method has its advantages and limitations and can be operated with the appropriate efficiency provided that the process parameters are properly selected to remove fluoride in the appropriate concentration range [26,31,122].

Large-scale industrial operations generate vast amounts of waste, the management of which can be a serious problem. An interesting possibility is to convert such waste into sorbents used for the water defluorination. Then, industrial waste becomes an adsorbent to remove fluoride from aqueous solutions [29]. Figure 1 shows selected types of industrial waste that are used as such adsorbents.

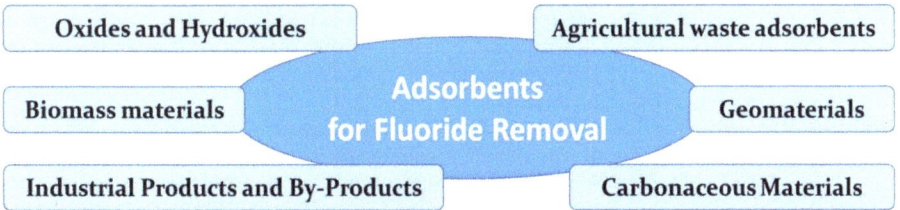

Figure 1. Selected types of industrial waste that are used as fluoride adsorbents.

Among the various methods of water defluorination (as mentioned earlier), adsorption is the most commonly used technique to remove fluoride. Fluorine is adsorbed on a barrier composed of a resin and some mineral particles. This method is efficient, simple, and cheap. These factors are especially important for developing countries [7,28,102]. The adsorbents may be also based on the biomass from plants, even being an agricultural waste, and several industrial by-products. These inexpensive materials help replace an expensive commercial adsorbent such as activated carbon, which again has a regeneration problem. Agricultural and industrial waste materials are available in massive amounts, Some of them are inexpensive and biodegradable, and thus environmentally friendly [123].

A wide range of adsorbents and their modifications were tested to remove fluoride from water [26]. These include activated carbon [83,124–128], activated alumina [129–132,135], bauxite [53,131,133,134,136–151], hematite [137,152–156], polymer resins [94,138,139,157], activated rice husk [125,140,141,158], brick powder [142], pumice stone [143,159–161], red earth, charcoal, brick, fly ash, serpentine [144,162–165], Moringa oleifera seed extracts [166], granular ceramics [167], chitin, chitosan and alginate [135,145–151,155,156,168–174], modified iron oxide/hydroxide [175–181], hydroxyapatite (HAP) [182–186], zirconium-modified materials and ceremonies [58,187–197], titanium adsorbent [198–200], schwertmannite [201], modified cellulose [202,203], clays [165,204–207], zeolite [57,208–213], and magnesium modified sorbent [106,118,202,214].

4. Industrial Waste, By-Product, and Biomass as Fluoride Adsorbents

Red mud is waste produced by the aluminium industry during alkaline processing, namely by the so-called Bayer process. The red sludge is strongly alkaline. The use of industrial wastes such as red sludge for defluorination will significantly reduce their volume for the problem of land removal, soil and groundwater contamination, and landscaping for alternative uses [53].

The removal of fluoride from water using red mud granular according to batch and column adsorption techniques is described by Tor et al. [215]. Cengeloglu et al. [165] have studied defluoridation by using red mud as such and acid-treated red mud by 5.5 M HCl for drinking purposes, and Wei et al. [216] have used modified red mud with $AlCl_3$ (MRMA) and further modified by heat-activated red mud (MRMAH) as an adsorbent for the removal of fluoride from water. Lv et al. [217] have investigated zirconium hydroxide modified red mud porous material to remove fluoride from aqueous solutions. Soni et al. [218] have studied red mud for defluoridation of water collected from the Sitapura Industrial Area, Jaipur (Rajasthan). The results of a study to remove fluoride from red mud by electrokinetic treatment and the feasibility of this technique were presented by Zhu et al. [219].

All authors reported promising results in removing fluoride. Waste mud was recently found as one of the most promising adsorbents due to its extremely low cost and wide availability. This waste is an untapped resource and, in some cases, presents serious disposal problems, so using waste sludge to remove contaminants is an important application.

The authors [220] tested three different forms of waste sludge for their fluoride removal efficiency: primary sludge, acid-treated sludge, and precipitated waste sludge [220]. The precipitated waste sludge showed a higher yield than the others [52].

Sujana et al. [221] investigated the defluorination limit of alum sludge, a waste product of the bauxite alum production process by adding sulfuric acid, which mainly contains aluminium oxide and titanium with a small number of undecomposed silicates. Nigussie et al. [222] investigated the removal of fluoride using the sludge formed during aluminium sulphate production (alum) from kaolin in the sulfuric acid process.

The potential of fluoride adsorption in drinking water treated with spent bleaching earth (SBE) was investigated by Mahramanlioglu et al. [223]. SBE is a solid waste generated during oil processing, as it contains mainly residual oil not removed by filter pressing. SBE applications were found very efficient [224] to adsorb fluorine from water in one of the Iran regions at concentrations ranging from 2.28 and 5.4 mg/L, pH 7, and processing time about 180 min [38].

Fly ash or coal ash, also known as UK Powdered Fuel Ash or Carbon Combustion Residue (CCR), is the product of coal combustion that consists of solid particles (fine particles of burnt fuel) that are driven from coal boilers along with exhaust fumes. The ash that falls to the bottom of the boiler combustion chamber (colloquially called the furnace) is called bottom ash. Singh et al. [53] have studied the defluoridation of groundwater of Agra city by means of fly ash (ATF).

The batch adsorption capacity of fly ash has been studied by Nemade [225]. He observed that fluoride adsorption decreased continuously between pH 2 to 12. Xue [226] observed that the high pH of the solution caused a slight turbidity of the filtered water, the effectiveness of defluorination increases with the increase of fluoride concentration in the inflow, and both the amount of sifted water and the effectiveness of the defluorination increase with an increasing temperature. Geethamani et al. [227] used calcium hydroxide-treated fly ash (CFA) to remove fluoride in a batch study. The removal of more than 80% was achieved with a 10 mg/L fluoride solution with an equilibrium contact time of 120 min and a dose of 3 g/L CFA. The maximum removal of fluoride was at pH 7 [53].

Ramesh et al. [228] investigated the ability to remove bottom ash fluoride in batch and column modes. Thus, 73.5% fluoride removal was achieved with a bottom ash dose of 70 mg/100 mL with an optimal contact time of 105 min. The maximum removal efficiency of 83.2% was observed at pH 6.

Zhang et al., in their work [229], characterized the mechanisms of the detoxification of water-soluble fluoride in bottom ash and the decomposition of fluorine during the combustion of spent potting material (SPL) in response to four calcium compounds $CaSiO_3$, CaO, $Ca(OH)_2$, and $CaCO_3$, which converted NaF into low toxicity compounds, with a conversion range at the level of 54.24–99.45%.

The cenosphere is a light, inert, hollow sphere made mainly of silica and aluminium oxide, filled with air or an inert gas, usually produced as a by-product of coal combustion in thermal power stations. Xu et al. [230] investigated fluoride removal using magnesium-loaded fly ash cenospheres (MLC) prepared by the wet impregnation of fly ash cenospheres with a magnesium chloride solution.

The removal of fluoride with aluminium hydroxide-coated rice husk ash was investigated Ganvir et al. [183]. Rice husk ash is obtained by burning rice husk ash and unshelled husk, the latter two being relatively cheap and massively produced materials. Mondal et al. [231] investigated the capacity of activated rice husk ash (ARHA) by washing and drying rice husk ash from a rice mill at 100 °C for 8 h in an electric furnace and further crushing into 250 μm particles. The fluoride adsorption capacity of such obtained adsorbent was 15.08 mg/g in the batch and 9.5 mg/g in the column test.

Aluminium Treated Bagasse Fly ash (ABF) treated with aluminium for drinking water defluorination with an initial fluorine concentration of 1–10 mg/L, with a sorbent dose range of 1–20 g/L at pH 6.0 were tested by Gupta et al. [232].

Jadhav et al. [233] used maize husk fly ash as an adsorbent for eliminating fluoride from water, with the efficiency reaching 86% at a pH value of 2 and reaction time about two hours.

Waste carbon slurry for fluoride removal was investigated by Gupta et al. [234]. This compound is obtained from fuel oil-based generators of the fertilizer industry. The maximum fluoride adsorption capacity was reported at a level of 4.861 mg/g, with a reaction time of about one hour and pH equal to 7–8.

The ability of the adsorbent produced from coal-mining waste to remove fluoride from an aqueous solution was investigated by CInarli et al. [235]. The optimal pH for the reaction was found at the level of 3.5. To the same goal, Kumari et al. [236] used shale (coal mine waste) as a native shell (NS) adsorbent and heat-activated shale (HAS) at various temperatures ranging from 350 °C to 550 °C.

Islam et al. [237] investigated the basic oxygen furnace slag, produced by the steel industry, to remove fluoride from water. Basic converter slag (BOFS) mainly contains 46.5% CaO, 16.7% iron oxide, 13.8% SiO_2, and some other components. The thermal activation of BOFS (TABOFS) by heating at 1000 °C for 24 h increased the porosity and surface area, leading to increased fluoride adsorption and resulting in fluoride removal at the level of 93% (in comparison with initial 70%).

Lai and Liu [238] used a spent catalyst (a by-product of the petrochemical industry) to remove fluoride from aquatic environments. This compound consists mainly of porous silica and alumina, and it is efficient enough to remove fluoride. Tsai and Lui [239] examined spent iron-coated catalyst by coating 0.1 and 0.5 M $Fe(NO_3)_3$ to remove fluoride from an aqueous solution. Fluoride adsorption decreased with an increasing pH. The fluoride adsorption reaction was endothermic, and the rate of reaction increased with temperature.

Bauxite is a basic source of such metals as aluminium and gallium. It is a sedimentary rock with a relatively high aluminium content. Das et al. [240] used a thermally activated titanium-rich bauxite (TRB) for the removal of the fluoride excess from drinking water. Lavecchia et al. [154] investigated bauxite with a high alumina content (81.5%) to remove fluoride from contaminated water. The percent removal of bauxite from fluoride in the pretest was 38.5%. Chaudhari [241] used bauxite to defluoridation water. It was observed that the optimal dose of the adsorbent was 1.8 g/50 mL, while the process took 90 min at the optimum pH of 6.0.

Bibi et al. used hydrated cement, brick dust, and marble flour to de-fluorine and remove arsenic from the water. The presence of co-anions did not significantly affect the effectiveness of arsenic and fluoride removal [53]. Kang et al. [242] investigated Cement Paste for removing fluoride as a low-cost solution. The cement paste was competitive with lime, the prevalent fluoride-removing agent [52].

Zhang et al. [186] investigated the possibility of removing fluoride using recycled phosphogypsum. The latter was applied in the form of HAP nanoparticles using microwave radiation technology [52].

Oguz used lightweight concrete (building material) [243] as an adsorbent to remove fluoride from water, and its effectiveness was tested. The maximum adsorption of fluoride took place at pH 6.9. Additionally, hydrated cement [244] and hardened alumina cement [245] were tested to remove fluoride from an aqueous solution. Various forms of apatite have been used to remove fluoride because it has shown good defluorination prospects, namely synthetic nano-hydroxyapatite (n-Hap), biogenic apatite, processed biogenic apatite, and geogenic apatite [246]. The fluoride adsorption was determined to decrease with increasing concentration levels and pH value. Ultrasonic and microwave treatment also increased the effectiveness of the fluoride removal process [247,248]. The influence of low molecular weight organic acids (LMWOA) on the defluoridation capacity of nano-hydroxyapatite (nHAP) from an aqueous solution was investigated [249]. Cellulose nanocomposites @ hydroxyapatite (HA) were prepared in $NaOH/thiourea/urea/H_2O$ by in situ hybridization [203]. Aluminium-modified hydroxyapatite (Al-HAP) was also used for defluoridation [250]. High-purity phosphogypsum (PG) nanoparticles were also used, showing an excellent fluoride adsorption capacity [54,186].

Waste clay brick (WCB) is a silicate solid waste, the recycling of which is of significant environmental and social importance. WCB is used in the production of concrete and mortar, as a raw material, or an additive for the production of secondary cement.

In recent years, more and more attention has been paid to the recycling of waste clay bricks, and the extension of their recyclable use has laid a solid foundation for improving its utility value [55,251].

Bleaching powder, also known as chlorinated lime (calcium oxychloride), is mainly composed of calcium hypochlorite. It is widely used as a disinfectant for drinking or swimming pool water and as a bleaching agent. The whitening powder generally has advantageous properties as an economical and viable replacement for other adsorbents for removing fluoride from an aqueous solution. In addition to being a disinfectant, it also acts as a defluorant. Kagne et al. [252] used a bleaching powder to remove fluoride, raising the removal ratio from to 90.6% [52].

Li Wang et al. [84] adopted the new calcium-containing calcite precipitating and assisted precipitating fluorspar to treat wastewater containing fluoride. Key parameters of the reaction were determined, such as reaction timing, the rate of the oscillation, the doses of hydrochloric acid and calcite, etc.

Chen et al. [253] developed a ceramic-based adsorbent for removing fluoride from an aqueous solution. The adsorbent showed sufficient mechanical resistance for long-term adsorption, as well as high efficiency. The same authors also reported results of batch tests of fluoride removal using a surface-modified granular ceramic with an Al-Fe complex [52].

Detailed information on the above-mentioned adsorbents is presented in Table 2.

Table 2. Detailed information on the adsorbents used for fluoride removal.

Adsorbent	Concentration Range (mg/L)	pH Range	Contact Time (min)	Model Used to Calculate Adsorption Capacity	Maximum Adsorption Capacity (mg/g)	Ref.
Waste mud	-	2–8	0–480	Langmuir and Freundlich	27.2	[220]
Red Mud	5–150	4.7	15–540	Freundlich	0.851	[215]
	5	4.7	360	Redlich–Peterson and Freundlich	0.644	[215]
	100–1000	5.5	120	Langmuir and Freundlich	3.12 and 6.29	[165]
Modified red mud with $AlCl_3$ (MRMA), heat activated red mud (MRMAH)	-	7–8		Langmuir	MRMA-68.07 MRMAH-91.28	[216]
Zirconium hydroxide modified red mud porous material Zr-modified RMPM	-	3	60	pseudo-second-order rate kinetics and pore diffusion models	0.6	[217]
Red mud	-	5.5	120	-		[218]
Alum sludge	-	5.5–6.5	-	-	5.35	[221]
Sludge produced during the manufacturing of aluminium sulphate (alum) from kaolin	10	3–8	-	-	332.5	[222]

Table 2. Cont.

Adsorbent	Concentration Range (mg/L)	pH Range	Contact Time (min)	Model Used to Calculate Adsorption Capacity	Maximum Adsorption Capacity (mg/g)	Ref.
Spent Bleach Earth (SBE)	-	3.5	-	-	7.75	[223]
Fly ash A and S	-	-	-	Freundlich	1.22 (A) 1.01 (S)	[226]
Calcium hydroxide treated fly ash (CFA)	10	7	120		10.86	[227]
Bottom ash	-	6	105	BDST	16.26	[228]
Magnesia-loaded fly ash cenospheres (MLC)	10	-	-	Thomas	5.884	[230]
aluminium-treated bagasse fly ash (ABF)	1–10	6	300	-	10	[232]
Maize husk fly ash	2.0 g/50 mL	2	120	Redlich-Peterson		[233]
Activated tea ash (AcTAP)		6	180	Langmuir	8.55	[231]
Waste carbon slurry obtained from fuel oil	15	7.58	120	Langmuir	4.861	[234]
Coal mining waste	-	3.5	-	Langmuir	15.67	[235]
Shale (coal mine waste) in the form of native shale (NS) and heat activated shale (HAS) at 350 °C, 450 °C and 550 °C	10-HAS550	3	24 h	Langmuir	0.358	[236]
Blast furnace slag generated from steel industry	10 mg/l	6–10	35	Langmuir	8.07	[237]
Spent catalyst (a by-product of petrochemical industry)	-	4	-		28	[238]
Iron coated spent catalyst	-	5.5–6.0	-	Langmuir	7.2–20.7	[239]
Thermally activated titanium rich bauxite (TRB)	10	5.5–6.5	-	Langmuir	3.8	[240]
High alumina (81.5%) content bauxite	-	-	-	Freundlich	3.125	[243]
Bauxite	10	6	90	Freundlicha, Langmuira Tempkina,	3	[241]
Hydrated cement (HC), brick powder (BP) marble powder (MP).	30	7 8 7	60	Langmuir	1.72 0.84 0.18	[254]
Bleaching powder	-	6–10	-	-	-	[244]

Table 2. Cont.

Adsorbent	Concentration Range (mg/L)	pH Range	Contact Time (min)	Model Used to Calculate Adsorption Capacity	Maximum Adsorption Capacity (mg/g)	Ref.
Rice husk ash, which was coated with aluminium hydroxide	10–60	7	60		15.08	[183]
Activated rice husk ash (ARHA)			100	Langmuir	0.402	[231]
Ceramic adsorbents consisting of Kanuma mud, with zeolite, starch, and $FeSO_4 \cdot 7H_2O$	20–100	4–11	0–48 h	pseudo-second-order	2.16	[253]
Porous granular ceramic adsorbents containing dispersed aluminium and iron oxides	10	4–9	48 h	Langmuir and Freundlich	1.79	[249]
Iron-impregnated granular ceramics		7, 4		Langmuir and Freundlich	-	[167]
Recycled phosphogypsum in a form of HAP nanoparticles		7		Langmuir-Freundlich	19.742–25 °C 26.108–35 °C 36.914–45 °C 40.818–55 °C	[186]
HAP-calcium phosphate based bioceramic	-	-	-	Langmuir and pseudo-second-order	32.57	[250]
HAP Apatitic tricalcium phosphate.	up to 20 up to 60	4.16 4		Langmuir Langmuir	13.88–25 °C 14.70–30 °C 15.15–37 °C	[118, 119]

5. Fluoride Wastes Removal in Industrial Processes

5.1. Industrial Production of Aluminium Fluoride

The mass production of aluminium fluoride forces significant amounts of silica gel being waste-contaminated with fluoride ions [24,255]. For example, a main fertilizer producer in Lithuania, a joint-stock company "Lifosa", generates approximately 15 thousand tons per year of the mentioned waste during the manufacture of 17 thousand tons of AlF3 [256]. AlF3 is formed in the reaction of neutralizing hexafluorosilicic acid with aluminium hydroxide. However, due to the strong bonding of fluoride ions to the crystal structure of the latter compound, the purification of silica gel waste is a challenge. As a result, the waste silica gel is mainly disposed in the landfill with no further treatment [25,26,255,257]. The long-term storage of this king of waste provokes many environmental problems, due to the fact of the leaching of fluoride into water [24,255]. According to the literature [258–263], the amount of toxic compounds may be reduced by removing them from waste or by reducing their mobility to the environment [11].

5.2. Industrial Waste from Semiconductor Factories

In recent years, the industrial production of electronic materials has contributed to an increase in the global concentration of fluoride and water pollution. The significant contributors to fluoride-contaminated wastewater are semiconductor manufacturers and industrial plants producing hydrofluoric acid, photovoltaic materials, plastics, and textiles [69].

It is assumed that almost 30% of waste produced in the semiconductor industry is of fluorine origin; thus, new treatment methods for this kind of waste are welcome, with one of them described in [65]. The authors converted fluoride waste to AlF3 by an aluminium treatment. AlF3 is then dissolved, and, at the same time, a calcium conditioner is added to replace the AlF3 with CaF2. The method is able to reduce the amount of fluorine contents by a factor of 75%. Moreover, the aluminium component is reusable, therefore the cost of the method is reasonably small [264].

Chemical vapor deposition (CVD) processes are widely used in the production of solar cells and include the deposition of crystalline silicon from chlorosilanes, iodides, bromides, and fluorides [265]. An undesirable side effect is the release of toxic SiO_2. The by-products of silicon film deposition consist of large amounts of SiO_2 powder, HF vapours, SiH_4, and PH_3 [266]. These by-products are usually transported to the factory's central scrubber or dust filter, and treatment produces large amounts of hazardous fluoride-containing sludges.

The effective and cheap treatment of fluorine-containing sludge resulting from CVD processes collected after cleaning the filter cartridge in a photovoltaic installation is located in southern Italy, as found by Zueva et al. [267].

In addition, the treatment of waste with alumina, magnesium sulphate, and lime was tested. These studies aimed to remove the F- content from the liquid phase of the sludge and examine the possibility of producing non-hazardous solid waste. Therefore, the toxicity characterization leaching (TCLP) procedure of the obtained solids was performed with and without thermal treatment. The best conditions for removing fluoride from liquid waste and converting the sludge into non-hazardous waste were related to water treatment with lime and magnesium sulphate.

Electrocoagulation coupled with flotation to treat semiconductor production wastewater was proposed by Hu et al. [77]. The fluoride ions were partially removed by precipitation with calcium in an electrolyser, to which sodium dodecyl sulphate was added to increase flotation. These treatments were effective in reducing fluoride and suspended solids in the wastewater. They lowered the concentration of fluoride from 806 mg/dm^3 to 5–6 mg/dm^3.

An original fluidization process to recover CaF_2 from a synthetic fluoride solution was developed by Aldaco et al. [268]. Granulated calcite and silica were used as seed materials to recover the calcium fluoride by crystallization in a fluidized bed reactor. The inlet concentration of fluoride was 250 mg/L, and the final fluoride conversion was 92%, with a CaF_2 content in the solid greater than 97% by weight. This process offers a good alternative for reducing solid waste and reusing calcium fluoride.

In the work of Shin et al. [269], more than 99 wt.% precipitated HF and silicon during the pre-treatment of the solution and recovered Na_2SiF_6 to commercial grade 98.2%. The remaining solution contained 279 g/L acetic acid, 513 g/L nitric acid, and some HF. It was extracted with 2 ethylhexyl alcohol. Acetic acid was removed from the organic phase with deionized water to give 96.3% acetic acid recovery.

The recycling of SiO_2-CaF_2 nanoparticle sludge recovered from the semiconductor industry wastewater treatment was investigated by Lee and Liu [270]. The dried and powdered sludge was replaced with 5 to 20 wt.% Portland cement in mortar. The compressive strength of the modified mortar was higher in comparison with the fresh cement mortar after three days of hardening. Moreover, the Toxicity Trait Leaching Procedure (TCLP) showed that no heavy metals were released from the modified mortars. In another study, similar results were obtained with a different deposit produced in the polishing operations of the IC industry. This sludge, consisting of hazardous compounds such as SiO_2, Al_2O_3, CaF_2, and unknown organic compounds, was used to replace 10 wt.% cement powder to produce concrete. The compressive strength was comparable to that of regular Portland cement, while the TCLP test did not detect any metal release [271]. In another study, Lee [270] investigated the addition of a PV sludge/fly ash slag mixture for the production of cement mortar. The optimal mixture, determined by the Taguchi method, was 20.9 wt.%. cement flour, 4.3% volatile slag, 3.4% PV sludge, and 71.4% sand. The optimally modified cement mortar showed an increased compressive strength from the fourth day of maturation, reaching the maximum value of 132% after 7 days in relation to the compressive strength of the mortar composed of fresh Portland cement.

As such, recycling sludge from the PV industry is essential to preventing their potential environmental hazards and helping to reduce the cement industry's carbon footprint and environmental impact.

One type of hazardous waste was the fluorine-containing sludge from the semiconductor industry, the safe treatment and disposal of which were ineffective. Da et al. [272] presented the research results on the assessment of the possibility of adding fluorine-containing sludge to cement clinker. The authors inform that the addition of 2.0% of the sludge significantly improved the flammability of the clinker and improved a formation of alite. However, increasing the amount of the sludge to 5.0% caused the profuse formation of interstitial phases and slowed down the formation of alite and belite. The presence of fluorite was high in the silicate phase, resulting in the accumulation of this compound mainly at the surface. The fluoride was immobilized by calcium, with the immobilization rates for fluorine, copper, zinc, and nickel reaching a level of 99.5%. A sludge addition did not cause any threats or side effects [272].

6. A New Concept(s) for the Production and Management of Fluoride Adsorbents

According to the review, there are many methods of removing fluoride, including using sorbents made from industrial waste, by-product, and biomass. Many sorbents and their application methods have been developed, dedicated to specific industrial processes, in which there are fluorides in the form of wastewater or waste. There are also known methods of using fluoride-containing wastes to produce new products used in many fields, including construction. However, further research on developing new solutions is still being carried out. One of the latest proposals is the production of composites from several types of waste, which, although they can be used alone as materials for removing fluorides, especially from water and sewage, must first be deeply processed; e.g., calcination or their direct use is associated with technical problems at the stage of separating the used sorbent from the treated liquid by means of filtration or sedimentation [273].

Paper sludge (PS) is generated as industrial waste in the process of recycling paper products, with the amount continuously increasing year by year [274]. PS mainly contains cellulose fibres (up to 50–60%) and inorganic fillers along with coating materials such as calcite, kaolinite, and talc [275]. The paper industry is of great importance to the natural environment due to the amount of PS produced and its disposal. A small part of PS waste is used in agriculture as a soil improver and fertilizer [276–280]. However, PS is mostly disposed in open landfills with no further treatment, which is a growing problem especially for highly developed countries. Recently, we observed some tries to use PS as an additive to cement [281], metakaolin for the production of ceramics and glass [282], fuel for energy recovery [283,284], and carbon adsorbent for removing organic pollutants [285], thus reducing the amount of PS disposed in landfills.

Takaaki Wajima et al. converted PS into an effective fluoride sequestrant by the process of calcination in a high temperature for several hours. They determined that PS fired at 800 °C shows highest fluorine absorption. The authors also pointed out the fact of the selective removal of fluoride in several solutions containing chlorides, nitrates, and sulphates [274].

Using paper slurries obtained in the creation of composites based on non-calcined sludge and post-soda lime is an interesting and innovative solution to remove fluoride from water and sewage [273]. Post-soda lime is a by-product formed in the process of separating the solid phase present in the still liquid during the production of soda by the Solvay method. The mixture mainly contains some calcium compounds ($CaCl_2$, $CaCO_3$, $CaSO_4$, $Ca(OH)_2$), magnesium and silica, sulphur, and aluminium. It is characterized by a very high hydration (up to 60%) and low particle size distribution (less than 2 μm). Unfortunately, such properties strongly limit the traditional usage of this waste material [286]. According to the invention, the above problems are solved by the method [287,288], where very fine soda ash, preferably from clarifiers, is applied to fibrous cellulosic support in the form of papermaking sludges. The result is a composite that is highly permeable to liquids and removes fluoride ions very efficiently, with the same amount of added fluoride precipitant being even twice as high as in the case of traditional materials used to remove fluorides (such as ground limestone and chalk). The content of cellulose fibres in the composite allows it to be shaped practically in any form, e.g., in the form of granules, pellets, flat membranes, plates, cylinders, etc., depending on the user's needs. They can also be placed in filter bags that are permeable to solutions, so they can be used repeatedly until the calcium compounds are entirely converted. After use, they can be easily removed from the treated solution. The used sorbent can be used analogously to paper sludge, for example as an additive/filler for building materials [36]. It was shown that, due to the calcium compounds used, the targeted material may be supplemented with hazardous mineral waste containing fluorine, e.g., in form of post-crystallization lye formed in the processing of fluosilicic acid or phosphogypsum. The approach improves the reuse of waste materials, as well as minimizes the usage of raw materials, contributing to the concept of a circular economy.

7. Conclusions

The negative impact of hazardous waste on the health of ecosystems, including humans, is becoming a rapidly growing global problem. The large amount of waste materials caused by the industry and human life is becoming a huge problem for both enterprises and local administrations, increasing the costs of everyday activities. Therefore, recycling and reusing industrial waste and by-products are of great importance. Many of them can be used to produce new construction materials after "solidification/stabilization" processes. Such materials may be used as admixtures or raw materials. In this case, an assessment of the leaching of the contaminants should be of particular concern, as well as the overall efficiency of the conversion process. Therefore, it is important to know the properties and conditions of use of fluoride binders.

Several materials have been proposed and tested as adsorbents towards efficient fluoride removal, taking into account also low processing costs and minimal side effects. The research in this area is undergoing, and the authors describe a high adsorption of fluoride-originated waste. However, the proposed methods usually depend on the particular pH and other process parameters that are difficult to achieve and maintain. Moreover, the adsorbents usually cannot be fully reused without costly regeneration. In addition, the competing ions show an affinity to the same active parts of the adsorbent, and the excess of some organic compounds delays the process balance.

In general, the concentration of fluoride ions can be reduced by a number of methods. Research on new methods for removing fluoride ions is still ongoing. At the same time, efforts are made to increase the efficiency of existing technologies. The removal of fluoride ions is a significant problem because they have a negative impact on human health [8]. Extensive research is required to develop and implement low-cost, sustainable hybrid technologies that can overcome the disadvantages of stand-alone processes.

Author Contributions: Conceptualization, M.O., I.R. and W.U.; investigation, M.O.; writing—original draft preparation, M.O. and I.R.; writing—review and editing, M.O., I.R. and W.U.; visualization, M.O.; supervision, W.U. All authors have read and agreed to the published version of the manuscript.

Funding: Project co-financed by the Regional Operational Programme for Pomorskie Voivodeship 2014–2020, Project no. RPPM.01.01.01-22-0040/18-00.

Institutional Review Board Statement: Not applicable.

Informed Consent Statement: Not applicable.

Data Availability Statement: Not applicable.

Conflicts of Interest: The authors declare no conflict of interest.

References

1. Gupta, N.; Yadav, K.K.; Kumar, V. A review on current status of municipal solid waste management in India. *J. Environ. Sci.* **2015**, *37*, 206–217. [CrossRef] [PubMed]
2. Melaré, A.V.D.S.; González, S.M.; Faceli, K.; Casadei, V. Technologies and decision support systems to aid solid-waste management: A systematic review. *Waste Manag.* **2017**, *59*, 567–584. [CrossRef] [PubMed]
3. Bing, X.; Bloemhof, J.M.; Ramos, T.R.P.; Barbosa-Povoa, A.; Wong, C.Y.; van der Vorst, J.G. Research challenges in municipal solid waste logistics management. *Waste Manag.* **2016**, *48*, 584–592. [CrossRef] [PubMed]
4. Hettiarachchi, H.; Meegoda, J.N.; Ryu, S. Organic Waste Buyback as a Viable Method to Enhance Sustainable Municipal Solid Waste Management in Developing Countries. *Int. J. Environ. Res. Public Health* **2018**, *15*, 2483. [CrossRef] [PubMed]
5. Ouda, O.K.M.; Raza, S.A.; Nizami, A.S.; Rehan, M.; Al-Waked, R.; Korres, N.E. Waste to energy potential: A case study of Saudi Arabia. *Renew. Sustain. Energy Rev.* **2016**, *61*, 328–340. [CrossRef]
6. Sadef, Y.; Nizami, A.-S.; Batool, S.A.; Chaudary, M.N.; Ouda, O.K.M.; Asam, Z.-U.; Habib, K.; Rehan, M.; Demirbas, A. Waste-to-energy and recycling value for developing integrated solid waste management plan in Lahore. *Energy Sources Part B Econ. Plan. Policy* **2016**, *11*, 569–579. [CrossRef]
7. Sawadogo, M.; Tanoh, S.T.; Sidibé, S.; Kpai, N.; Tankoano, I. Cleaner production in Burkina Faso: Case study of fuel briquettes made from cashew industry waste. *J. Clean. Prod.* **2018**, *195*, 1047–1056. [CrossRef]
8. Ghisolfi, V.; Chaves, G.D.L.D.; Siman, R.R.; Xavier, L.H. System dynamics applied to closed loop supply chains of desktops and laptops in Brazil: A perspective for social inclusion of waste pickers. *Waste Manag.* **2017**, *60*, 14–31. [CrossRef]
9. Matter, A.; Ahsan, M.; Marbach, M.; Zurbrügg, C. Impacts of policy and market incentives for solid waste recycling in Dhaka, Bangladesh. *Waste Manag.* **2015**, *39*, 321–328. [CrossRef]
10. Ferronato, N.; Torretta, V. Waste Mismanagement in Developing Countries: A Review of Global Issues. *Int. J. Environ. Res. Public Health* **2019**, *16*, 1060. [CrossRef]
11. Rudelis, V.; Dambrauskas, T.; Grineviciene, A.; Baltakys, K. The Prospective Approach for the Reduction of Fluoride Ions Mobility in Industrial Waste by Creating Products of Commercial Value. *Sustainability* **2019**, *11*, 634. [CrossRef]
12. Angin, D. Utilization of activated carbon produced from fruit juice industry solid waste for the adsorption of Yellow 18 from aqueous solutions. *Bioresour. Technol.* **2014**, *168*, 259–266. [CrossRef] [PubMed]
13. Bujak, J.W. Thermal utilization (treatment) of plastic waste. *Energy* **2015**, *90*, 1468–1477. [CrossRef]
14. Moh, Y. Solid waste management transformation and future challenges of source separation and recycling practice in Malaysia. *Resour. Conserv. Recycl.* **2017**, *116*, 1–14. [CrossRef]
15. Shen, C.-W.; Tran, P.P.; Ly, P.T.M. Chemical Waste Management in the U.S. Semiconductor Industry. *Sustainability* **2018**, *10*, 1545. [CrossRef]
16. Stonys, R.; Kuznetsov, D.; Krasnikovs, A.; Škamat, J.; Baltakys, K.; Antonovič, V.; Černašėjus, O. Reuse of ultrafine mineral wool production waste in the manufacture of refractory concrete. *J. Environ. Manag.* **2016**, *176*, 149–156. [CrossRef]
17. Yilmaz, O.; Kara, B.Y.; Yetis, U. Hazardous waste management system design under population and environmental impact considerations. *J. Environ. Manag.* **2017**, *203*, 720–731. [CrossRef]
18. Woźniak, J.; Pactwa, K. Overview of Polish Mining Wastes with Circular Economy Model and Its Comparison with Other Wastes. *Sustainability* **2018**, *10*, 3994. [CrossRef]
19. Siddique, R. Utilization of Industrial By-products in Concrete. *Procedia Eng.* **2014**, *95*, 335–347. [CrossRef]
20. Dyachenko, A.; Petlin, I.; Malyutin, L. The Research of Sulfuric Acidic Recycling of Aluminum Industry Fluorine-containing Waste. *Procedia Chem.* **2014**, *11*, 10–14. [CrossRef]
21. Li, Y.; Zhang, H.; Zhang, Z.; Shao, L.; He, P. Treatment and resource recovery from inorganic fluoride-containing waste produced by the pesticide industry. *J. Environ. Sci.* **2015**, *31*, 21–29. [CrossRef] [PubMed]
22. de Beer, M.; Doucet, F.; Maree, J.; Liebenberg, L. Synthesis of high-purity precipitated calcium carbonate during the process of recovery of elemental sulphur from gypsum waste. *Waste Manag.* **2015**, *46*, 619–627. [CrossRef] [PubMed]
23. Baek, C.; Seo, J.; Choi, M.; Cho, J.; Ahn, J.; Cho, K. Utilization of CFBC Fly Ash as a Binder to Produce In-Furnace Desulfurization Sorbent. *Sustainability* **2018**, *10*, 4854. [CrossRef]
24. Vaičiukynienė, D.; Vaitkevičius, V.; Kantautas, A.; Sasnauskas, V. Utilization of by-product waste silica in concrete-based materials. *Mater. Res.* **2012**, *15*, 561–567. [CrossRef]
25. Iljina, A.; Baltakys, K.; Baltakys, M.; Siauciunas, R. Neutralization and removal of compounds containing fluoride ions from waste silica gel. *Rom. J. Mater.* **2014**, *44*, 265–271.
26. Iljina, A.; Baltakys, K.; Bankauskaite, A.; Eisinas, A.; Kitrys, S. The stability of formed CaF_2 and its influence on the thermal behavior of C–S–H in CaO–silica gel waste–H_2O system. *J. Therm. Anal.* **2016**, *127*, 221–228. [CrossRef]
27. Vaičiukynienė, D.; Kantautas, A.; Vaitkevičius, V.; Jakevičius, L.; Rudžionis, Ž.; Paškevičius, M. Effects of ultrasonic treatment on zeolite NaA synthesized from by-product silica. *Ultrason. Sonochem.* **2015**, *27*, 515–521. [CrossRef]
28. Baltakys, K.; Iljina, A.; Bankauskaite, A. Thermal properties and application of silica gel waste contaminated with F− ions for C-S-H synthesis. *J. Therm. Anal.* **2015**, *121*, 145–154. [CrossRef]
29. Hydrothermal Synthesis of Calcium Sulfoaluminate–Belite Cement from Industrial Waste Materials—Advances in Engineering. Available online: https://advanceseng.com/hydrothermal-synthesis-calcium-sulfoaluminate-belite-cement-industrial-waste-materials/ (accessed on 9 January 2022).

30. Shen, Z.; Jin, F.; O'Connor, D.; Hou, D. Solidification/Stabilization for Soil Remediation: An Old Technology with New Vitality. *Environ. Sci. Technol.* **2019**, *53*, 11615–11617. [CrossRef]
31. Ma, W.; Chen, D.; Pan, M.; Gu, T.; Zhong, L.; Chen, G.; Yan, B.; Cheng, Z. Performance of chemical chelating agent stabilization and cement solidification on heavy metals in MSWI fly ash: A comparative study. *J. Environ. Manag.* **2019**, *247*, 169–177. [CrossRef]
32. Chen, W.; Wang, F.; Li, Z.; Li, Q. A comprehensive evaluation of the treatment of lead in MSWI fly ash by the combined cement solidification and phosphate stabilization process. *Waste Manag.* **2020**, *114*, 107–114. [CrossRef] [PubMed]
33. Feng, Y.-S.; Du, Y.-J.; Zhou, A.; Zhang, M.; Li, J.-S.; Zhou, S.-J.; Xia, W.-Y. Geoenvironmental properties of industrially contaminated site soil solidified/stabilized with a sustainable by-product-based binder. *Sci. Total Environ.* **2020**, *765*, 142778. [CrossRef] [PubMed]
34. Zhang, W.-L.; Zhao, L.-Y.; Yuan, Z.-J.; Li, D.-Q.; Morrison, L. Assessment of the long-term leaching characteristics of cement-slag stabilized/solidified contaminated sediment. *Chemosphere* **2020**, *267*, 128926. [CrossRef]
35. Solidification/Stabilization-ITRC. Available online: https://itrcweb.org/itrcwebsite/teams/projects/solidificationstabilization (accessed on 2 March 2022).
36. Kizinievic, O.; Kizinievic, V.; Trambitski, Y.; Voisniene, V. Application of paper sludge and clay in manufacture of composite materials: Properties and biological susceptibility. *J. Build. Eng.* **2022**, *48*, 104003. [CrossRef]
37. Jayarathne, T.; Stockwell, C.E.; Yokelson, R.J.; Nakao, S.; Stone, E.A. Emissions of Fine Particle Fluoride from Biomass Burning. *Environ. Sci. Technol.* **2014**, *48*, 12636–12644. [CrossRef] [PubMed]
38. Nayak, B.; Samant, A.; Patel, R.; Misra, P.K. Comprehensive Understanding of the Kinetics and Mechanism of Fluoride Removal over a Potent Nanocrystalline Hydroxyapatite Surface. *ACS Omega* **2017**, *2*, 8118–8128. [CrossRef] [PubMed]
39. Bodzek, M.; Konieczny, K. Open Access (CC BY-NC 4K.0) Fluorki w środowisku wodnym-zagrożenia i metody usuwania Fluorine in the Water Environment—Hazards and Removal Methods. *Eng. Prot. Environ.* **2018**, *21*, 113–141. [CrossRef]
40. Tang, W.; Kovalsky, P.; He, D.; Waite, T.D. Fluoride and nitrate removal from brackish groundwaters by batch-mode capacitive deionization. *Water Res.* **2015**, *84*, 342–349. [CrossRef]
41. Wang, L.; Sun, N.; Wang, Z.; Han, H.; Yang, Y.; Liu, R.; Hu, Y.; Tang, H.; Sun, W. Self-assembly of mixed dodecylamine–dodecanol molecules at the air/water interface based on large-scale molecular dynamics. *J. Mol. Liq.* **2019**, *276*, 867–874. [CrossRef]
42. Claveau-Mallet, D.; Wallace, S.; Comeau, Y. Removal of phosphorus, fluoride and metals from a gypsum mining leachate using steel slag filters. *Water Res.* **2013**, *47*, 1512–1520. [CrossRef]
43. Wang, Y.; Zhu, H.; Jiang, X.; Lv, G.; Yan, J. Study on the evolution and transformation of Cl during Co-incineration of a mixture of rectification residue and raw meal of a cement kiln. *Waste Manag.* **2019**, *84*, 112–118. [CrossRef] [PubMed]
44. Guth, S.; Hüser, S.; Roth, A.; Degen, G.; Diel, P.; Edlund, K.; Eisenbrand, G.; Engel, K.-H.; Epe, B.; Grune, T.; et al. Toxicity of fluoride: Critical evaluation of evidence for human developmental neurotoxicity in epidemiological studies, animal experiments and in vitro analyses. *Arch. Toxicol.* **2020**, *94*, 1375–1415. [CrossRef] [PubMed]
45. World Health Organization. *Guidelines for Drinking-Water Quality 3rd Edition Incorporating the First and Second Addenda*; Recommendations Geneva 2008 WHO Library Cataloguing-in-Publication Data; World Health Organization: Geneva, Switzerland, 2008; Volume 1.
46. Karthikeyan, G.; Shunmugasundarraj, A. Isopleth mapping and in-situ fluoride dependence on water quality in the krishnagiri block of tamil nadu in south india. *Res. Rep.* **2000**, *33*, 121–127.
47. Rao, N.S. Groundwater quality: Focus on fluoride concentration in rural parts of Guntur district, Andhra Pradesh, India. *Hydrol. Sci. J.* **2003**, *48*, 835–847. [CrossRef]
48. Viswanathan, G.; Jaswanth, A.; Gopalakrishnan, S.; Ilango, S.S.; Aditya, G. Determining the optimal fluoride concentration in drinking water for fluoride endemic regions in South India. *Sci. Total Environ.* **2009**, *407*, 5298–5307. [CrossRef]
49. Abdelgawad, A.; Watanabe, K.; Takeuchi, S.; Mizuno, T. The origin of fluoride-rich groundwater in Mizunami area, Japan—Mineralogy and geochemistry implications. *Eng. Geol.* **2009**, *108*, 76–85. [CrossRef]
50. Rafique, T.; Naseem, S.; Bhanger, M.I.; Usmani, T.H. Fluoride ion contamination in the groundwater of Mithi sub-district, the Thar Desert, Pakistan. *Environ. Earth Sci.* **2008**, *56*, 317–326. [CrossRef]
51. Meenakshi; Maheshwari, R. Fluoride in drinking water and its removal. *J. Hazard. Mater.* **2006**, *137*, 456–463. [CrossRef]
52. Habuda-Stanić, M.; Ravančić, M.E.; Flanagan, A. A Review on Adsorption of Fluoride from Aqueous Solution. *Materials* **2014**, *7*, 6317–6366. [CrossRef]
53. Waghmare, S.S.; Arfin, T. Fluoride removal by industrial, agricultural and biomass wastes as adsorbents: Review. *J. Adv. Res. Innov. Ideas Educ.* **2015**, *1*, 628–653.
54. Tomar, V.; Kumar, D. A critical study on efficiency of different materials for fluoride removal from aqueous media. *Chem. Cent. J.* **2013**, *7*, 51. [CrossRef] [PubMed]
55. Kumar, P.S.; Suganya, S.; Srinivas, S.; Priyadharshini, S.; Karthika, M.; Karishma Sri, R.; Lichtfouse, E. Treatment of fluoride-contaminated water. A review. *Environ. Chem. Lett.* **2019**, *17*, 1707–1726. [CrossRef]
56. Waghmare, S.S.; Arfin, T. Fluoride Removal by Clays, Geomaterials, Minerals, Low Cost Materials and Zeolites by Ad-sorption: A Review. *Int. J. Sci. Eng. Technol. Res.* **2015**, *4*, 3663–3676.
57. Waghmare, S.S.; Arfin, T. Fluoride Removal from Water by various techniques: Review. *IJISET—Int. J. Innov. Sci. Eng. Technol.* **2015**, *2*, 560–571.

58. Rao, N. Fluoride and Environment—A Review. In Proceedings of the 3rd International Conference on Environment and Health, Chennai, India, 15–17 December 2003; pp. 386–399.
59. Schlesinger, W.H.; Klein, E.M.; Vengosh, A. Global Biogeochemical Cycle of Fluorine. *Glob. Biogeochem. Cycles* **2020**, *34*, e2020GB006722. [CrossRef]
60. He, J.; Yang, Y.; Wu, Z.; Xie, C.; Zhang, K.; Kong, L.; Liu, J. Review of fluoride removal from water environment by adsorption. *J. Environ. Chem. Eng.* **2020**, *8*, 104516. [CrossRef]
61. Bhatnagar, A.; Kumar, E.; Sillanpää, M. Fluoride removal from water by adsorption—A review. *Chem. Eng. J.* **2011**, *171*, 811–840. [CrossRef]
62. Wang, L.; Sun, N.; Tang, H.; Sun, W. A Review on Comprehensive Utilization of Red Mud and Prospect Analysis. *Minerals* **2019**, *9*, 362. [CrossRef]
63. Arora, H.C.; Chattopadhya, S.N. A study on the effluent disposal of superphosphate fertilizer factory. *Environ. Health* **1994**, *16*, 140–150.
64. Mourad, N.; Sharshar, T.; Elnimr, T.; Mousa, M. Radioactivity and fluoride contamination derived from a phosphate fertilizer plant in Egypt. *Appl. Radiat. Isot. Incl. Data Instrum. Methods Use Agric. Ind. Med.* **2009**, *67*, 1259–1268. [CrossRef]
65. Shen, J.; Schaefer, A. Removal of fluoride and uranium by nanofiltration and reverse osmosis: A review. *Chemosphere* **2014**, *117*, 679–691. [CrossRef] [PubMed]
66. Fan, C.-S.; Li, K.-C. Production of insulating glass ceramics from thin film transistor-liquid crystal display (TFT-LCD) waste glass and calcium fluoride sludge. *J. Clean. Prod.* **2013**, *57*, 335–341. [CrossRef]
67. Ponsot, I.; Falcone, R.; Bernardo, E. Stabilization of fluorine-containing industrial waste by production of sintered glass-ceramics. *Ceram. Int.* **2013**, *39*, 6907–6915. [CrossRef]
68. Sujana, M.G.; Thakur, R.S.; Das, S.N.; Rao, S.B. Defluorination of Waste Water. *Asian J. Chem.* **1997**, *9*, 561–570.
69. Shen, F.; Chen, X.; Gao, P.; Chen, G. Electrochemical removal of fluoride ions from industrial wastewater. *Chem. Eng. Sci.* **2003**, *58*, 987–993. [CrossRef]
70. Blagojevic, S.; Jakovljevic, M.; Radulovic, M. Content of fluorine in soils in the vicinity of aluminium plant in Podgorica. *J. Agric. Sci.* **2002**, *47*, 1–8. [CrossRef]
71. Paulson, E.G. Reducing fluoride in industrial wastewater. *Chem. Eng.* **1997**, *84*, 89–94.
72. Khatibikamal, V.; Torabian, A.; Janpoor, F.; Hoshyaripour, G. Fluoride removal from industrial wastewater using electrocoagulation and its adsorption kinetics. *J. Hazard. Mater.* **2010**, *179*, 276–280. [CrossRef]
73. Drouiche, N.; Aoudj, S.; Hecini, M.; Ghaffour, N.; Lounici, H.; Mameri, N. Study on the treatment of photovoltaic wastewater using electrocoagulation: Fluoride removal with aluminium electrodes—Characteristics of products. *J. Hazard. Mater.* **2009**, *169*, 65–69. [CrossRef]
74. Drouiche, N.; Djouadi-Belkada, F.; Ouslimane, T.; Kefaifi, A.; Fathi, J.; Ahmetovic, E. Photovoltaic solar cells industry wastewater treatment. *Desalination Water Treat.* **2013**, *51*, 5965–5973. [CrossRef]
75. Huang, C. Precipitate flotation of fluoride-containing wastewater from a semiconductor manufacturer. *Water Res.* **1999**, *33*, 3403–3412. [CrossRef]
76. Paudyal, H.; Pangeni, B.; Inoue, K.; Kawakita, H.; Ohto, K.; Alam, S. Removal of Fluoride from Aqueous Solution by Using Porous Resins Containing Hydrated Oxide of Cerium(IV) and Zirconium(IV). *J. Chem. Eng. Jpn.* **2012**, *45*, 331–336. [CrossRef]
77. Hu, C.-Y.; Lo, S.; Kuan, W.; Lee, Y. Removal of fluoride from semiconductor wastewater by electrocoagulation–flotation. *Water Res.* **2005**, *39*, 895–901. [CrossRef] [PubMed]
78. Warmadewanthi, B.; Liu, J.C. Selective separation of phosphate and fluoride from semiconductor wastewater. *Water Sci. Technol.* **2009**, *59*, 2047–2053. [CrossRef]
79. Neuwahl, F.; Cusano, G.; Benavides, J.G.; Holbrook, S.; Roudier, S. *Best Available Techniques (BAT) Reference Document for Waste Incineration*; European Union: Luxembourg, 2006.
80. Ghosh, A.; Mukherjee, K.; Ghosh, S.K.; Saha, B. Sources and toxicity of fluoride in the environment. *Res. Chem. Intermed.* **2012**, *39*, 2881–2915. [CrossRef]
81. Dreveton, A. Overview of the Fluorochemicals Industrial Sectors. *Procedia Eng.* **2016**, *138*, 240–247. [CrossRef]
82. Chlebna-Sokol, D. *Wpływ Ponadoptymalnych Stężeń Fluorków w Wodzie Pitnej na Rozwój Biologiczny i Stan Zdrowia Dzieci w Wieku Szkolnym*; Polska Akademia Nauk: Łódź, Poland, 1995.
83. Ndiaye, P.; Moulin, P.; Dominguez, L.; Millet, J.; Charbit, F. Removal of fluoride from electronic industrial effluent by RO membrane separation. *Desalination* **2005**, *173*, 25–32. [CrossRef]
84. Wang, L.; Zhang, Y.; Sun, N.; Sun, W.; Hu, Y.; Tang, H. Precipitation Methods Using Calcium-Containing Ores for Fluoride Removal in Wastewater. *Minerals* **2019**, *9*, 511. [CrossRef]
85. Liu, Y.; Fan, Q.; Wang, S.; Liu, Y.; Zhou, A.; Fan, L. Adsorptive removal of fluoride from aqueous solutions using Al-humic acid-La aerogel composites. *Chem. Eng. J.* **2016**, *306*, 174–185. [CrossRef]
86. Zeng, G.; Ling, B.; Li, Z.; Luo, S.; Xu, X.; Guan, Q. Fluorine removal and calcium fluoride recovery from rare-earth smelting wastewater using fluidized bed crystallization process. *J. Hazard. Mater.* **2019**, *373*, 313–320. [CrossRef]
87. Raghav, S.; Nehra, S.; Kumar, D. Adsorptive removal studies of fluoride in aqueous system by bimetallic oxide incorporated in cellulose. *Process. Saf. Environ. Prot. Trans. Inst. Chem. Eng. Part B* **2019**, *127*, 211–225. [CrossRef]

88. Chigondo, M.; Paumo, H.K.; Bhaumik, M.; Pillay, K.; Maity, A. Hydrous CeO_2-Fe_3O_4 decorated polyaniline fibers nanocomposite for effective defluoridation of drinking water. *J. Colloid Interface Sci.* 2018, *532*, 500–516. [CrossRef] [PubMed]
89. Sarkar, M.; Banerjee, A.; Pramanick, P.P.; Sarkar, A.R. Use of laterite for the removal of fluoride from contaminated drinking water. *J. Colloid Interface Sci.* 2006, *302*, 432–441. [CrossRef] [PubMed]
90. Oguz, E. Adsorption of fluoride on gas concrete materials. *J. Hazard. Mater.* 2005, *117*, 227–233. [CrossRef]
91. Alagumuthu, G.; Veeraputhiran, V.; Venkataraman, R. Adsorption Isotherms on Fluoride Removal: Batch Techniques. *Arch. Appl. Sci. Res.* 2010, *2*, 170–185. Available online: https://www.academia.edu/2555273/Adsorption_Isotherms_on_Fluoride_Removal_Batch_Techniques (accessed on 9 January 2022).
92. Chaudhary, M.; Maiti, A. Defluoridation by highly efficient calcium hydroxide nanorods from synthetic and industrial wastewater. *Colloids Surf. A Physicochem. Eng. Asp.* 2019, *561*, 79–88. [CrossRef]
93. Venditti, F.; Cuomo, F.; Giansalvo, G.; Giustini, M.; Cinelli, G.; Lopez, F. Fluorides decontamination by means of Aluminum polychloride based commercial coagulant. *J. Water Process. Eng.* 2018, *26*, 182–186. [CrossRef]
94. Tolkou, A.K.; Mitrakas, M.; Katsoyiannis, I.A.; Ernst, M.; Zouboulis, A.I. Fluoride removal from water by composite Al/Fe/Si/Mg pre-polymerized coagulants: Characterization and application. *Chemosphere* 2019, *231*, 528–537. [CrossRef]
95. Turner, B.D.; Binning, P.; Stipp, S.L.S. Fluoride Removal by Calcite: Evidence for Fluorite Precipitation and Surface Adsorption. *Environ. Sci. Technol.* 2005, *39*, 9561–9568. [CrossRef]
96. El-Gohary, F.; Tawfik, A.; Mahmoud, U. Comparative study between chemical coagulation/precipitation (C/P) versus coagulation/dissolved air flotation (C/DAF) for pre-treatment of personal care products (PCPs) wastewater. *Desalination* 2010, *252*, 106–112. [CrossRef]
97. Saha, S. Treatment of aqueous effluent for fluoride removal. *Water Res.* 1993, *27*, 1347–1350. [CrossRef]
98. Reardon, E.J.; Wang, Y. A Limestone Reactor for Fluoride Removal from Wastewaters. *Environ. Sci. Technol.* 2000, *34*, 3247–3253. [CrossRef]
99. Gong, W.-X.; Qu, J.-H.; Liu, R.-P.; Lan, H.-C. Effect of aluminum fluoride complexation on fluoride removal by coagulation. *Colloids Surf. A Physicochem. Eng. Asp.* 2012, *395*, 88–93. [CrossRef]
100. Herath, H.M.A.S.; Kawakami, T.; Tafu, M. Repeated Heat Regeneration of Bone Char for Sustainable Use in Fluoride Removal from Drinking Water. *Healthcare* 2018, *6*, 143. [CrossRef]
101. Asimeng, B.O.; Fianko, J.R.; Kaufmann, E.E.; Tiburu, E.K.; Hayford, C.F.; Anani, P.A.; Dzikunu, O.K. Preparation and characterization of hydroxyapatite from *Achatina achatina* snail shells: Effect of carbonate substitution and trace elements on defluoridation of water. *J. Asian Ceram. Soc.* 2018, *6*, 205–212. [CrossRef]
102. Kodama, H.; Kabay, N. Reactivity of inorganic anion exchanger BiPbO2(NO3) with fluoride ions in solution. *Solid State Ion.* 2001, *141-142*, 603–607. [CrossRef]
103. Chubar, N.; Samanidou, V.; Kouts, V.; Gallios, G.; Kanibolotsky, V.; Strelko, V.; Zhuravlev, I. Adsorption of fluoride, chloride, bromide, and bromate ions on a novel ion exchanger. *J. Colloid Interface Sci.* 2005, *291*, 67–74. [CrossRef]
104. Hänninen, K.; Kaukonen, A.M.; Murtomäki, L.; Hirvonen, J.T. Mechanistic evaluation of factors affecting compound loading into ion-exchange fibers. *Eur. J. Pharm. Sci. Off. J. Eur. Fed. Pharm. Sci.* 2007, *31*, 306–317. [CrossRef]
105. Ruixia, L.; Jinlong, G.; Hongxiao, T. Adsorption of Fluoride, Phosphate, and Arsenate Ions on a New Type of Ion Exchange Fiber. *J. Colloid Interface Sci.* 2002, *248*, 268–274. [CrossRef]
106. Meenakshi, S.; Viswanathan, N. Identification of selective ion-exchange resin for fluoride sorption. *J. Colloid Interface Sci.* 2007, *308*, 438–450. [CrossRef]
107. Paudyal, H.; Pangeni, B.; Inoue, K.; Kawakita, H.; Ohto, K.; Ghimire, K.N.; Alam, S. Preparation of novel alginate based anion exchanger from *Ulva japonica* and its application for the removal of trace concentrations of fluoride from water. *Bioresour. Technol.* 2013, *148*, 221–227. [CrossRef] [PubMed]
108. Guo, Q.; Tian, J. Removal of fluoride and arsenate from aqueous solution by hydrocalumite via precipitation and anion exchange. *Chem. Eng. J.* 2013, *231*, 121–131. [CrossRef]
109. Hichour, M.; Persin, F.; Molénat, J.; Sandeaux, J.; Gavach, C. Fluoride removal from diluted solutions by Donnan dialysis with anion-exchange membranes. *Desalination* 1999, *122*, 53–62. [CrossRef]
110. Amor, Z.; Malki, S.; Taky, M.; Bariou, B.; Mameri, N.; Elmidaoui, A. Optimization of fluoride removal from brackish water by electrodialysis. *Desalination* 1998, *120*, 263–271. [CrossRef]
111. Ahamad, K.U.; Mahanta, A.; Ahmed, S. Removal of Fluoride from Groundwater by Adsorption onto Brick Powder–Alum–Calcium-Infused Adsorbent. *Adv. Waste Manag.* 2019, 231–242. [CrossRef]
112. Hashim, K.S.; Shaw, A.; Al Khaddar, R.; Pedrola, M.O.; Phipps, D. Defluoridation of drinking water using a new flow column-electrocoagulation reactor (FCER)—Experimental, statistical, and economic approach. *J. Environ. Manag.* 2017, *197*, 80–88. [CrossRef]
113. Cui, H.; Qian, Y.; An, H.; Sun, C.; Zhai, J.; Li, Q. Electrochemical removal of fluoride from water by PAOA-modified carbon felt electrodes in a continuous flow reactor. *Water Res.* 2012, *46*, 3943–3950. [CrossRef]
114. Lin, J.-Y.; Raharjo, A.; Hsu, L.-H.; Shih, Y.-J.; Huang, Y.-H. Electrocoagulation of tetrafluoroborate (BF4−) and the derived boron and fluorine using aluminum electrodes. *Water Res.* 2019, *155*, 362–371. [CrossRef]
115. Tahaikt, M.; Achary, I.; Sahli, M.M.; Amor, Z.; Taky, M.; Alami, A.; Boughriba, A.; Hafsi, M.; Elmidaoui, A. Defluoridation of Moroccan groundwater by electrodialysis: Continuous operation. *Desalination* 2006, *189*, 215–220. [CrossRef]

116. Renuka, P.; Pushpanjali, K. Review on Defluoridation Techniques of Water. *Int. J. Eng. Sci.* **2013**, *2*, 86–94.
117. Sahli, M.M.; Annouar, S.; Tahaikt, M.; Mountadar, M.; Soufiane, A.; Elmidaoui, A. Fluoride removal for underground brackish water by adsorption on the natural chitosan and by electrodialysis. *Desalination* **2007**, *212*, 37–45. [CrossRef]
118. Zuo, Q.; Chen, X.; Li, W.; Chen, G. Combined electrocoagulation and electroflotation for removal of fluoride from drinking water. *J. Hazard. Mater.* **2008**, *159*, 452–457. [CrossRef] [PubMed]
119. Ergun, E.; Tor, A.; Cengeloglu, Y.; Kocak, I. Electrodialytic removal of fluoride from water: Effects of process parameters and accompanying anions. *Sep. Purif. Technol.* **2008**, *64*, 147–153. [CrossRef]
120. Kabay, N.; Arar, Ö.; Samatya, S.; Yüksel, Ü.; Yüksel, M. Separation of fluoride from aqueous solution by electrodialysis: Effect of process parameters and other ionic species. *J. Hazard. Mater.* **2008**, *153*, 107–113. [CrossRef] [PubMed]
121. Mameri, N.; Lounici, H.; Belhocine, D.; Grib, H.; Piron, D.; Yahiat, Y. Defluoridation of Sahara water by small plant electrocoagulation using bipolar aluminium electrodes. *Sep. Purif. Technol.* **2001**, *24*, 113–119. [CrossRef]
122. Arar, O.; Yavuz, E.; Yuksel, U.; Kabay, N. Separation of Low Concentration of Fluoride from Water by Electrodialysis (ED) in the Presence of Chloride and Sulfate Ions. *Sep. Sci. Technol.* **2009**, *44*, 1562–1573. [CrossRef]
123. Un, U.T.; Koparal, A.S.; Ogutveren, U.B. Fluoride removal from water and wastewater with a bach cylindrical electrode using electrocoagulation. *Chem. Eng. J.* **2013**, *223*, 110–115. [CrossRef]
124. Karabelas, A.; Yiantsios, S.; Metaxiotou, Z.; Andritsos, N.; Akiskalos, A.; Vlachopoulos, G.; Stavroulias, S. Water and materials recovery from fertilizer industry acidic effluents by membrane processes. *Desalination* **2001**, *138*, 93–102. [CrossRef]
125. Sehn, P. Fluoride removal with extra low energy reverse osmosis membranes: Three years of large scale field experience in Finland. *Desalination* **2008**, *223*, 73–84. [CrossRef]
126. Guo, L.; Hunt, B.J.; Santschi, P.H. Ultrafiltration behavior of major ions (Na, Ca, Mg, F, Cl, and SO_4) in natural waters. *Water Res.* **2001**, *35*, 1500–1508. [CrossRef]
127. Lhassani, A.; Rumeau, M.; Benjelloun, D.; Pontie, M. Selective demineralization of water by nanofiltration Application to the defluorination of brackish water. *Water Res.* **2001**, *35*, 3260–3264. [CrossRef]
128. Hu, K.; Dickson, J.M. Nanofiltration membrane performance on fluoride removal from water. *J. Membr. Sci.* **2006**, *279*, 529–538. [CrossRef]
129. Malaisamy, R.; Talla-Nwafo, A.; Jones, K.L. Polyelectrolyte modification of nanofiltration membrane for selective removal of monovalent anions. *Sep. Purif. Technol.* **2011**, *77*, 367–374. [CrossRef]
130. Ghosh, D.; Sinha, M.; Purkait, M. A comparative analysis of low-cost ceramic membrane preparation for effective fluoride removal using hybrid technique. *Desalination* **2013**, *327*, 2–13. [CrossRef]
131. Chakrabortty, S.; Roy, M.; Pal, P. Removal of fluoride from contaminated groundwater by cross flow nanofiltration: Transport modeling and economic evaluation. *Desalination* **2013**, *313*, 115–124. [CrossRef]
132. Yadav, K.K.; Kumar, S.; Pham, Q.B.; Gupta, N.; Rezania, S.; Kamyab, H.; Yadav, S.; Vymazal, J.; Kumar, V.; Tri, D.Q.; et al. Fluoride contamination, health problems and remediation methods in Asian groundwater: A comprehensive review. *Ecotoxicol. Environ. Saf.* **2019**, *182*, 109362. [CrossRef]
133. Jeihanipour, A.; Shen, J.; Abbt-Braun, G.; Huber, S.A.; Mkongo, G.; Schäfer, A.I. Seasonal variation of organic matter characteristics and fluoride concentration in the Maji ya Chai River (Tanzania): Impact on treatability by nanofiltration/reverse osmosis. *Sci. Total Environ.* **2018**, *637-638*, 1209–1220. [CrossRef]
134. Grzegorzek, M.; Majewska-Nowak, K. The use of micellar-enhanced ultrafiltration (MEUF) for fluoride removal from aqueous solutions. *Sep. Purif. Technol.* **2018**, *195*, 1–11. [CrossRef]
135. Sequeira, E.A.T.; Miranda, V.M.; Solache-Ríos, M.; Hernández, I.L. Aluminum and lanthanum effects in natural materials on the adsorption of fluoride ions. *J. Fluor. Chem.* **2013**, *148*, 6–13. [CrossRef]
136. Saxena, A.; Patel, A. Role of Bioremediation as a Low-Cost Adsorbent for Excessive Fluoride Removal in Groundwater. In *Handbook of Environmental Materials Management*; Springer: Berlin/Heidelberg, Germany, 2018; pp. 1–32. [CrossRef]
137. Mohan, D.; Singh, K.P.; Singh, V.K. Wastewater treatment using low cost activated carbons derived from agricultural byproducts—A case study. *J. Hazard. Mater.* **2008**, *152*, 1045–1053. [CrossRef]
138. Alagumuthu, G.; Rajan, M. Kinetic and equilibrium studies on fluoride removal by zirconium (IV): Impregnated groundnut shell carbon. *Chem. Ind.* **2010**, *64*, 295–304. [CrossRef]
139. Alagumuthu, G.; Veeraputhiran, V.; Venkataraman, R. Fluoride sorption using *Cynodon dactylon* based activated carbon. *Chem. Ind.* **2011**, *65*, 23–35. [CrossRef]
140. Alagumuthu, G.; Rajan, M. Equilibrium and kinetics of adsorption of fluoride onto zirconium impregnated cashew nut shell carbon. *Chem. Eng. J.* **2010**, *158*, 451–457. [CrossRef]
141. Daifullah, A.A.M.; Yakout, S.M.; A Elreefy, S. Adsorption of fluoride in aqueous solutions using $KMnO_4$-modified activated carbon derived from steam pyrolysis of rice straw. *J. Hazard. Mater.* **2007**, *147*, 633–643. [CrossRef] [PubMed]
142. Hernández-Montoya, V.; Ramírez-Montoya, L.A.; Bonilla-Petriciolet, A.; Montes-Moran, M.A. Optimizing the removal of fluoride from water using new carbons obtained by modification of nut shell with a calcium solution from egg shell. *Biochem. Eng. J.* **2012**, *62*, 1–7. [CrossRef]
143. Meenakshi, S. *Studies on Defluoridation of Water with a Few Adsorbents and Development of an Indigenous Defluoridation Unit for Do-mestic Use*; The Gandhigram Rural Institute: Tamil Nadu, India, 1992.

144. Malay, D.K.; Salim, A.J. Salim, Comparative Study of Batch Adsorption of Fluoride Using Commercial and Natural Adsorbent. *Res. J. Chem. Sci.* **2011**, *1*, 68–75.
145. Viswanathan, N.; Meenakshi, S. Role of metal ion incorporation in ion exchange resin on the selectivity of fluoride. *J. Hazard. Mater.* **2009**, *162*, 920–930. [CrossRef]
146. Viswanathan, N.; Meenakshi, S. Effect of metal ion loaded in a resin towards fluoride retention. *J. Fluor. Chem.* **2008**, *129*, 645–653. [CrossRef]
147. Vardhan, C.V.; Karthikeyan, J. Removal of Fluoride from Water Using Low-Cost Materials. In Proceedings of the Fifteenth International Water Technology Conference, IWTC-15, Alexandria, Egypt, 28–30 May 2011.
148. Coetzee, P.; Coetzee, L.; Puka, R.; Mubenga, S. Characterisation of selected South African clays for defluoridation of natural waters. *Water SA* **2004**, *29*, 331–338. [CrossRef]
149. Yadav, A.K.; Kaushik, C.P.; Haritash, A.K.; Kansal, A.; Rani, N. Defluoridation of groundwater using brick powder as an adsorbent. *J. Hazard. Mater.* **2006**, *128*, 289–293. [CrossRef]
150. Malakootian, M.; Moosazadeh, M.; Yousefi, N.; Fatehizadeh, A. Fluoride removal from aqueous solution by pumice: Case study on Kuhbonan water. *Afr. J. Environ. Sci. Technol.* **2011**, *5*, 299–306. [CrossRef]
151. Chidambaram, S.; Ramanathan, A.; Vasudevan, S. Fluoride removal studies in water using natural materials: Technical note. *Water SA* **2004**, *29*, 339–344. [CrossRef]
152. Maliyekkal, S.M.; Shukla, S.; Philip, L.; Nambi, I.M. Enhanced fluoride removal from drinking water by magnesia-amended activated alumina granules. *Chem. Eng. J.* **2008**, *140*, 183–192. [CrossRef]
153. Tripathy, S.S.; Raichur, A. Abatement of fluoride from water using manganese dioxide-coated activated alumina. *J. Hazard. Mater.* **2008**, *153*, 1043–1051. [CrossRef] [PubMed]
154. Lavecchia, R.; Medici, F.; Piga, L.; Rinaldi, G.; Zuorro, A. Fluoride removal from water by adsorption on a high-alumina content bauxite. *Chem. Eng. Trans.* **2012**, *26*, 225–230. [CrossRef]
155. Teutli-Sequeira, A.; Solache-Rios, M.; Balderas-Hernández, P. Modification Effects of Hematite with Aluminum Hydroxide on the Removal of Fluoride Ions from Water. *Water Air Soil Pollut.* **2011**, *223*, 319–327. [CrossRef]
156. Shan, Y.; Guo, H. Fluoride adsorption on modified natural siderite: Optimization and performance. *Chem. Eng. J.* **2013**, *223*, 183–191. [CrossRef]
157. Sajidu, S.; Kayira, C.; Masamba, W.; Mwatseteza, J. Defluoridation of Groundwater Using Raw Bauxite: Rural Domestic Defluoridation Technology. *Environ. Nat. Resour. Res.* **2012**, *2*, 1. [CrossRef]
158. Goswami, D.; Das, A.K. Removal of fluoride from drinking water using a modified fly ash adsorbent. *J. Sci. Ind. Res.* **2006**, *65*, 77–79.
159. Sundaram, C.S.; Viswanathan, N.; Meenakshi, S. Defluoridation chemistry of synthetic hydroxyapatite at nano scale: Equilibrium and kinetic studies. *J. Hazard. Mater.* **2008**, *155*, 206–215. [CrossRef]
160. Sundaram, C.S.; Viswanathan, N.; Meenakshi, S. Uptake of fluoride by nano-hydroxyapatite/chitosan, a bioinorganic composite. *Bioresour. Technol.* **2008**, *99*, 8226–8230. [CrossRef] [PubMed]
161. Salifu, A.; Petrusevski, B.; Ghebremichael, K.; Modestus, L.; Buamah, R.; Aubry, C.; Amy, G. Aluminum (hydr)oxide coated pumice for fluoride removal from drinking water: Synthesis, equilibrium, kinetics and mechanism. *Chem. Eng. J.* **2013**, *228*, 63–74. [CrossRef]
162. Tembhurkar, A.R.; Dongre, S. Studies on fluoride removal using adsorption process. *J. Environ. Sci. Eng. Technol.* **2006**, *48*, 151–156.
163. Nath, S.K.; Dutta, R.K. Fluoride removal from water using crushed limestone. *Indian J. Chem. Technol.* **2010**, *17*, 120–125.
164. Bhargava, D.; Killedar, D. Fluoride adsorption on fishbone charcoal through a moving media adsorber. *Water Res.* **1992**, *26*, 781–788. [CrossRef]
165. Çengeloglu, Y. Removal of fluoride from aqueous solution by using red mud. *Sep. Purif. Technol.* **2002**, *28*, 81–86. [CrossRef]
166. Asgari, G.; Roshani, B.; Ghanizadeh, G. The investigation of kinetic and isotherm of fluoride adsorption onto functionalize pumice stone. *J. Hazard. Mater.* **2012**, *217–218*, 123–132. [CrossRef]
167. Chen, N.; Zhang, Z.; Feng, C.; Li, M.; Zhu, D.; Sugiura, N. Studies on fluoride adsorption of iron-impregnated granular ceramics from aqueous solution. *Mater. Chem. Phys.* **2011**, *125*, 293–298. [CrossRef]
168. Kamble, S.P.; Jagtap, S.; Labhsetwar, N.K.; Thakare, D.; Godfrey, S.; Devotta, S.; Rayalu, S.S. Defluoridation of drinking water using chitin, chitosan and lanthanum-modified chitosan. *Chem. Eng. J.* **2007**, *129*, 173–180. [CrossRef]
169. Viswanathan, N.; Meenakshi, S. Development of chitosan supported zirconium(IV) tungstophosphate composite for fluoride removal. *J. Hazard. Mater.* **2010**, *176*, 459–465. [CrossRef]
170. Jagtap, S.; Thakre, D.; Wanjari, S.; Kamble, S.; Labhsetwar, N.; Rayalu, S. New modified chitosan-based adsorbent for defluoridation of water. *J. Colloid Interface Sci.* **2009**, *332*, 280–290. [CrossRef] [PubMed]
171. Sujana, M.; Mishra, A.; Acharya, B. Hydrous ferric oxide doped alginate beads for fluoride removal: Adsorption kinetics and equilibrium studies. *Appl. Surf. Sci.* **2013**, *270*, 767–776. [CrossRef]
172. Liang, P.; Zhang, Y.; Wang, D.; Xu, Y.; Luo, L. Preparation of mixed rare earths modified chitosan for fluoride adsorption. *J. Rare Earths* **2013**, *31*, 817–822. [CrossRef]
173. Davila-Rodriguez, J.L.; Escobar-Barrios, V.; Rangel-Mendez, J.R. Removal of fluoride from drinking water by a chitin-based biocomposite in fixed-bed columns. *J. Fluor. Chem.* **2012**, *140*, 99–103. [CrossRef]

174. Swain, S.K.; Patnaik, T.; Patnaik, P.; Jha, U.; Dey, R. Development of new alginate entrapped Fe(III)–Zr(IV) binary mixed oxide for removal of fluoride from water bodies. *Chem. Eng. J.* **2013**, *215-216*, 763–771. [CrossRef]
175. Shams, M.; Nabizadeh Nodehi, R.; Hadi Dehghani, M.; Younesian, M.; Hossein Mahvia, A. Efficiency of granular ferric hydroxide (GFH) for removal of fluoride from water. *Fluoride* **2010**, *43*, 61–66.
176. Chai, L.; Wang, Y.; Zhao, N.; Yang, W.; You, X. Sulfate-doped Fe_3O_4/Al_2O_3 nanoparticles as a novel adsorbent for fluoride removal from drinking water. *Water Res.* **2013**, *47*, 4040–4049. [CrossRef]
177. Liu, R.; Gong, W.; Lan, H.; Yang, T.; Liu, H.; Qu, J. Simultaneous removal of arsenate and fluoride by iron and aluminum binary oxide: Competitive adsorption effects. *Sep. Purif. Technol.* **2012**, *92*, 100–105. [CrossRef]
178. García-Sánchez, J.; Solache-Ríos, M.; Miranda, V.M.; Morelos, C.S. Removal of fluoride ions from drinking water and fluoride solutions by aluminum modified iron oxides in a column system. *J. Colloid Interface Sci.* **2013**, *407*, 410–415. [CrossRef]
179. Kang, D.; Yu, X.; Tong, S.; Ge, M.; Zuo, J.; Cao, C.; Song, W. Performance and mechanism of Mg/Fe layered double hydroxides for fluoride and arsenate removal from aqueous solution. *Chem. Eng. J.* **2013**, *228*, 731–740. [CrossRef]
180. Wu, H.-X.; Wang, T.-J.; Chen, L.; Jin, Y.; Zhang, Y.; Dou, X.-M. Granulation of Fe–Al–Ce hydroxide nano-adsorbent by immobilization in porous polyvinyl alcohol for fluoride removal in drinking water. *Powder Technol.* **2011**, *209*, 92–97. [CrossRef]
181. Dou, X.; Zhang, Y.; Wang, H.; Wang, T.; Wang, Y. Performance of granular zirconium–iron oxide in the removal of fluoride from drinking water. *Water Res.* **2011**, *45*, 3571–3578. [CrossRef] [PubMed]
182. Viswanathan, N.; Prabhu, S.M.; Meenakshi, S. Development of amine functionalized co-polymeric resins for selective fluoride sorption. *J. Fluor. Chem.* **2013**, *153*, 143–150. [CrossRef]
183. Ganvir, V.; Das, K. Removal of fluoride from drinking water using aluminum hydroxide coated rice husk ash. *J. Hazard. Mater.* **2011**, *185*, 1287–1294. [CrossRef]
184. Mourabet, M.; El Rhilassi, A.; El Boujaady, H.; Bennani-Ziatni, M.; El Hamri, R.; Taitai, A. Removal of fluoride from aqueous solution by adsorption on hydroxyapatite (HAp) using response surface methodology. *J. Saudi Chem. Soc.* **2015**, *19*, 603–615. [CrossRef]
185. Mourabet, M.; El Rhilassi, A.; El Boujaady, H.; Bennani-Ziatni, M.; El Hamri, R.; Taitai, A. Removal of fluoride from aqueous solution by adsorption on *Apatitic tricalcium* phosphate using Box–Behnken design and desirability function. *Appl. Surf. Sci.* **2012**, *258*, 4402–4410. [CrossRef]
186. Zhang, D.; Luo, H.; Zheng, L.; Wang, K.; Li, H.; Wang, Y.; Feng, H. Utilization of waste phosphogypsum to prepare hydroxyapatite nanoparticles and its application towards removal of fluoride from aqueous solution. *J. Hazard. Mater.* **2012**, *241–242*, 418–426. [CrossRef]
187. Dou, X.; Mohan, D.; Pittman, C.U.; Yang, S. Remediating fluoride from water using hydrous zirconium oxide. *Chem. Eng. J.* **2012**, *198-199*, 236–245. [CrossRef]
188. Swain, S.K.; Patnaik, T.; Singh, V.; Jha, U.; Patel, R.; Dey, R. Kinetics, equilibrium and thermodynamic aspects of removal of fluoride from drinking water using meso-structured zirconium phosphate. *Chem. Eng. J.* **2011**, *171*, 1218–1226. [CrossRef]
189. Poursaberi, T.; Hassanisadi, M.; Torkestani, K.; Zare, M. Development of zirconium (IV)-metalloporphyrin grafted Fe_3O_4 nanoparticles for efficient fluoride removal. *Chem. Eng. J.* **2012**, *189-190*, 117–125. [CrossRef]
190. Swain, S.K.; Mishra, S.; Patnaik, T.; Patel, R.; Jha, U.; Dey, R. Fluoride removal performance of a new hybrid sorbent of Zr(IV)–ethylenediamine. *Chem. Eng. J.* **2012**, *184*, 72–81. [CrossRef]
191. Koilraj, P.; Kannan, S. Aqueous fluoride removal using ZnCr layered double hydroxides and their polymeric composites: Batch and column studies. *Chem. Eng. J.* **2013**, *234*, 406–415. [CrossRef]
192. Wang, J.; Xu, W.; Chen, L.; Jia, Y.; Wang, L.; Huang, X.-J.; Liu, J. Excellent fluoride removal performance by CeO_2–ZrO_2 nanocages in water environment. *Chem. Eng. J.* **2013**, *231*, 198–205. [CrossRef]
193. Chen, L.; Wang, T.-J.; Wu, H.-X.; Jin, Y.; Zhang, Y.; Dou, X. Optimization of a Fe–Al–Ce nano-adsorbent granulation process that used spray coating in a fluidized bed for fluoride removal from drinking water. *Powder Technol.* **2011**, *206*, 291–296. [CrossRef]
194. Zhao, B.; Zhang, Y.; Dou, X.; Wu, X.; Yang, M. Granulation of Fe–Al–Ce trimetal hydroxide as a fluoride adsorbent using the extrusion method. *Chem. Eng. J.* **2012**, *185–186*, 211–218. [CrossRef]
195. Sivasankar, V.; Murugesh, S.; Rajkumar, S.; Darchen, A. Cerium dispersed in carbon (CeDC) and its adsorption behavior: A first example of tailored adsorbent for fluoride removal from drinking water. *Chem. Eng. J.* **2013**, *214*, 45–54. [CrossRef]
196. Mandal, S.; Tripathy, S.; Padhi, T.; Sahu, M.K.; Patel, R.K. Removal efficiency of fluoride by novel Mg-Cr-Cl layered double hydroxide by batch process from water. *J. Environ. Sci.* **2013**, *25*, 993–1000. [CrossRef]
197. Zhang, T.; Li, Q.; Xiao, H.; Mei, Z.; Lu, H.; Zhou, Y. Enhanced fluoride removal from water by non-thermal plasma modified CeO_2/Mg–Fe layered double hydroxides. *Appl. Clay Sci.* **2013**, *72*, 117–123. [CrossRef]
198. Wajima, T.; Umeta, Y.; Narita, S.; Sugawara, K. Adsorption behavior of fluoride ions using a titanium hydroxide-derived adsorbent. *Desalination* **2009**, *249*, 323–330. [CrossRef]
199. Chen, L.; He, B.-Y.; He, S.; Wang, T.-J.; Su, C.-L.; Jin, Y. Fe—Ti oxide nano-adsorbent synthesized by co-precipitation for fluoride removal from drinking water and its adsorption mechanism. *Powder Technol.* **2012**, *227*, 3–8. [CrossRef]
200. Babaeivelni, K.; Khodadoust, A.P. Adsorption of fluoride onto crystalline titanium dioxide: Effect of pH, ionic strength, and co-existing ions. *J. Colloid Interface Sci.* **2013**, *394*, 419–427. [CrossRef] [PubMed]
201. Eskandarpour, A.; Onyango, M.S.; Ochieng, A.; Asai, S. Removal of fluoride ions from aqueous solution at low pH using schwertmannite. *J. Hazard. Mater.* **2008**, *152*, 571–579. [CrossRef] [PubMed]

202. Zhao, Y.; Li, X.; Liu, L.; Chen, F. Fluoride removal by Fe(III)-loaded ligand exchange cotton cellulose adsorbent from drinking water. *Carbohydr. Polym.* **2008**, *72*, 144–150. [CrossRef]
203. Yu, X.; Tong, S.; Ge, M.; Zuo, J. Removal of fluoride from drinking water by cellulose@hydroxyapatite nanocomposites. *Carbohydr. Polym.* **2013**, *92*, 269–275. [CrossRef]
204. Gogoi, P.K.; Baruah, R. Fluoride removal from water by adsorption on acid activated kaolinite clay. *Indian J. Chem. Technol.* **2008**, *15*, 500–503.
205. Meenakshi, S.; Sundaram, C.S.; Sukumar, R. Enhanced fluoride sorption by mechanochemically activated kaolinites. *J. Hazard. Mater.* **2008**, *153*, 164–172. [CrossRef]
206. Guo, Q.; Reardon, E.J. Fluoride removal from water by meixnerite and its calcination product. *Appl. Clay Sci.* **2012**, *56*, 7–15. [CrossRef]
207. Suzuki, T.; Nakamura, A.; Niinae, M.; Nakata, H.; Fujii, H.; Tasaka, Y. Immobilization of fluoride in artificially contaminated kaolinite by the addition of commercial-grade magnesium oxide. *Chem. Eng. J.* **2013**, *233*, 176–184. [CrossRef]
208. Lebedynets, M.; Sprynskyy, M.; Sakhnyuk, I.; Zbytniewski, R.; Golembiewski, R.; Buszewski, B. Adsorption of Ammonium Ions onto a Natural Zeolite: Transcarpathian Clinoptilolite. *Adsorpt. Sci. Technol.* **2016**, *22*, 731–741. [CrossRef]
209. Erdem, E.; Karapinar, N.; Donat, R. The removal of heavy metal cations by natural zeolites. *J. Colloid Interface Sci.* **2004**, *280*, 309–314. [CrossRef]
210. Rahmani, A.; Nouri, J.; Ghadiri, S.K.; Mahvi, A.; Zare, M.R. Adsorption of fluoride from water by Al^{3+} and Fe^{3+} pretreated natural Iranian zeolites. *Int. J. Environ. Res.* **2010**, *4*, 607–614. [CrossRef]
211. Wang, S.; Peng, Y. Natural zeolites as effective adsorbents in water and wastewater treatment. *Chem. Eng. J.* **2010**, *156*, 11–24. [CrossRef]
212. Sun, Y.; Fang, Q.; Dong, J.; Cheng, X.; Xu, J. Removal of fluoride from drinking water by natural stilbite zeolite modified with Fe(III). *Desalination* **2011**, *277*, 121–127. [CrossRef]
213. Gómez-Hortigüela, L.; Pérez-Pariente, J.; García, R.; Chebude, Y.; Díaz, I. Natural zeolites from Ethiopia for elimination of fluoride from drinking water. *Sep. Purif. Technol.* **2013**, *120*, 224–229. [CrossRef]
214. Sasaki, K.; Fukumoto, N.; Moriyama, S.; Yu, Q.; Hirajima, T. Chemical regeneration of magnesium oxide used as a sorbent for fluoride. *Sep. Purif. Technol.* **2012**, *98*, 24–30. [CrossRef]
215. Tor, A.; Danaoglu, N.; Arslan, G.; Cengeloglu, Y. Removal of fluoride from water by using granular red mud: Batch and column studies. *J. Hazard. Mater.* **2009**, *164*, 271–278. [CrossRef]
216. Ning, W.; Zhao-Kun, L.; Jun, W.; Li, S.; Ying, Z.; Jie-Wei, W. Preparation of Modified Red Mud with Aluminum and Its Adsorption Characteristics on Fluoride Removal. *Chin. J. Inorg. Chem.* **2009**, *25*, 849–854.
217. Lv, G.; Wu, L.; Liao, L.; Zhang, Y.; Li, Z. Preparation and characterization of red mud sintered porous materials for water defluoridation. *Appl. Clay Sci.* **2013**, *74*, 95–101. [CrossRef]
218. Tor, A.; Danaoglu, N.; Arslan, G.; Cengeloglu, Y. Removal of Fluoride From Drinking Water Using Red Mud Introduction. *Int. J. Sci. Technol. Res.* **2013**, *2*, 120–122.
219. Zhu, S.; Zhu, D.; Wang, X. Removal of fluorine from red mud (bauxite residue) by electrokinetics. *Electrochim. Acta* **2017**, *242*, 300–306. [CrossRef]
220. Kemer, B.; Ozdes, D.; Gundogdu, A.; Bulut, V.N.; Duran, C.; Soylak, M. Removal of fluoride ions from aqueous solution by waste mud. *J. Hazard. Mater.* **2009**, *168*, 888–894. [CrossRef] [PubMed]
221. Sujana, M.; Thakur, R.; Rao, S. Removal of Fluoride from Aqueous Solution by Using Alum Sludge. *J. Colloid Interface Sci.* **1998**, *206*, 94–101. [CrossRef] [PubMed]
222. Nigussie, W.; Zewge, F.; Chandravanshi, B. Removal of excess fluoride from water using waste residue from alum manufacturing process. *J. Hazard. Mater.* **2007**, *147*, 954–963. [CrossRef] [PubMed]
223. Mahramanlioglu, M.; Kizilcikli, I.; Bicer, I.O. Adsorption of fluoride from aqueous solution by acid treated spent bleaching earth. *J. Fluor. Chem.* **2002**, *115*, 41–47. [CrossRef]
224. Malakootian, M.; Fatehizadeh, A.; Yousefi, N.; Ahmadian, M.; Moosazadeh, M. Fluoride removal using Regenerated Spent Bleaching Earth (RSBE) from groundwater: Case study on Kuhbonan water. *Desalination* **2011**, *277*, 244–249. [CrossRef]
225. Nemade, P.D.; Vasudeva Rao, A.; Alappat, B.J. Removal of fluorides from water using low cost adsorbents. *Water Supply* **2002**, *2*, 311–317. [CrossRef]
226. Xue, Y.; Hou, H.; Zhu, S.; Zha, J. Utilization of municipal solid waste incineration ash in stone mastic asphalt mixture: Pavement performance and environmental impact. *Constr. Build. Mater.* **2009**, *23*, 989–996. [CrossRef]
227. Geethamani, C.; Ramesh, S.; Gandhimathi, R.; Nidheesh, P. Alkali-treated fly ash for the removal of fluoride from aqueous solutions. *Desalin. Water Treat.* **2013**, *52*, 3466–3476. [CrossRef]
228. Ramesh, S.; Gandhimathi, R.; Nidheesh, P.; Taywade, M. Batch and Column Operations for the Removal of Fluoride from Aqueous Solution Using Bottom Ash. *Environ. Res. Eng. Manag.* **2012**, *60*, 12–20. [CrossRef]
229. Zhang, G.; Sun, G.; Chen, Z.; Evrendilek, F.; Liu, J. Water-soluble fluorine detoxification mechanisms of spent potlining incineration in response to calcium compounds. *Environ. Pollut.* **2020**, *266*, 115420. [CrossRef]
230. Xu, X.; Li, Q.; Cui, H.; Pang, J.; An, H.; Wang, W.; Zhai, J. Column-mode fluoride removal from aqueous solution by magnesia-loaded fly ash cenospheres. *Environ. Technol.* **2012**, *33*, 1409–1415. [CrossRef] [PubMed]

231. Mondal, N.K.; Bhaumik, R.; Baur, T.; Das, B.; Roy, P.; Datta, P.R.A.J.K. Studies on Defluoridation of Water by Tea Ash: An Unconventional Biosorbent. *Chem. Sci. Trans.* **2012**, *1*, 239–256. [CrossRef]
232. Gupta, N.; Gupta, V.; Singh, A.P.; Singh, R.P. Defluoridation of Groundwater using Low Cost Adsorbent like Bagasse Dust, Aluminium Treated Bagasse Flyash, Bone Powder and Shell Powder. *Bonfring Int. J. Ind. Eng. Manag. Sci.* **2014**, *4*, 72–75. [CrossRef]
233. Jadhav, A.S.; Jadhav, M.V. Use of Maize Husk Fly Ash as an Adsorbent for Removal of Fluoride from Water. *Int. J. Recent Dev. Eng. Technol.* **2014**, *2*, 2. Available online: www.ijrdet.com (accessed on 25 July 2021).
234. Gupta, V.K.; Ali, I.; Saini, V.K. Defluoridation of wastewaters using waste carbon slurry. *Water Res.* **2007**, *41*, 3307–3316. [CrossRef]
235. Cinarli, A.; Bicer, O.; Mahramanlioglu, M. Removal of fluoride using the adsorbents produced from mining waste. *Fresenius Environ. Bull.* **2005**, *14*, 520–525.
236. Kumari, M.; Adhikari, K.; Dutta, S. Fluoride removal using shale: A mine waste. In *India Water Week-Efficient Water Management: Challenges and Opportunities*; Government of India, Ministry of Water Resources: New Delhi, India, 2013; p. 157.
237. Islam, M.; Patel, R. Thermal activation of basic oxygen furnace slag and evaluation of its fluoride removal efficiency. *Chem. Eng. J.* **2011**, *169*, 68–77. [CrossRef]
238. Lai, Y.D.; Liu, J.C. Fluoride Removal from Water with Spent Catalyst. *Sep. Sci. Technol.* **1996**, *31*, 2791–2803. [CrossRef]
239. Tsai, C.Y.; Liu, J.C. Fluoride removal from water with iron-coated spent catalyst. *Chinese J. Environ. Eng.* **1999**, *9*, 107–114.
240. Das, N.; Pattanaik, P.; Das, R. Defluoridation of drinking water using activated titanium rich bauxite. *J. Colloid Interface Sci.* **2005**, *292*, 1–10. [CrossRef]
241. Chaudhari, V.S.; Sasane, V.V. Investigation of Optimum Operating Parameters for Removal of Fluoride Using Naturally Available Geomaterial. *Int. J. Eng. Res. Technol.* **2014**, *3*, 2123–2128.
242. Kang, W.-H.; Kim, E.-I.; Park, J.-Y. Fluoride removal capacity of cement paste. *Desalination* **2007**, *202*, 38–44. [CrossRef]
243. Oguz, E. Equilibrium isotherms and kinetics studies for the sorption of fluoride on light weight concrete materials. *Colloids Surf. A Physicochem. Eng. Asp.* **2007**, *295*, 258–263. [CrossRef]
244. Kagne, S.; Jagtap, S.; Dhawade, P.; Kamble, S.; Devotta, S.; Rayalu, S. Hydrated cement: A promising adsorbent for the removal of fluoride from aqueous solution. *J. Hazard. Mater.* **2008**, *154*, 88–95. [CrossRef] [PubMed]
245. Ayoob, S.; Gupta, A. Insights into isotherm making in the sorptive removal of fluoride from drinking water. *J. Hazard. Mater.* **2008**, *152*, 976–985. [CrossRef] [PubMed]
246. Gao, S.; Cui, J.; Wei, Z. Study on the fluoride adsorption of various apatite materials in aqueous solution. *J. Fluor. Chem.* **2009**, *130*, 1035–1041. [CrossRef]
247. Gao, S.; Sun, R.; Wei, Z.; Zhao, H.; Li, H.; Hu, F.; Gao, S.; Sun, R. Size-dependent defluoridation ultrasonic and microwave combined technique. *J. Hazard. Mater.* **2011**, *185*, 29–37.
248. Poinern, G.E.J.; Ghosh, M.K.; Ng, Y.-J.; Issa, T.B.; Anand, S.; Singh, P. Defluoridation behavior of nanostructured hydroxyapatite synthesized through an ultrasonic and microwave combined technique. *J. Hazard. Mater.* **2011**, *185*, 29–37. [CrossRef]
249. Wang, Y.; Chen, N.; Wei, W.; Cui, J.; Wei, Z. Enhanced adsorption of fluoride from aqueous solution onto nanosized hydroxyapatite by low-molecular-weight organic acids. *Desalination* **2011**, *276*, 161–168. [CrossRef]
250. Nie, Y.; Hu, C.; Kong, C. Enhanced fluoride adsorption using Al (III) modified calcium hydroxyapatite. *J. Hazard. Mater.* **2012**, *233–234*, 194–199. [CrossRef]
251. Cheng, H. Reuse Research Progress on Waste Clay Brick. *Procedia Environ. Sci.* **2016**, *31*, 218–226. [CrossRef]
252. Kagne, S.; Jagtap, S.; Thakare, D.; Devotta, S.; Rayalu, S.S. Bleaching powder: A versatile adsorbent for the removal of fluoride from aqueous solution. *Desalination* **2009**, *243*, 22–31. [CrossRef]
253. Chen, N.; Zhang, Z.; Feng, C.; Li, M.; Zhu, D.; Chen, R.; Sugiura, N. An excellent fluoride sorption behavior of ceramic adsorbent. *J. Hazard. Mater.* **2009**, *183*, 460–465. [CrossRef] [PubMed]
254. Bibi, S.; Farooqi, A.; Hussain, K.; Haider, N. Evaluation of industrial based adsorbents for simultaneous removal of arsenic and fluoride from drinking water. *J. Clean. Prod.* **2015**, *87*, 882–896. [CrossRef]
255. Krysztafkiewicz, A.; Rager, B.; Maik, M. Silica recovery from waste obtained in hydrofluoric acid and aluminum fluoride production from fluosilicic acid. *J. Hazard. Mater.* **1996**, *48*, 31–49. [CrossRef]
256. Alusilica | Alufluor. Available online: https://www.alufluor.com/alusilica/ (accessed on 1 February 2018).
257. Şener, Ş.; Sener, E.; Karagüzel, R. Solid waste disposal site selection with GIS and AHP methodology: A case study in Senirkent–Uluborlu (Isparta) Basin, Turkey. *Environ. Monit. Assess.* **2010**, *173*, 533–554. [CrossRef]
258. Kaminskas, R.; Kubiliūtė, R. Artificial pozzolana from silica gel waste–clay–limestone composite. *Adv. Cem. Res.* **2014**, *26*, 155–168. [CrossRef]
259. Li, X.; He, S.; Feng, C.; Zhu, Y.; Pang, Y.; Hou, J.; Luo, K.; Liao, X. Non-Competitive and Competitive Adsorption of Pb^{2+}, Cd^{2+} and Zn^{2+} Ions onto SDS in Process of Micellar-Enhanced Ultrafiltration. *Sustainability* **2018**, *10*, 92. [CrossRef]
260. Zhu, Y.; Du, X.; Gao, C.; Yu, Z. Adsorption Behavior of Inorganic and Organic Phosphate by Iron Manganese Plaques on Reed Roots in Wetlands. *Sustainability* **2018**, *10*, 4578. [CrossRef]
261. Chen, L.; Chen, Q.; Rao, P.; Yan, L.; Shakib, A.; Shen, G. Formulating and Optimizing a Novel Biochar-Based Fertilizer for Simultaneous Slow-Release of Nitrogen and Immobilization of Cadmium. *Sustainability* **2018**, *10*, 2740. [CrossRef]
262. Zhang, X.; Wang, X.-Q.; Wang, D.-F. Immobilization of Heavy Metals in Sewage Sludge during Land Application Process in China: A Review. *Sustainability* **2017**, *9*, 2020. [CrossRef]

263. Kim, J.; Tae, S.; Kim, R. Theoretical Study on the Production of Environment-Friendly Recycled Cement Using Inorganic Construction Wastes as Secondary Materials in South Korea. *Sustainability* **2018**, *10*, 4449. [CrossRef]
264. Kurosaki, H. Reduction of Fluorine-containing Industrial Waste Using Aluminum-solubility Method. *Oki Tech. Rev.* **1998**, *63*, 53–56.
265. Adhikari, S.; Kayastha, M.S.; Ghimire, D.C.; Aryal, H.R.; Adhikary, S.; Takeuchi, T.; Murakami, K.; Kawashimo, Y.; Uchida, H.; Wakita, K.; et al. Improved Photovoltaic Properties of Heterojunction Carbon Based Solar Cell. *J. Surf. Eng. Mater. Adv. Technol.* **2013**, *3*, 178–183. [CrossRef]
266. Seshan, H. *Handbook of Thin—Film Deposition Processes and Techniques*, 2nd ed.; Noyes Publications: Park Ridge, NJ, USA, 2002.
267. Zueva, S.B.; BIrloaga, I.; Ferella, F.; Baturina, E.V.; Corradini, V.; Veglio, F. Mitigation of Fluorine-Containing Waste Resulting from Chemical Vapour Deposition Used In Manufacturing Of Silicon Solar Cells. *Processes* **2021**, *9*, 1745. [CrossRef]
268. Aldaco, R.; Garea, A.; Irabien, A. Calcium fluoride recovery from fluoride wastewater in a fluidized bed reactor. *Water Res.* **2007**, *41*, 810–818. [CrossRef]
269. Shin, C.-H.; Kim, J.-Y.; Kim, J.-Y.; Kim, H.-S.; Lee, H.-S.; Mohapatra, D.; Ahn, J.-W.; Ahn, J.-G.; Bae, W. A solvent extraction approach to recover acetic acid from mixed waste acids produced during semiconductor wafer process. *J. Hazard. Mater.* **2009**, *162*, 1278–1284. [CrossRef]
270. Lee, T.-C.; Liu, F.-J. Recovery of hazardous semiconductor-industry sludge as a useful resource. *J. Hazard. Mater.* **2009**, *165*, 359–365. [CrossRef]
271. Lee, T.-C.; Lin, K.-L.; Su, X.-W.; Lin, K.-K. Recycling CMP sludge as a resource in concrete. *Constr. Build. Mater.* **2012**, *30*, 243–251. [CrossRef]
272. Da, Y.; He, T.; Shi, C.; Wang, M.; Feng, Y. Potential of preparing cement clinker by adding the fluorine-containing sludge into raw meal. *J. Hazard. Mater.* **2021**, *403*, 123692. [CrossRef]
273. Olejarczyk, M.; Urbaniak, W.; Rykowska, I. Wapno posodowe jako składnik sorbentów jonów fluorkowych. In Proceedings of the II Ogólnopolska Przyrodnicza Konferencja Naukowa "Mater naturae", Online, 11 December 2020; pp. 41–42, ISBN 978-83-66261-98-3.
274. Wajima, T.; Rakovan, J.F. Removal of fluoride ions using calcined paper sludge. *J. Therm. Anal.* **2013**, *113*, 1027–1035. [CrossRef]
275. Frías, M.; García, R.; Vigil, R.; Ferreiro, S. Calcination of art paper sludge waste for the use as a supplementary cementing material. *Appl. Clay Sci.* **2008**, *42*, 189–193. [CrossRef]
276. Henry, C.L. Nitrogen Dynamics of Pulp and Paper Sludge Amendment to Forest Soils. *Water Sci. Technol.* **1991**, *24*, 417–425. [CrossRef]
277. Tripepi, R.R.; Zhang, X.; Campbell, A.G. Use of Raw and Composted Paper Sludge as a Soil Additive or Mulch for Cottonwood Plants. *Compos. Sci. Util.* **1996**, *4*, 26–36. [CrossRef]
278. Dell'Abate, M.T.; Benedetti, A.; Sequi, P. Thermal Methods of Organic Matter Maturation Monitoring During a Composting Process. *J. Therm. Anal.* **2000**, *61*, 389–396. [CrossRef]
279. Barriga, S.; Méndez, A.; Cámara, J.; Guerrero, F.; Gascó, G. Agricultural valorisation of de-inking paper sludge as organic amendment in different soils: Thermal study. *J. Therm. Anal. Calorim.* **2010**, *99*, 981–986. [CrossRef]
280. Méndez, A.; Barriga, S.; Guerrero, F.; Gascó, G. Thermal analysis of growing media obtained from mixtures of paper mill waste materials and sewage sludge. *J. Therm. Anal.* **2010**, *104*, 213–221. [CrossRef]
281. Rodríguez, O.; Frías, M.; de Rojas, M.I.S. Influence of the calcined paper sludge on the development of hydration heat in blended cement mortars. *J. Therm. Anal.* **2008**, *92*, 865–871. [CrossRef]
282. Melo, C.R.; Angioletto, E.; Riella, H.G.; Peterson, M.; Rocha, M.R.; Melo, A.R.; Silva, L.; Strugale, S. Production of metakaolin from industrial cellulose waste. *J. Therm. Anal.* **2012**, *109*, 1341–1345. [CrossRef]
283. Król, D.; Poskrobko, S. Waste and fuels from waste Part, I. Analysis of thermal decomposition. *J. Therm. Anal. Calorim.* **2012**, *109*, 619–628. [CrossRef]
284. Taş, S.; Yürüm, Y. Co-firing of biomass with coals: Part 2. Thermogravimetric kinetic analysis of co-combustion of fir (*Abies bornmulleriana*) wood with Beypazari lignite. *J. Therm. Anal. Calorim.* **2012**, *107*, 293–298. [CrossRef]
285. Méndez, A.; Barriga, S.; Saa, A.; Gascó, G. Removal of malachite green by adsorbents from paper industry waste materials: Thermal analysis. *J. Therm. Anal. Calorim.* **2010**, *99*, 993–998. [CrossRef]
286. Olejarczyk, M.; Urbaniak, W.; Rykowska, I. Reclamation materials based on post-soda lime. In *Practical Aspects of Remediation, Recultivation and Revitalization*; Kołwzan, B., Bukowski, Z., Eds.; Publishing House UKW: Bydgoszcz, Poland, 2022.
287. Waciński, W.; Olejarczyk, M.; Urbaniak, W.; Rykowska, I. Sorbent, especially for removing aqueous solutions of ions in the form of sparingly soluble salts and how it is maintained. *Pol. Pat. Appl.* **2022**, 440956.
288. Waciński, W.; Olejarczyk, M.; Urbaniak, W.; Rykowska, I. Method of removing fluoride ions from contaminated waters, especially sewage. *Pol. Pat. Appl.* **2022**, 440957.

Review

Influence of Rock Dust Additives as Fine Aggregate Replacement on Properties of Cement Composites—A Review

Magdalena Dobiszewska [1,*], Orlando Bagcal [2], Ahmet Beycioğlu [3], Dimitrios Goulias [4], Fuat Köksal [5], Maciej Niedostatkiewicz [6] and Hüsamettin Ürünveren [3]

1 Faculty of Civil and Environmental Engineering and Architecture, Bydgoszcz University of Science and Technology, 85-796 Bydgoszcz, Poland
2 School of Engineering, Construction Science and Management, Tarleton State University, Stephenville, TX 76402, USA; bagcal@tarleton.edu
3 Department of Civil Engineering, Adana Alparslan Türkeş Science and Technology University, Adana 01250, Turkey; abeycioglu@gmail.com (A.B.); hurunveren@atu.edu.tr (H.Ü.)
4 Department of Civil and Environmental Engineering, University of Maryland, College Park, MD 20742, USA; dgoulias@umd.edu
5 Department of Civil Engineering, Yozgat Bozot University, Yozgat 66900, Turkey; fuat.koksal@yobu.edu.tr
6 Faculty of Civil and Environmental Engineering, Gdańsk University of Technology, 80-233 Gdańsk, Poland; maciej.niedostatkiewicz@pg.edu.pl
* Correspondence: magdalena.dobiszewska@pbs.edu.pl

Citation: Dobiszewska, M.; Bagcal, O.; Beycioğlu, A.; Goulias, D.; Köksal, F.; Niedostatkiewicz, M.; Ürünveren, H. Influence of Rock Dust Additives as Fine Aggregate Replacement on Properties of Cement Composites—A Review. *Materials* 2022, *15*, 2947. https://doi.org/10.3390/ma15082947

Academic Editor: Malgorzata Ulewicz

Received: 4 March 2022
Accepted: 15 April 2022
Published: 18 April 2022

Publisher's Note: MDPI stays neutral with regard to jurisdictional claims in published maps and institutional affiliations.

Copyright: © 2022 by the authors. Licensee MDPI, Basel, Switzerland. This article is an open access article distributed under the terms and conditions of the Creative Commons Attribution (CC BY) license (https://creativecommons.org/licenses/by/4.0/).

Abstract: Concrete production consumes enormous amounts of fossil fuels, raw materials, and is energy intensive. Therefore, scientific research is being conducted worldwide regarding the possibility of using by-products in the production of concrete. The objective is not only to identify substitutes for cement clinker, but also to identify materials that can be used as aggregate in mortar and concrete productions. Among the potential alternative materials that can be used in cement composite production is rock dust of different geological origin. However, some adversarial effects may be encountered when using rock dust regarding the properties and durability of mortars and concrete. Therefore, comprehensive research is needed to evaluate the adequacy of rock dust use in cementitious composite production. This paper presents a comprehensive review of the scientific findings from past studies concerning the use of various geological origins of rock dust in the production of mortars and concrete. The influence of rock dust as a replacement of fine aggregates on cementitious composites was analyzed and evaluated. In this assessment and review, fresh concrete and mortar properties, i.e., workability, segregation, and bleeding, mechanical properties, and the durability of hardened concrete and mortar were considered.

Keywords: concrete; mortar; waste management; rock dust; concrete strength; concrete durability

1. Introduction

The industrialization and advancement of society generate large quantities of waste, creating a significant impact on the environment. Worldwide, waste generation has increased greatly in recent years and shows no signs of slowing down, other than the temporary effects of the COVID19 pandemic in construction. Each year, approximately 2.01 billion tons of municipal solid waste are generated worldwide, at least 33% of which is not well managed in terms of being environmentally friendly [1]. This means that the daily waste generation per capita worldwide ranges widely from 0.11 and 4.54 kg, with an average of 0.74 kg [1]. By 2050, the amount of municipal solid waste generated worldwide is expected to increase by approximately 70% and reach 3.4 billion metric tons. Less than 20% of waste is recycled each year, with vast amounts still sent to hazardous open landfill sites, thus posing a significant threat to the environment [1]. Coal combustion residues, such as blast furnace slag and rock dust of different geological origins and solid wastes produced in various industrial processes and mining sectors, constitute a waste group of their own. These industrial and mining wastes are complex and complicate the task of safe disposal and/or environmentally sound use in terms of quality and quantity. In most industrialized countries, waste disposal has become a cause for concern due to limited site conditions and stringent environmental standards. Thus, there is a pressing demand for authorities and agencies to ensure sufficient waste treatment and disposal services in order to attain better efficiency in waste management, particularly focusing on the reuse of waste materials.

Industrial and economic development has intensified activities in the construction industry, demanding an increase in the production of building materials. It is undeniable that concrete is the most extensively and generally used construction material worldwide. The continuous rapid growth of urban areas and infrastructure has led to an increase in demand, and this puts excessive pressure on the concrete industry for the production of large quantities of concrete to meet that demand. According to data compiled by the U.S. Geological Survey in 2019, the yearly global production of Portland cement was about 4.1 billion mt and it is forecasted to increase to 5.8 billion mt in 2050 [2–4]. Assuming that an average of 350 kg of cement is used per cubic meter of ordinary concrete, it can be estimated that the annual production of concrete in the world amounts to about 12 billion mt, which leads to an annual global average consumption rate of about 1.6 tons of concrete per person. The production of such an amount of concrete also requires 9 billion mt of aggregates and 2.2 billion mt of fresh water. Environmental constraints considerably decrease the scale of the natural deposits, which are used for the manufacture of cement and natural aggregates. The use of approximately 40% of the world's resources, such as water, fine and coarse aggregates, and wood, is the responsibility of the construction industry [5]. Not only over-exploitation and finite natural resources, but also the growing increase in the amount of various industrial waste and the lack of storage and landfill space have led to the development of extensive research throughout the years for assessing the potential use of these wastes in building materials production. Sustainable development principles, which is expressed as the rational management of non-renewable resources and the substituted use of these resources with recycling wastes, are compatible with such efforts.

There is a sort of material that can be utilized not only as a substitute for cement clinker, but also as a substitute for natural raw materials for the production of building materials [6–9]. In conjunction with the rapid increase in concrete production, the demands for natural aggregates are also increasing. To meet the increasing demand for aggregates, natural river sand, regarded as the most appropriate and commonly used fine aggregate in the production of mortar and concrete, is comprehensively exploited. This has led towards the uncontrollable exploitation of natural aggregate and serious environmental and economic concerns [10–12]. The mining of river sand has a very harmful influence on the environment, such as river flow, erosion levels, and aquatic habitats. Dredging a riverbed can destroy not only riverbanks, but also the habitat occupied by the bottom-dwelling organisms. The water may become cloudy due to the sediment that will form during the dredging operations, the fish may drown due to the sediment that will form, and the sunlight from which the aquatic vegetation feeds may be blocked [13–16]. Due to the massive depletion of river sand and strict environmental requirements, there is a shortage of sand for building materials production in many countries around the world [15,17–21]. Furthermore, there is currently a deficiency of good quality natural sand that may be used in concrete production in many regions of the world [15,17,18,22]. In some countries, preventive restrictions on the extraction of river sand have been introduced in order to protect valuable natural areas [13–16]. Considering that natural sand is about 35% of the concrete volume, combined with the increased demand from construction, this implies a serious shortage. Such shortage creates challenges for the concrete industry to identify alternative solutions. As mentioned earlier, rock dust may be a promising alternative for fine aggregate in mortars and concrete. These inert fillers, which may be composed of rock of different geological origin from the grinding process, can be used to enhance both particle size distribution and packing density. The optimization of the cementitious and aggregate blended systems has become the art of maximizing the use of various by-products and their positive synergetic effects [23].

The following sections of the manuscript present and synthesize the findings of various studies when rock dust is to be used as a replacement of fine aggregate in mortars and concrete.

2. Rock Dust Characteristics

Aggregate is defined by the European standard EN 12620 [24] as granular materials of natural, manufactured, and recycled origin used in the construction industry. Similar definitions are provided from various ASTM standards [25,26]. The aggregate grains are originally part of the parent rock and are divided into fine fractions by either natural factors (i.e., by weathering and abrasion) or artificially by mechanical grinding and crushing of the rock. Thus, many properties of aggregates, namely chemical and mineral composition, petrographic characteristics, density, water absorption, and strength, are dependent on the bedrock. On the other hand, other characteristics of the aggregates, such as shape, grain size, and surface texture, depend primarily on the technique used to crush the bedrock. The name natural aggregate covers all mineral aggregates that come from deposits, i.e., gravel and sand (fine aggregate) and pebbles obtained from loose rock materials, as well as crushed aggregates produced from mechanically treated rocks.

Approximately 4 billion mt of aggregates are produced and consumed in Europe and almost 91% of these aggregates are obtained from natural deposits [27]. In the US, the estimated annual output of construction aggregates produced for consumption was around 2.5 billion mt in 2020, with an estimated increase of 3–5% per year [2,28]. Crushed aggregates are mainly produced from igneous (basalt, melaphyre, diabase, porphyry, gabbro, and granite), metamorphic (amphibolite, gneiss, serpentinite), and sedimentary rocks (limestones, dolomites, sandstones, greywackes,). Large amounts of waste material in the form of rock dust are generated during the extraction and mechanical treatment of rocks, and then as a result of their categorization. Rock dust is also obtained during the aggregate production process for asphalt mixes. Waste dust accounts for around 5% of the aggregate mass used in the production of asphalt mixes. This means about 5000 tons of waste dust are produced annually in an average size asphalt mixture plant. Similar dusty waste is generated in dimensional stone factories, where mainly granite and marble are processed. They are used for paving stones, floors, cladding panels, tombstones, monuments, and statues. About 68 million tons of rock are processed annually in the stone industry around the world [29]. Countries where over a million tonnes of stone are processed annually include Italy, Portugal, Greece, France, Turkey, USA, Brazil, South African, India, and China [29]. When cutting, grinding, and polishing the rock blocks, water is used to cool and moisten the saws and polishing equipment. As a result of such a processing, semi-liquid sludge is formed as a waste in the amount of about 20–30% [30–36]. This waste is collected in settling tanks and then stored in the pulp form in landfills [18]. Part of the water contained in the pulp penetrates into the ground and paves the way for fine dust particles, which fill the voids and gaps in the ground. This causes significant soil permeability reduction, which negatively affects the soil fertility and groundwater level [18,37,38]. Part of the water is evaporated, and then dried dust is carried by the wind to the atmosphere, posing a threat to people and the environment [29,39].

According to the European standard EN 12620 [24], natural mineral dust is a fraction of aggregates with grain sizes smaller than 0.063 mm. Dust in crushed aggregate is generated from the crushing of the bedrock. On the other hand, uncrushed natural aggregate may contain dust resulting from natural weathering processes, as well as clays. These dusts can coat the surface of aggregate grains, which reduces the adhesion of the cement paste to the aggregate grains, resulting in a decrease of concrete strength. Moreover, clay grains may be adsorbed on the cement grain surface and create a water-impermeable coating, which delays hydration [29]. Additionally, due to the propensity of clay minerals to swelling due to the presence of water, the volume stability of mortars and concretes is influenced. Therefore, clay grains are not desirable as aggregate in concrete. Similar definitions and guidelines are provided by the American Concrete Institute, ACI, and the American Society for Testing and Materials, ASTM, for aggregates to be used in concrete, and in terms of classifications that are based on bulk density (i.e., unit weight) [25], mineralogical composition [40], and particle shape [26]. The National Stone, Sand, and Gravel Association, NSSGA [41], provides further guidance on the characteristics, physical properties, and mineralogical composition of rock dust.

Standards adopted by agencies in various countries contain guidelines regarding the limit content of dust in aggregate to be used in the production of concrete. The European standard, EN 12620 [24], presents the categories of maximum dust contents to be used by aggregate producers. The total dust content in the fine aggregate is considered to be harmless if it less than 3% by weight of the aggregate. The content of grains smaller than 75 μm in coarse aggregate cannot exceed 4%, according to the British standard BS-EN 12620 [42], whereas the content of fine aggregate depends on the concrete application and may not exceed 10–14%. On the other hand, the American guidelines specified in ASTM C33 [25] limit the maximum content of dust to 3% in the aggregate used for the production of concrete exposed to abrasion and 5% for other concretes. The aggregate must be washed before used in concrete in the case of excess on the permissible dust content. The fine-grained material obtained in this way is a dusty waste. It can be pointed out that the classification of dust as a deleterious additive in concrete solely on the basis of its grain size is incorrect [13,29]. As indicated earlier, mineral dusts that adhere to the surface of aggregate particles are not desirable in concrete. However, as shown later in this paper, when they are added to concrete or mortar, they can be beneficial in terms of the properties of hardened composites [18,33,43–50].

The characteristics of mineral dusts presented in this review were limited to dusts from limestone, marble, granite, and basalt rocks, which represent the most rock dust in concrete production and thus examined on past studies. Marble is formed by the metamorphism of limestone and dolomite over a wide range of pressures and temperatures. In the process of grain recrystallization, carbonate sedimentary rocks are transformed into crystalline rocks. Thus, petrographically, marble is a limestone. However, in published studies on the use of rock dust in cement composites, a distinction is made between limestone dust and marble dust. Therefore, in this paper, it was also decided to keep such a division. The oxide composition of rock dusts is presented in Table 1.

Table 1. Chemical composition of limestone [22,35,44,46,51,52], marble [30,35,38,48,53], granite [14,54], and basalt dust [46,48,55–57].

Oxide Composition	Limestone Dust	Marble Dust	Granite Dust	Basalt Dust
		[%]		
SiO_2	0.22–12.90	0.18–6.01	51.98–85.50	44.59–56.33
CaO	42.30–56.09	40.73–83.22	1.82–5.90	6.42–12.80
Al_2O_3	0.18–2.70	0.29–0.73	2.10–16.30	5.76–20.70
Fe_2O_3	0.11–2.00	0.05–0.80	0.40–27.89	4.14–17.73
MgO	0.20–9.64	0.23–15.21	0.58–2.50	2.99–8.73
Na_2O	0.01–0.54	0.06–2.44	2.02–3.69	0.84–4.11
K_2O	0.03–0.60	0.05–1.80	2.99–4.12	0.35–1.62
SO_3	0.01–0.88	0.08–0.56	0.05–1.80	0.02–1.10

Natural rock dust from rock fragmentation has a rough surface, sharp edges, and irregular shapes. Examples of the texture of limestone, marble, and basalt rock dusts observed under a scanning electron microscope are shown in Figure 1.

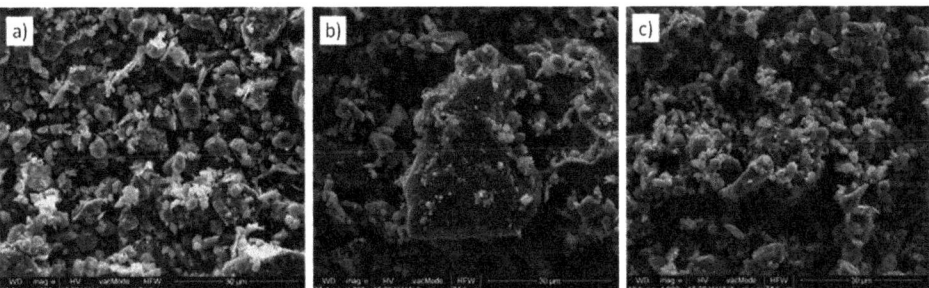

Figure 1. Microscopic images of rock dust: (**a**) limestone, (**b**) marble, (**c**) basalt.

A question arises whether, in light of the applicable standards, rock dust can be used and classified as additive for mortars and concrete. In the case of fillers or fine aggregates, their suitability for use in concrete is determined on the basis of EN 12620 [24] and EN 13055-1 [58] standards. Rock dust can therefore be considered as type I concrete additive, i.e., chemically inert mineral fillers. According to the EN 12620 [24] standard, the filler is the aggregates, the majority of which pass through a sieve of 0.063 mm and provide specific features by their addition to the construction materials. However, the European standards do not explicitly specify the permissible content of filler aggregates in mortars and concrete mixes. Nevertheless, concrete specification should provide the type and amount of this additive and the rock dust origin.

Thus, it should be stated that, considering the applicable standards, the addition of dusts as mineral fillers in the composition of mortars and concretes is determined by these cement composites properties, which may not be affected by the addition of dusts.

3. Fresh Concrete and Mortar Properties

The use of stone dusts as filler in concrete has a significant effect on workability. Fine filler additives in appropriate quantity improve the workability of cementitious materials and may not increase the water requirement [59,60]. However, for a constant w/c ratio, when too much powder content is used, more water is necessary in order to wet the grains surface, resulting in reduced mixing water and consequently poor workability [54,57,61]. On the other hand, crusher dust consumes more water than sand because of its rough texture. Thus, it causes a reduction in workability [62]. Hameed and Sekar [63] stated that 50% replacement of marble dust with river sand improves the workability of mortar. Janakiram and Murahari [64] investigated the workability of concrete where quarry dust and marble dust were used in various proportions of 0%, 10%, 20%, 30%, 40%, 50%, and 60% instead of natural sand. They reported that the workability decreased for all replacement percentages and a reduction of 28.57% was obtained for the 60% replacement. Idrees and Faiz [65] used marble powder and quarry dust in concrete as a replacement of sand at the percentages of 12.5% and 25% separately, and 25%, and 50% as combined replacement in equal proportions. They reported that marble powder negatively affects the workability while quarry dust improves it. They also indicated that the use of quarry dust increased slump while marble powder reduced it. It was concluded that the combined replacement of marble powder and quarry dust moderately improved the workability of concrete. Other authors also observed that adding marble powder into concrete or mortar shows a reduction in workability [30,38,66–68]. Several studies reported in the literature examined the influence of limestone powders in terms of concrete workability [69–74]. Some studies concluded that limestone powder decreases workability of concrete [70,72,74], while others reported improvements in workability [69]. Filler, dilution, and morphological effects of limestone powder play a role on the flowability of concrete [71]. Dobiszewska et al. [56] analysed the workability of concrete by replacing 5%, 10%, 20%, and 30% of sand (by mass) with basalt powder. It was concluded that workability decreased because of the much greater specific surface area of the basalt powder in comparison to the sand.

Segregation and bleeding in cement-based materials are two effects related to the loss of homogeneity. Segregation is observed as the settlement of aggregates within the mortars and concrete. Bleeding is associated with excess water rising to the surface of a highly fluid concrete mixture. Bleeding and segregation can be controlled by using well graded aggregates, finer cement, proper water to cement ratio, entraining agents, and mineral additives [75]. Uniform mixing is also important in reducing the propensity to bleeding and segregation. The use of fine granulated materials reduces bleeding and segregation by creating a longer path for water to rise to the surface, blocking the pores and improving the cohesion of the mix [59,75,76]. There are many studies on the effects of quarry dust, such as marble dust, granite dust, crushed rock dust, and limestone powder, on the bleeding and segregation of cement-based composites [77–81]. In all the studies examined, it was emphasized that the use of non-pozzolanic fillers in mortar or concrete mixtures increased bleeding and segregation resistance. Danish and Mohan Ganesh [78] reported that a reduction of 65.2% in the bleeding resistance of self-compacting concrete (SCC) was obtained by using marble powder. Schankoski et al. [82] indicated that no bleeding occurred in the quarry dust pastes which had a lower viscosity than those with limestone fillers. It was mentioned that bleeding is prevented due to the higher surface area and longer shaped quarry dust particles [82]. Elyamany et al. [79] conducted a study on the effects of various pozzolanic, such as silica fume and metakaolin, and non-pozzolanic fillers, such as limestone powder, granite dust, and marble dust, on the bleeding and segregation of self-compacting concrete. It was concluded that marble and granite powders showed better bleeding resistance compared to other filler types used. It was also concluded that a significant effect on bleeding was observed with a filler content of 15.0%. Nguyen et al. [80] emphasized that SCC would have sufficient bleeding resistance if 30% of dolomite powder was replaced with pozzolanic fillers.

Further, air content is a very important ingredient for cementitious materials because it directly affects the mechanical and durability properties cementitious materials. Cementitious based composites contain two types of air, namely entrapped and entrained. Entrapped air occurs naturally in the mix during mixing operations. These voids are convoluted and interrelated. On the other hand, entrained air is formed by the addition of air entrained admixture into the mixture. Those voids are spherical in form and independent from each other. In a conventional concrete (non-air entrained concrete) with a suitable mixture and sufficient compaction, the air content is around 1.5–2%. Air content can be increased up to 4–8% by using air entrained admixture for the improvement of freeze-thaw resistance for cold weather concreting [59,75]. The factors affecting the air content of concrete can be listed as follows: water and cement contents, maximum size and grading of aggregate, mixing and compaction of concrete, temperature of concrete, admixtures (mineral and chemical), and the use of fillers (stone powder, rock dust etc.). As a replacement of sand, rock powders and quarry dusts are the most preferable filler materials as ultrafine aggregates filling the voids to control or decrease the air content of cementitious materials. The use of very fine materials, with larger specific surface area than cement and in adequate quantities, reduce the air content of concrete [76,83–85]. However, the use of excessive rock dust particles in relation to the voids between cement and sand particles has a reducing effect on pore filling, resulting in an increase in air content due to a reduction in packing density [86–89]. Therefore, the optimum substitution of fine rock powder into cementitious composites is an important consideration.

It is seen that there are conflicting interpretations in the literature on workability, but in general, the use of fillers affects the workability negatively by effectively changing the water/cement in the concrete. For this reason, when stone dust is used, preliminary tests must be performed, and water/cement adjustment must take into account the stone dust used instead of the fixed water-cement ratio. In addition, it is seen that there is a consensus in the literature that the effect of stone dust on segregation and bleeding resistance is very significant. It is also seen that the use of stone dust reduces the air content by filling the voids in the concrete. Although this situation seems positive in terms of reducing the permeability of concrete, it should also be investigated in terms of durability problems, such as freezing and thawing.

4. Hardened Mortar and Concrete Properties

4.1. Compressive Strength

The introduction of stone powder, which partially replaces fine aggregate, affects the properties of mortar and concrete physically and mechanically. The relative compressive strength of mortar and concrete with rock powder addition at a different curing age are shown in Figures 2–7. Soroka and Stern [90] noticed that the addition of rock dust powder positively affects the cement mortars mechanical properties. They observed that as the amount of dust used sand replacement and the fineness of the dust increases, so does the strength of the mortar. Similar conclusions were obtained with studies examining the usage of rock dust as a partial replacement for sand in mortars and concrete: limestone dust [44,46,51,55,91,92], marble dust [12,30,33,34,61,67,76,86,92–94], granite dust [14,17,18,43,54,95], and basalt powder [46,55–57,96].

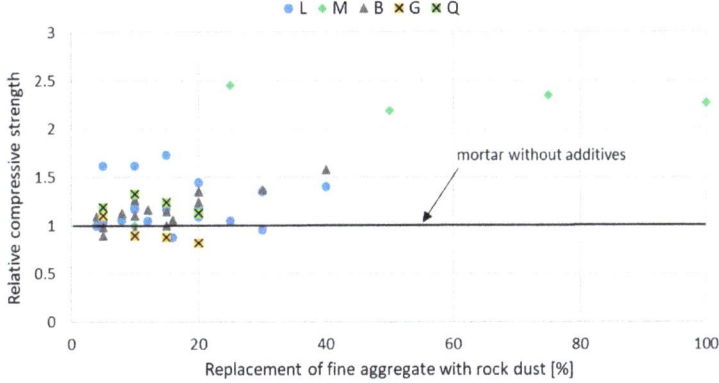

Figure 2. The effect of replacing fine aggregate with rock dust on the compressive strength of mortar after 7 days of curing: limestone dust (L) [46,55,95,97], marble dust (M) [67,76], basalt dust (B) [46,55,57,96], granite dust (G) [95], quartz dust (Q) [95].

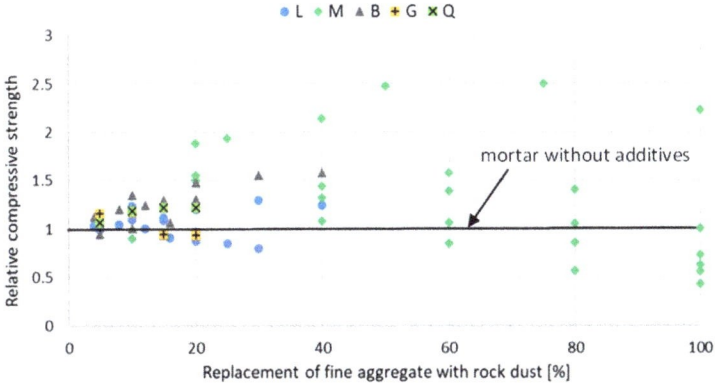

Figure 3. The effect of replacing fine aggregate with rock dust on the compressive strength of mortar after 28 days of curing: limestone dust (L) [46,55,95,97], marble dust (M) [63,67,76,86], basalt dust (B) [46,55,57,96], granite dust (G) [95], quartz dust (Q) [95].

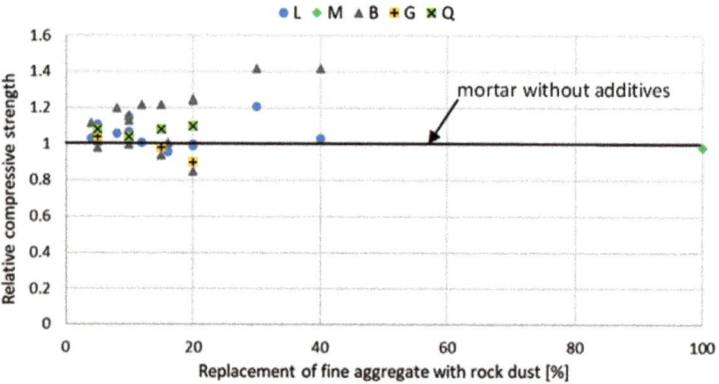

Figure 4. The effect of replacing fine aggregate with rock dust on the compressive strength of mortar after 90 days of curing: limestone dust (L) [46,55,95], marble dust (M) [63], basalt dust (B) [46,55,57,96], granite dust (G) [95], quartz dust (Q) [95].

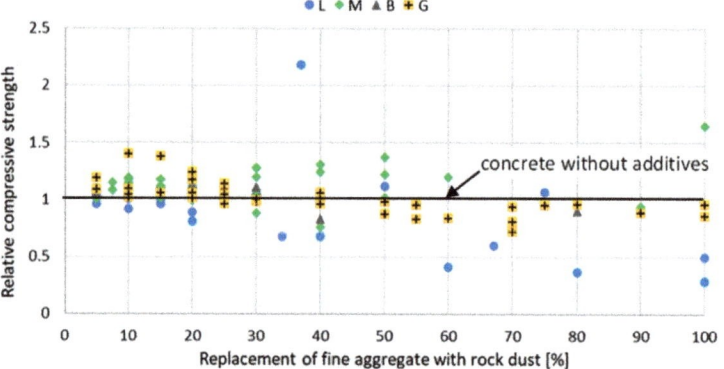

Figure 5. The effect of replacing fine aggregate with rock dust on the compressive strength of concrete after 7 days of curing: limestone dust (L) [13,33,91,92,98,99], marble dust (M) [12,30,33,34,61,93,94,100–103], basalt dust (B) [56,104,105], granite dust (G) [54,106–110].

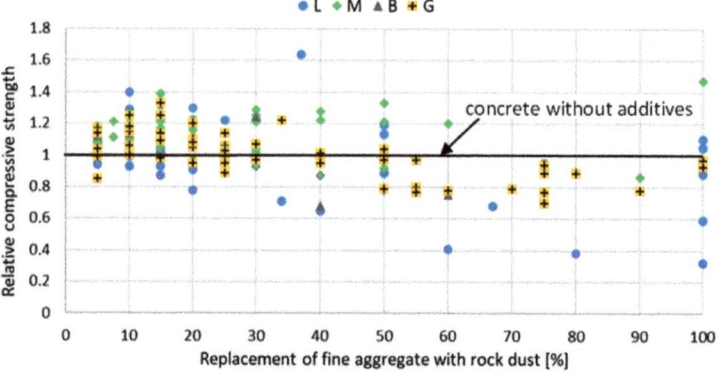

Figure 6. The effect of replacing fine aggregate with rock dust on the compressive strength of concrete after 28 days of curing: limestone dust (L) [13,22,33,44,91,92,98,99,111], marble dust (M) [12,30,33,34, 61,93,94,100–103], basalt dust (B) [56,104,105], granite dust (G) [14,17,18,54,106–110].

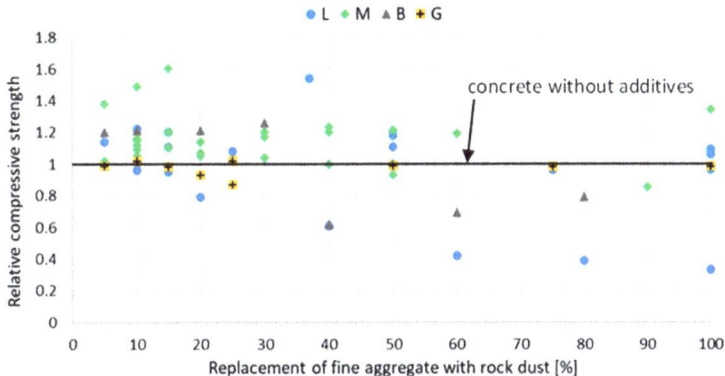

Figure 7. The effect of replacing fine aggregate with rock dust on the compressive strength of concrete after 90 days of curing: limestone dust (L) [22,33,92,98,99,111], marble dust (M) [12,33,34,93,94,100,102], basalt dust (B) [56,104,105], granite dust (G) [17,54,109].

The filler role of stone powder is primarily responsible for the improvement in strength in cement composites with rock dust addition. As mentioned earlier, the process of the heteronucleation of cement clinker hydrates on dust particles mechanically improves the cement matrix microstructure and interfacial transition zone [12,30,46,56,67,99,112–115]. The chemical composition of the parent rock and the rock type from which the powder comes has a minor impact on the rock dust operation mechanism in this situation [46,116]. Much more important and dominant influence is the fineness of the rock dust. Nevertheless, it should be noticed that the rock dust specific surface area affects to a greater extent the mortar and concrete mechanical properties when rock dust is used as a partial cement substitution rather than as fine aggregate replacement. As observed earlier, cement substitution with inert additives of finer particles size and greater fineness compared to cement particles results in the increase of hydration products nucleation sites. Hydration products crystallize on cement particles as well as on the rock powder surface, which contributes to an increase of the rate and hydration degree of cement clinker. This leads to an increase in the content of C-S-H phase, decrease of cement paste porosity, and therefore to an increase in cement matrix strength, particularly in the early hydration process [117]. When cement is replaced with rock dust of larger particles diameter than cement grains, it results in a reduction of the specific surface area where hydrates can crystallize. This action leads to lower hydration rate and degree of hydration and lower early strength at early ages. The addition of stone dust as sand substitution does not affect cement content, and thus the nucleation centers specific surface area increases in each case. Therefore, in this situation, it is irrelevant if the fineness of stone dust is larger or smaller than cement.

Rock dust is an inert filler and thus contributes in filling a greater range of the intergranular free space in cement composites. This results in the densification of the cement matrix, which leads to lower porosity and therefore higher strength [12,95,102,103,106,118]. Roy et al. [119] discussed the particular role of particle packing in achieving optimal mortar and concrete properties. The more regular cement grains dispersion, and therefore the faster hydration of the clinker phases of cement, occurs by the addition of microfiller grain [54,118,120].

Uchikawa et. al. [98] indicated that the increase in the strength of concrete by substitution of fine aggregate with rock dust is achieved by the increase in the density of the hardened concrete structure due to the generation of the C-S-H phase during the pozzolanic reaction, in addition to the mineral powder's filling property. Abdelaziz et al. [55] observed the same in which compressive strength of mortar increases with basalt dust addition. Such an effect was attributed to the filler effects as well as to basalt pozzolanic activity. The reaction result of active silica and alumina ions in basalt with the calcium hydroxide CH in the cement pore solution, an additional amount of C-S-H phase is formed. In turn, this results in the increase of cementitious matrix density and strength improvement. Other researchers reached similar conclusions in regard to the slightly greater strength of mortars when basalt powder additive was used compared to mortars with other rock dust [46,55,121].

In the case of limestone and marble powder, the development in concrete's compressive strength is connected with the physical and chemical effects of dust. The dominant reason is due to the physical filler effect of mineral powder. This results in filling the spaces between the cement grains and refining the pore structure, enhancing the concrete matrix microstructure, and thereby a strength increase. The chemical effect of limestone and marble powder concerns the reaction of calcium carbonate $CaCO_3$ and alite C3A available in the cement [53] and results in the formation of calcium carboaluminate hydrates [63,92]. This increases the degree of hydration reactions and reduces porosity especially at early ages of hydration [122–124]. Thus, it contributes to an increase in early age strength [125,126].

However, with a certain rock dust content replacing fine aggregate, decreases in the mortar and concrete strength were observed [12,61,63,92,97,107,127,128]. Fine dust particles feature a large surface area and therefore need more water for moistening the grain surface. However, when the w/c ratio is kept constant, the increase in dust content leads to the reduction of available water necessary for hydration of cement clinker phases, poor workability, and thus poor compactness and a decrease in compressive strength [54,57,61]. Alyamac and Aydin [94] observed that high dust content leads to an improper grain-size distribution. This results in larger free space between particles and therefore strength reduction. When rock dust particles are in excess of the voids between cement and sand particles, then the particles push each other apart, leading to a reduction in packing density, and thus a reduction in compressive strength [86,88,89]. Hence, the pore filling effect is being downplayed. Once the optimum substitution level is reached, any higher amount of rock dust increases the surface area of particles instead of filling up the voids. The increase in the surface area requires an excess amount of cement to bind dust particles as well as aggregate grains. When the cement content is constant, a strength reduction is observed at higher rock powder inclusion [107]. Additionally, the presence of excessive permeable voids accelerates crack propagation and crack connectivity, thus, resulting in strength reduction at higher substitution rates of sand with rock powder [86,106]. Knop et al. [117] confirmed that high amount of very fine dust particles leads to the agglomeration as a consequence of the inter-particle interaction which generated massive grain aggregate formation with a diameter exceeding even 100 μm. As a result, it decreases the effective specific surface area and causes lower particle packing density, which directly affects strength reduction.

In general, the addition of rock dust has a positive effect on the strength as it fills the concrete voids and reduces the porosity. In addition, when the water/cement ratio is not modified according to the added stone powder, since the water required for cement hydration is used by rock dust, it can significantly reduce the compressive strength as well as the workability of the concrete.

4.2. Tensile and Flexural Strength

Few studies have addressed the impact on tensile and flexural strength of cement composites with rock dust addition as a fine aggregate. Overall, an increase in the rock powder replacing fine aggregate, an increase in tensile and flexural strength were observed [30,36,86,89,92,129–131]. As in the case of compressive strength, the increase in tensile and flexural strength is related mainly by the filler action of fine rock dust particles. The fine rock dust particles fill the voids in the cement matrix, and therefore a denser microstructure contributes to an increase in strength properties [89,132]. Some studies pointed out that the effects of rough surface texture and irregular shape of rock dust particles are the most significant parameters in increasing flexural and tensile strength of cement-based materials [133]. Such rock powder particle properties might improve the adherence of the aggregate phase to the cement paste, resulting in better bonding on the crack route created throughout the split tensile and flexural strength testing. This enhances the strength properties [12,92,106,107]. The decrease in porosity and improvement of the strength of both the cement paste matrix and the interfacial transition zone might be ascribed to the improvement in bond strength [115,134,135].

However, above a certain amount of stone dust replacing sand (i.e., about 30%), a decrease in h is observed. The increased fine content may also increase the pores in the concrete, which explains the flexural strength reduction [103,136].

Results similar to compressive strength in tensile and flexural strength appear in the literature, but the most important issue here is the rough surface texture and irregular shape of rock dust used.

5. Concrete and Mortar Durability

Galetakis and Soultana [39], as well as many other authors, have asserted that permeability is one of the most important factors characterizing the durability of concrete. The permeability of concrete is often measured based on its resistance to allow the penetration and movement of aggressive substances within its mass. The published research results indicate that concrete with the addition of different mineralogical origin rock waste demonstrated lower water permeability as compared with conventional concrete [33,35,92,137,138]. A study on the use of marble waste as coarse aggregate replacement conducted by Ulubeyli et al. [139] found that marble waste acted as a filler, reducing the gaps within the hardened concrete, thus providing a less porous structure of concrete. It can be stated that water permeability depends primarily on the capillary pores volume. However, Kurdowski [118] concluded that permeability is determined not only by the total porosity, but also to the distribution, tortuosity, shape of pores, as well as their size and continuity. The study conducted by Holly et al. [140] supported this concept by demonstrating a remarkable impact of the interconnectivity of cement paste pores and the pore size distribution on permeability. Menadi [22] observed a reduction in the water penetration depth with an increase in limestone powder content. This is the effect of the improvement of pore structure in the interfacial transition zone. The increase in concrete water permeability with the increase in limestone powder substitution level was also confirmed by Celik et al. [91]. The decrease in the permeability of cement matrix with the addition of rock dust is generally related to the filler effect, i.e., physical rock dust interaction. In addition, fine particles of rock dust block the continuity of capillary pores, which leads to the reduction of the capillary rise of water as well as permeability [29,91,140]. Dobiszewska et al. [56] observed the phenomenon of heteronucleation on the surface of rock dust particles. This phenomenon increases the production of crystallization nuclei, which leads to the densification of the cement paste and has a significant impact on the permeability reduction of the cement matrix when rock dust is added. The addition of rock powder accelerates cement hydration. It can be argued that hydrates fill free space between cement and dust particles, which directly contributes to the capillary pore content reduction and breaking of its continuity.

As mentioned earlier, the water absorption of concrete affects concrete durability. The ability of water absorption depends mainly on the distribution, size, shape, and tortuosity of pores, as well as their continuity [52]. Studies conducted by Almeida et al. [29] as well as and Celik and Marar [91] confirm that adding powdered limestone as a partial replacement for fine aggregate reduces concrete absorption. This is a consequence of the reduction of the pore content and the disruption of their continuity. The beneficial effects of rock powder on reducing water absorption of concrete were also confirmed by Alyamac and Aydin [94], Gameiro et al. [45], and Ulubeyli et al. [139], where marble dust was used as a partial fine aggregate substitute. Hameed et al. [137] further observed that adding of marble beyond 15–20% resulted in an increased water requirement in the concrete mixture due to the very high-specific surface area of the marble waste. This finding strengthened the results of previous studies conducted by Tasdemir [141], Gesoglu et al. [35], as well as Tsivilis et al. [52] indicating that the addition of rock powder of larger specific surface area than cement particles results in a reduction of porosity. The consequence of this is a lower absorption of concrete and greater resistance to the aggressive media penetration. However, some studies where quarry rock dust additives were used as a fine aggregate replacement indicated an increase of water absorption when a higher percentage of aggregate were substituted, resulting in a higher level of pores [91,142,143].

Further, the dissolution of compounds or chemical reactions between concrete and substance constituents occurs due to a chemical attack [144]. The most destructive agents that caused concrete deterioration are chlorides. Chloride ions penetrate concrete and replace hydroxide ions in cement hydrates during leaching. This leads to a lower pH of pore solution and, as a consequence, to the gradual disintegration of cement matrix. The resistance of concrete to the penetration of chloride ions is closely related to the concrete permeability and porosity. The ability of ion diffusion depends significantly on pore structure, the content of gel, and capillary pores. The effective diffusion coefficient decreases with the increase of gel pore contents and the disruption of capillary pore continuity [118]. As indicated earlier, heteronucleation on the rock dust particles surface leads to the increase in C-S-H phase content, and as a result to the densification of the cement matrix and change in pore size and structure [56]. The increase of fine pores content and break in continuity of capillary pores with the increase in C-S-H phase content is also observed. Thus, it results in a reduction of the rate of ion diffusion. The positive effect of limestone powder addition on the reduction of chloride ion permeability in concrete was noticed by Li et al. [145]. The enhancement in chloride resistance of concrete was observed also in the case of using granite powder as a partial replacement of fine aggregate [17]. In contrast, the conclusions made by Kepniak et al. [138] concerning the influence of the substitution for fine aggregate with limestone powder on concrete resistance to chloride corrosion observed an increase in the chloride ion concentration, and at the same time a reduction of total porosity with the increase of limestone powder addition. This indicates the faster chloride ion penetration which was confirmed by determination of the effective diffusion coefficient of chloride ions. Menadi et al. [22] have come to similar conclusions where the resistance to chloride ion penetration and gas permeability of concrete decrease with limestone powder increase, whereas water permeability is reduced. A negative effect of the influence of granite powder on chloride resistance of concrete was also observed by Vijayalakshmi et al. [54]. The presented results show that the concrete chloride permeability is proportional to the substitution rate, and the penetration rate increases with an increase in granite powder share. Vijayalakshmi et al. [54] stated that increase in the permeability of chloride ions is attributed to poor compaction, which results in higher porosity and a discontinuous pore system. This leads also to an increase in the carbonation depth value of the concrete with the increase in granite powder waste substitution.

Kępniak et al. [138] noticed an increase in the sulfate attack resistance of concrete with an increase in limestone powder amount, despite stated lower chloride resistance, as mentioned earlier. The authors noticed that, with an increase in limestone powder substitution level, the capillary pores content increases, in spite of the total porosity reduction. This favors the increase in the rate of chloride ion diffusion in concrete. The effect of a faster filling of smaller capillary pores with corrosion products prevents the further migration of sulphate ions from the solution, which results in the inhibition of the sulphate degradation process. The improvement of the mortar sulphate resistance as an effect of the incorporation of limestone powder was confirmed by Li et al. [145]. The decrease in sulphate resistance of the concrete with granite powder addition was noticed by Vijayalakshmi et al. [54]. This was caused by the contamination of granite powder with kerosene, diesel, and wax, which has been used during the process of sawing and polishing granite rock. In a study conducted by Inlangovana et al. [142], it was found that using quarry rock dust as fine aggregate increased concrete durability when compared to conventional concrete exposed to sulfate and acid action. As is known, the durability of concrete is directly related to the void structure and permeability of the concrete. Studies generally mention that more impermeable concrete can be produced thanks to the gap-filling effect of stone dust, but it is also seen that the materials used as fillers plays a much more effective role if it is finer-grained than cement, even if they are used as a fine aggregate substitution.

6. Conclusions

Concrete production is associated with environmental concerns since it consumes large amounts of raw materials, energy, and labor. Thus, worldwide there is an urgent demand to use by-products in building material production. In addition to materials that can be used as aggregate in the production of mortar and concrete, materials that can be substituted with cement clinker are also sought. The potential alternative materials that can be used in cement composites production as fine aggregate substitution include rock dust of different geological origin. The management of this waste is currently a serious problem for producers of mineral aggregates, asphalt mixture plants, and dimension stone industry. This indicates that more research concerning the management and utilization of these waste products in cement composites production is required. The review of past studies in this area synthesized in this manuscript provided the following valuable conclusions that can be considered in further studies.

1. The addition of rock powder significantly affects fresh concrete and mortar properties. The substitution for fine aggregate with rock dust leads generally to a significant decrease in workability. The much greater specific surface area of rock dust compared to fine aggregate results in a significant increase in water required by wet the particle surfaces, and thus poor workability. The solution to this problem is to use high water reducing admixtures to improve the workability of concrete. Therefore, there is a need to conduct research concerning the analysis of the influence of admixtures on concrete workability when the rock dust is used for fine aggregate substitution. As rock powder is very fine material, its addition leads to a reduction in bleeding and segregation. This is mainly the result of mix cohesion improvement by fine particles of rock dust and water retention enhancement.

2. Improved mechanical properties of cement composites are due to the use of rock powder as a partial replacement for fine aggregate. The most important and dominant mechanism of beneficial rock dust interaction is connected with the filler effect, i.e., physical interaction. The space between the cement and aggregate grains is filled with very small particles of stone powder, which results in reducing the cement matrix porosity. With the addition of stone dust, the number of large capillary pores decreases and the content of small pores increases, which leads to sealing in the microstructure of the hardened cement paste and, accordingly, to a less permeable structure. As a result, cement composites with rock dust additive feature higher strength. Aside from the physical influence of stone dust on the cement matrix microstructure, other phenomena also occur. The rock dust grain surface is mainly the active center, which leads to the improvement of the properties and durability of cement composites from which heteronuclei of the C-S-H phase are formed. The heteronucleation on rock dust particles is much more favored by the fineness than geological origin of rock powder. As mentioned earlier, basalt dusts have some pozzolanic activity, which results in the increase of cement matrix density and thus strength improvement. In the case of using rock dust for fine aggregate substitution, the dominate role in property improvement is played by the filler effect, while the rock origin from which stone powder comes is less of importance. That is because analysed rock dust is in any case much finer than fine aggregate and possesses the greater specific surface area. The optimum fine aggregate replacement is about 20–30% and it depends more on rock dust fineness than its geological origin. With such a substitution level, an approximately 30% increase in mortar and concrete strength is observed.

3. Reported results confirmed the positive effect of rock dust on concrete with an increase in the permeability and decrease in water absorption. Generally, the outcome is a result of the densification of cement matrix with fine rock powder particles, i.e., the filler effect. However, there were some contradictions regarding the influence of rock dust on permeability of concrete mainly to chloride ions. This depends on the finesses of the rock dust particles as compared to capillary pore and substitution level of fine aggregate with rock powder. In the case of sulphate attack, the addition of stone

powder leads mainly to an improvement of sulphate resistance. Undoubtedly, further research is necessary to analyse effect of rock dust on cement composites durability, especially regarding chloride and sulphate corrosion, carbonation, and freeze-thaw resistance. Profound analysis concerning the influence of the fineness of rock dust on the penetration of chloride and sulphate ions is needed.

4. Rock dust utilization in cement composite production requires the development of concrete design methods that allow to determine the optimal dust content in terms of obtaining the desired properties of both fluid concrete mix properties as well as hardened properties. Profound analysis is necessary to establish the optimum ratio for fine aggregate substitution with regard to the fineness of rock dust and addition of water reducing admixtures.

5. Rock dust, which is currently considered as a by-product, can be used as a partial replacement for fine aggregates or even cement in cement mortars and concrete production. The utilization of rock dust waste is technically, economically, and ecologically justified and addresses the principle of sustainable development as it allows to reduce the consumption and dependency of natural resources for the production of cement composites and to manage the waste effectively.

As a result, considering the extensive studies in the literature, it can be concluded that rock dust is an environmentally friendly material that contributes economically to the mixture of cement-based materials. The use of rock dust for fine aggregate replacement at a certain amount in cement-based composites improves many fresh and hardened state properties. Therefore, rock dust should be taken into account in the optimum mix design of cement-based composites. Most of the studies in the literature also mentioned that, in addition to improving the properties of concretes, using stone dust in concrete led to the consumption of by-products, thus providing a twofold benefit.

When the results are evaluated for future studies, it is recommended that more research should be conducted on evaluating the usage of rock dust in high-performance concrete production and self-compacting concrete production, besides reactive powder concrete. The use of rock dust in the production process of cement composites requires the development of concrete design methods that allow the determination of the optimal rock dust content in terms of obtaining the desired properties of both the concrete mix and hardened concrete. If a careful analysis of the literature is performed, another important issue comes to the fore for future studies. Generally, the particle size distribution within the stone dust itself has not been taken into account by researchers. As known, it can be encountered in some cases that the particle size distribution in some intervals forms a significant part of the heap compared to other grain intervals. This situation directly affects many important parameters, such as water requirement, workability, gap-filling ability, etc., in concrete containing stone dust. For this reason, specifying the particle size distribution of these powder materials in studies instead of just calling them a material under 150 microns is recommended for future studies. A detailed analysis of particles size distribution can help to better interpret the results of the use of stone powder, as this affects the internal structure and many related properties.

Author Contributions: M.D.: conceptualization, formal analysis, investigation, writing—original draft preparation, writing—review and editing, supervision; O.B.: formal analysis, investigation, writing—original draft preparation, writing—review and editing; A.B.: formal analysis, investigation, writing—original draft preparation, writing—review and editing; D.G.: formal analysis, writing—review and editing; F.K.: formal analysis, writing—review and editing; M.N.: formal analysis, writing—review and editing; H.Ü.: writing—review and editing. All authors have read and agreed to the published version of the manuscript.

Funding: This research received no external funding.

Institutional Review Board Statement: Not applicable.

Informed Consent Statement: Not applicable.

Data Availability Statement: The study did not report any data.

Acknowledgments: This article has been supported by the Polish National Agency for Academic Exchange under Grant No. PPI/APM/2019/1/00003.

Conflicts of Interest: The authors declare no conflict of interest. The funders had no role in the design of the study; in the collection, analyses, or interpretation of data; in the writing of the manuscript, or in the decision to publish the results.

References

1. Kaza, S.; Yao, L.; Bhada-Tata, P.; Van Woerden, F. *What a Waste 2.0: A Global Snapshot of Solid Waste Management to 2050*; Urban Development Series; World Bank Group: Washington, DC, USA, 2018.
2. U.S. Geological Survey. Cement—Mineral Commodity Summaries. In *Cement Statistics and Information—Annual Publications*; U.S. Department of the Interior, U.S. Geological Survey: Reston, VI, USA, 2020.
3. Adewumi, A.A.; Mohd Ariffin, M.A.; Maslehuddin, M.; Yusuf, M.O.; Ismail, M.; Al-Sodani, K.A.A. Influence of Silica Modulus and Curing Temperature on the Strength of Alkali-Activated Volcanic Ash and Limestone Powder Mortar. *Materials* 2021, *14*, 5204. [CrossRef] [PubMed]
4. Reiterman, P.; Jaskulski, R.; Kubissa, W.; Holčapek, O.; Keppert, M. Assessment of Rational Design of Self-Compacting Concrete Incorporating Fly Ash and Limestone Powder in Terms of Long-Term Durability. *Materials* 2020, *13*, 2863. [CrossRef] [PubMed]
5. Mateus, R.; Neiva, S.; Bragança, L.; Mendonça, P.; Macieira, M. Sustainability Assessment of an Innovative Lightweight Building Technology for Partition Walls—Comparison with Conventional Technologies. *Build. Environ.* 2013, *67*, 147–159. [CrossRef]
6. Ryu, H.-S.; Kim, D.-M.; Shin, S.-H.; Kim, W.-K.; Lim, S.-M.; Park, W.-J. Properties of Cement Mortar Using Limestone Sludge Powder Modified with Recycled Acetic Acid. *Sustainability* 2019, *11*, 879. [CrossRef]
7. Luo, W.; Wang, H.; Li, X.; Wang, X.; Wu, Z.; Zhang, Y.; Lian, X.; Li, X. Mechanical Properties of Reactive Powder Concrete with Coal Gangue as Sand Replacement. *Materials* 2022, *15*, 1807. [CrossRef] [PubMed]
8. Karakurt, C.; Dumangöz, M. Rheological and Durability Properties of Self-Compacting Concrete Produced Using Marble Dust and Blast Furnace Slag. *Materials* 2022, *15*, 1795. [CrossRef] [PubMed]
9. Sosoi, G.; Abid, C.; Barbuta, M.; Burlacu, A.; Balan, M.C.; Branoaea, M.; Vizitiu, R.S.; Rigollet, F. Experimental Investigation on Mechanical and Thermal Properties of Concrete Using Waste Materials as an Aggregate Substitution. *Materials* 2022, *15*, 1728. [CrossRef]
10. Sankh, A.C.; Biradar, P.; Naghathan, S.J.; Manjunath, B. Recent Trends in Replacement of Natural Sand with Different Alternatives. *IOSR J. Mech. Civ. Eng.* 2014, 59–66.
11. Rashad, A.M. A Preliminary Study on the Effect of Fine Aggregate Replacement with Metakaolin on Strength and Abrasion Resistance of Concrete. *Constr. Build. Mater.* 2013, *44*, 487–495. [CrossRef]
12. Vardhan, K.; Siddique, R.; Goyal, S. Strength, Permeation and Micro-Structural Characteristics of Concrete Incorporating Waste Marble. *Constr. Build. Mater.* 2019, *203*, 45–55. [CrossRef]
13. Almeida, N.; Branco, F.; Santos, J.R. Recycling of Stone Slurry in Industrial Activities: Application to Concrete Mixtures. *Build. Environ.* 2007, *42*, 810–819. [CrossRef]
14. Arulraj, G.P.; Adin, A.; Kannan, T.S. Granite Powder Concrete. *IRACST–Eng. Sci. Technol. Int. J. ESTIJ* 2013, *3*, 193–198.
15. Li, H.; Huang, F.; Cheng, G.; Xie, Y.; Tan, Y.; Li, L.; Yi, Z. Effect of Granite Dust on Mechanical and Some Durability Properties of Manufactured Sand Concrete. *Constr. Build. Mater.* 2016, *109*, 41–46. [CrossRef]
16. Felixkala, T. Effect of Granite Powder on Strength Properties of Concrete. *Int. J. Eng. Sci.* 2013, *2*, 36–50.
17. Arivumangai, A.; Felixkala, T. Strength and Durability Properties of Granite Powder Concrete. *J. Civ. Eng. Res.* 2014, *4*, 1–6. [CrossRef]
18. Divakar, Y.; Manjunath, S.; Aswath, M.U. Experimental Investigation on Behaviour of Concrete with the Use of Granite Fines. *Int. J. Adv. Eng. Res. Stud.* 2012, *I*, 84–87.
19. Martirena, J.F.; Day, R.L.; Middendorf, B.; Gehrke, M.; Martínez, L.; Dopico, J.M. Lime-Pozzolan Binder as a Very Fine Mineral Admixture in Concrete. In Proceedings of the International Symposium on Ultra High Performance Concrete, Kassel, Germany, 13 September 2004; Schmidt, M., Fehling, E., Geisenhanslüke, C., Eds.; Kassel University Press GmbH: Kassel, Germany; Volume 3, pp. 117–131.
20. Jain, A.; Gupta, R.; Chaudhary, S. Performance of Self-Compacting Concrete Comprising Granite Cutting Waste as Fine Aggregate. *Constr. Build. Mater.* 2019, *221*, 539–552. [CrossRef]
21. Shon, C.-S.; Tugelbayev, A.; Shaimakhanov, R.; Karatay, N.; Zhang, D.; Kim, J.R. Use of Off-ASTM Class F Fly Ash and Waste Limestone Powder in Mortar Mixtures Containing Waste Glass Sand. *Sustainability* 2022, *14*, 75. [CrossRef]
22. Menadi, B.; Kenai, S.; Khatib, J.; Aït-Mokhtar, A. Strength and Durability of Concrete Incorporating Crushed Limestone Sand. *Constr. Build. Mater.* 2009, *23*, 625–633. [CrossRef]
23. Youness, D.; Mechaymech, A.; Al Wardany, R. Flow Assessment and Development towards Sustainable Self-Consolidating Concrete Using Blended Basalt and Limestone-Cement Systems. *J. Clean. Prod.* 2021, *283*, 124582. [CrossRef]
24. *EN 12620:2013*; Aggregates for Concrete. European Committee for Standardization (CEN): Brussels, Belgium, 2013.

25. *ASTM C33/C33M-13*; Standard Specification for Concrete Aggregates. American Society for Testing and Materials: West Conshohocken, PA, USA, 2018.
26. *ACI Bulletin E1-07*; ACI Aggregates for Concrete. American Concrete Institute: Farmington Hills, MI, USA, 2007.
27. Kozioł, W.; Ciepliński, A.; Machniak, Ł.; Borcz, A. Kruszywa w Budownictwie. Cz. 1. Kruszywa Naturalne. *Nowocz. Bud. Inżynieryjne* **2015**, *4*, 98–100.
28. Aggregates Production. Rock Products 2020. p. 6. Available online: https://rockproducts.com/2020/07/06/aggregates-production-4 (accessed on 15 March 2022).
29. Almeida, N.; Branco, F.; de Brito, J.; Santos, J.R. High-Performance Concrete with Recycled Stone Slurry. *Cem. Concr. Res.* **2007**, *37*, 210–220. [CrossRef]
30. Aliabdo, A.A.; Abd Elmoaty, A.E.M.; Auda, E.M. Re-Use of Waste Marble Dust in the Production of Cement and Concrete. *Constr. Build. Mater.* **2014**, *50*, 28–41. [CrossRef]
31. Aruntaş, H.Y.; Gürü, M.; Dayı, M.; Tekin, İ. Utilization of Waste Marble Dust as an Additive in Cement Production. *Mater. Des.* **2010**, *31*, 4039–4042. [CrossRef]
32. Bacarji, E.; Toledo Filho, R.D.; Koenders, E.A.B.; Figueiredo, E.P.; Lopes, J.L.M.P. Sustainability Perspective of Marble and Granite Residues as Concrete Fillers. *Constr. Build. Mater.* **2013**, *45*, 1–10. [CrossRef]
33. Binici, H.; Kaplan, H.; Yilmaz, S. Influence of Marble and Limestone Dusts as Additives on Some Mechanical Properties of Concrete. *Sci. Res. Essay* **2007**, *2*, 372–379.
34. Demirel, B. The Effects of Waste Marble Dust Applying as a Fi Ne Sand on the Mechanical Properties of Concrete. *Cem. Wapno Beton* **2010**, *5*, 259–267.
35. Gesoğlu, M.; Güneyisi, E.; Kocabağ, M.E.; Bayram, V.; Mermerdaş, K. Fresh and Hardened Characteristics of Self Compacting Concretes Made with Combined Use of Marble Powder, Limestone Filler, and Fly Ash. *Constr. Build. Mater.* **2012**, *37*, 160–170. [CrossRef]
36. Keleştemur, O.; Arıcı, E.; Yıldız, S.; Gökçer, B. Performance Evaluation of Cement Mortars Containing Marble Dust and Glass Fiber Exposed to High Temperature by Using Taguchi Method. *Constr. Build. Mater.* **2014**, *60*, 17–24. [CrossRef]
37. Rana, A.; Kalla, P.; Verma, H.K.; Mohnot, J.K. Recycling of Dimensional Stone Waste in Concrete: A Review. *J. Clean. Prod.* **2016**, *135*, 312–331. [CrossRef]
38. Vardhan, K.; Goyal, S.; Siddique, R.; Singh, M. Mechanical Properties and Microstructural Analysis of Cement Mortar Incorporating Marble Powder as Partial Replacement of Cement. *Constr. Build. Mater.* **2015**, *96*, 615–621. [CrossRef]
39. Galetakis, M.; Soultana, A. A Review on the Utilisation of Quarry and Ornamental Stone Industry Fine By-Products in the Construction Sector. *Constr. Build. Mater.* **2016**, *102*, 769–781. [CrossRef]
40. *ASTM C294-12*; Standard Descriptive Nomenclature for Constituents of Concrete Aggregates. American Society for Testing and Materials: West Conshohocken, PA, USA, 2017.
41. *The Aggregates Handbook*, 2nd ed.; The National Stone, Sand and Gravel Association: Alexandria, VA, USA, 2013; ISBN 9780988995000.
42. *BS EN 12620:2013*; Aggregates for Concrete. British Standards Institution (BSI): London, UK, 2013.
43. Chiranjeevi Reddy, K.; Yaswanth Kumar, Y.; Poornima, P. Experimental Study on Concrete with Waste Granite Powder as an Admixture. *Int. J. Eng. Res. Appl.* **2015**, *5*, 87–93.
44. Eren, Ö.; Marar, K. Effects of Limestone Crusher Dust and Steel Fibers on Concrete. *Constr. Build. Mater.* **2009**, *23*, 981–988. [CrossRef]
45. Gameiro, F.; de Brito, J.; Correia da Silva, D. Durability Performance of Structural Concrete Containing Fine Aggregates from Waste Generated by Marble Quarrying Industry. *Eng. Struct.* **2014**, *59*, 654–662. [CrossRef]
46. Soroka, I.; Setter, N. The Effect of Fillers on Strength of Cement Mortars. *Cem. Concr. Res.* **1977**, *7*, 449–456. [CrossRef]
47. Uysal, M. Self-Compacting Concrete Incorporating Filler Additives: Performance at High Temperatures. *Constr. Build. Mater.* **2012**, *26*, 701–706. [CrossRef]
48. Uysal, M.; Sumer, M. Performance of Self-Compacting Concrete Containing Different Mineral Admixtures. *Constr. Build. Mater.* **2011**, *25*, 4112–4120. [CrossRef]
49. Uysal, M.; Yilmaz, K. Effect of Mineral Admixtures on Properties of Self-Compacting Concrete. *Cem. Concr. Compos.* **2011**, *33*, 771–776. [CrossRef]
50. Uysal, M.; Yilmaz, K.; Ipek, M. The Effect of Mineral Admixtures on Mechanical Properties, Chloride Ion Permeability and Impermeability of Self-Compacting Concrete. *Constr. Build. Mater.* **2012**, *27*, 263–270. [CrossRef]
51. Topçu, İ.B.; Uğurlu, A. Effect of the Use of Mineral Filler on the Properties of Concrete. *Cem. Concr. Res.* **2003**, *33*, 1071–1075. [CrossRef]
52. Tsivilis, S.; Tsantilas, J.; Kakali, G.; Chaniotakis, E.; Sakellariou, A. The Permeability of Portland Limestone Cement Concrete. *Cem. Concr. Res.* **2003**, *33*, 1465–1471. [CrossRef]
53. Ergün, A. Effects of the Usage of Diatomite and Waste Marble Powder as Partial Replacement of Cement on the Mechanical Properties of Concrete. *Constr. Build. Mater.* **2011**, *25*, 806–812. [CrossRef]
54. Vijayalakshmi, M.; Sekar, A.S.S.; Ganesh Prabhu, G. Strength and Durability Properties of Concrete Made with Granite Industry Waste. *Constr. Build. Mater.* **2013**, *46*, 1–7. [CrossRef]

55. Abdelaziz, M.A.; El-Aleem, S.A.; Menshawy, W.M. Effect of Fine Materials in Local Quarry Dusts of Limestone and Basalt on the Properties of Portland Cement Pastes and Mortars. *Int. J. Eng. Res. Technol.* **2014**, *3*, 1038–1056.
56. Dobiszewska, M.; Schindler, A.K.; Pichór, W. Mechanical Properties and Interfacial Transition Zone Microstructure of Concrete with Waste Basalt Powder Addition. *Constr. Build. Mater.* **2018**, *177*, 222–229. [CrossRef]
57. Kmecová, V.; Štefunková, Z. Effect of Basalt Powder on Workability and Initial Strength of Cement Mortar. *Int. J. Civ. Eng. Archit.* **2014**, *1*, 260–267.
58. *EN 13055:2016*; Lightweight Aggregates. European Committee for Standardization (CEN): Brussels, Belgium, 2016.
59. Shetty, M.S. *Concrete Technology: Theory and Practice*; S. Chand & Co., Ltd.: Ram Nagar, India, 2005; ISBN 9788121900034.
60. Khan, M.I.; Usman, M.; Rizwan, S.A.; Hanif, A. Self-Consolidating Lightweight Concrete Incorporating Limestone Powder and Fly Ash as Supplementary Cementing Material. *Materials* **2019**, *12*, 3050. [CrossRef]
61. Anitha Selvasofia, S.D.; Dinesh, A.; Sarath Babu, V. Investigation of Waste Marble Powder in the Development of Sustainable Concrete. *Mater. Today Proc.* **2021**, *44*, 4223–4226. [CrossRef]
62. Mir, A.H. Improved Concrete Properties Using Quarry Dust as Replacement for Natural Sand. *Int. J. Eng. Res. Dev.* **2015**, *11*, 46–52.
63. Hameed, M.; Sekar, A. Properties of Green Concrete Containing Quarry Rock Dust and Marble Sludge Powder as Fine Aggregate. *J. Eng. Appl. Sci.* **2009**, *4*, 83–89.
64. Janakiram, N.; Murahari Krishna, P. Partial Replacement of Fine Aggregates with Quarry Dust and Marble Powder in Concrete. *Int. J. Eng. Trends Appl. IJETA* **2018**, *5*, 57–70.
65. Idrees, M.; Faiz, A. Utilization of Waste Quarry Dust and Marble Powder in Concrete. In Proceedings of the Fifth International Conference on Sustainable Construction Materials and Technologies (SCMT5), London, UK, 14–17 July 2019; Volume 1, pp. 302–315.
66. Ghani, A.; Ali, Z.; Khan, F.A.; Shah, S.R.; Khan, S.W.; Rashid, M. Experimental Study on the Behavior of Waste Marble Powder as Partial Replacement of Sand in Concrete. *SN Appl. Sci.* **2020**, *2*, 1554. [CrossRef]
67. Khyaliya, R.K.; Kabeer, K.I.S.A.; Vyas, A.K. Evaluation of Strength and Durability of Lean Mortar Mixes Containing Marble Waste. *Constr. Build. Mater.* **2017**, *147*, 598–607. [CrossRef]
68. Sutcu, M.; Alptekin, H.; Erdoğmuş, E.; Er, Y.; Gencel, O. Characteristics of Fired Clay Bricks with Waste Marble Powder Addition as Building Materials. *Constr. Build. Mater.* **2015**, *82*, 1–8. [CrossRef]
69. Rizwan, S.; Bier, T. Blends of Limestone Powder and Fly-Ash Enhance the Response of Self-Compacting Mortars. *Constr. Build. Mater.* **2011**, *27*, 398–403. [CrossRef]
70. Sua-iam, G.; Makul, N. Use of Limestone Powder during Incorporation of Pb-Containing Cathode Ray Tube Waste in Self-Compacting Concrete. *J. Environ. Manag.* **2013**, *128*, 931–940. [CrossRef]
71. Wang, D.; Shi, C.; Farzadnia, N.; Shi, Z.; Jia, H.; Ou, Z. A Review on Use of Limestone Powder in Cement-Based Materials: Mechanism, Hydration and Microstructures. *Constr. Build. Mater.* **2018**, *181*, 659–672. [CrossRef]
72. Derabla, R.; Benmalek, M.L. Characterization of Heat-Treated Self-Compacting Concrete Containing Mineral Admixtures at Early Age and in the Long Term. *Constr. Build. Mater.* **2014**, *66*, 787–794. [CrossRef]
73. Yahia, A.; Tanimura, M.; Shimoyama, Y. Rheological Properties of Highly Flowable Mortar Containing Limestone Filler-Effect of Powder Content and W/C Ratio. *Cem. Concr. Res.* **2005**, *35*, 532–539. [CrossRef]
74. Kanazawa, K.; Yamada, K.; Sogo, S. Properties of Low-Heat Generating Concrete Containing Large Volumes of Blast-Furnace Slag and Fly Ash. *Int. Concr. Abstr. Portal* **1992**, *132*, 97–118. [CrossRef]
75. Duggal, S.K. *Building Materials*, 4th ed.; New Age International Pvt. Ltd.: New Delhi, India, 2010; ISBN 9788122433791.
76. Corinaldesi, V.; Moriconi, G.; Naik, T.R. Characterization of Marble Powder for Its Use in Mortar and Concrete. *Constr. Build. Mater.* **2010**, *24*, 113–117. [CrossRef]
77. Venkata Sairam Kumar, N.; Ram, K.S.S. Performance of Concrete at Elevated Temperatures Made with Crushed Rock Dust as Filler Material. *Mater. Today Proc.* **2019**, *18*, 2270–2278. [CrossRef]
78. Danish, P.; Mohan Ganesh, G. Study on Influence of Metakaolin and Waste Marble Powder on Self-Compacting Concrete—A State of the Art Review. *Mater. Today Proc.* **2021**, *44*, 1428–1436. [CrossRef]
79. Elyamany, H.E.; Abd Elmoaty, A.E.M.; Mohamed, B. Effect of Filler Types on Physical, Mechanical and Microstructure of Self Compacting Concrete and Flow-Able Concrete. *Alex. Eng. J.* **2014**, *53*, 295–307. [CrossRef]
80. Nguyen, H.-A.; Chang, T.-P.; Shih, J.-Y.; Suryadi Djayaprabha, H. Enhancement of Low-Cement Self-Compacting Concrete with Dolomite Powder. *Constr. Build. Mater.* **2018**, *161*, 539–546. [CrossRef]
81. Paralada, S. Use of Granite Waste as Powder in SCC. *Int. Res. J. Eng. Technol.* **2016**, *3*, 1129–1135.
82. Schankoski, R.A.; Pilar, R.; Prudêncio, L.R.; Ferron, R.D. Evaluation of Fresh Cement Pastes Containing Quarry By-Product Powders. *Constr. Build. Mater.* **2017**, *133*, 234–242. [CrossRef]
83. Kalcheff, I.V. *Portland Cement Concrete with Stone Sand*; National Crushed Stone Association: Washington, DC, USA, 1977; p. 20.
84. Popovics, S. *Concrete Materials. Properties, Specifications and Testing*, 2nd ed.; Elsevier Science: Norwich, UK, 1992; ISBN 9780815516552.
85. Gonzalez, M.; Irassar, E.F. Effect of Limestone Filler on the Sulfate Resistance of Low C3A Portland Cement. *Cem. Concr. Res.* **1998**, *28*, 1655–1667. [CrossRef]

86. Kabeer, K.I.S.A.; Vyas, A.K. Utilization of Marble Powder as Fine Aggregate in Mortar Mixes. *Constr. Build. Mater.* **2018**, *165*, 321–332. [CrossRef]
87. De Weerdt, K.; Justnes, H.; Kjellsen, K.; Sellevold, E. Fly Ash-Limestone Ternary Composite Cements: Synergy Effect at 28 Days. *Nord. Concr. Res.* **2010**, *42*, 51–70.
88. Kwan, A.K.H.; McKinley, M. Effects of Limestone Fines on Water Film Thickness, Paste Film Thickness and Performance of Mortar. *Powder Technol.* **2014**, *261*, 33–41. [CrossRef]
89. Kabeer, K.I.S.A.; Vyas, A.K. Experimental Investigation on Utilization of Dried Marble Slurry as Fine Aggregate in Lean Masonry Mortars. *J. Build. Eng.* **2019**, *23*, 185–192. [CrossRef]
90. Soroka, I.; Stern, N. Calcareous Fillers and the Compressive Strength of Portland Cement. *Cem. Concr. Res.* **1976**, *6*, 367–376. [CrossRef]
91. Celik, T.; Marar, K. Effects of Crushed Stone Dust on Some Properties of Concrete. *Cem. Concr. Res.* **1996**, *26*, 1121–1130. [CrossRef]
92. Omar, O.M.; Abd Elhameed, G.D.; Sherif, M.A.; Mohamadien, H.A. Influence of Limestone Waste as Partial Replacement Material for Sand and Marble Powder in Concrete Properties. *HBRC J.* **2012**, *8*, 193–203. [CrossRef]
93. Ashish, D.K. Feasibility of Waste Marble Powder in Concrete as Partial Substitution of Cement and Sand Amalgam for Sustainable Growth. *J. Build. Eng.* **2018**, *15*, 236–242. [CrossRef]
94. Alyamaç, K.E.; Aydin, A.B. Concrete Properties Containing Fine Aggregate Marble Powder. *KSCE J. Civ. Eng.* **2015**, *19*, 2208–2216. [CrossRef]
95. Bonavetti, V.L.; Irassar, E.F. The Effect of Stone Dust Content in Sand. *Cem. Concr. Res.* **1994**, *24*, 580–590. [CrossRef]
96. Dobiszewska, M.; Barnes, R.W. Properties of Mortar Made with Basalt Powder as Sand Replacement. *ACI Mater. J.* **2020**, *117*, 3–9. [CrossRef]
97. Benabed, B.; Hamza, S.; Eddine, B.A.S.; Lakhdar, A.; Hadj, K.E.; Said, K. Effect of Limestone Powder as a Partial Replacement of Crushed Quarry Sand on Properties of Self-Compacting Repair Mortars. *J. Build. Mater. Struct.* **2016**, *3*, 15–30. [CrossRef]
98. Uchikawa, H.; Hanehara, S.; Hirao, H. Influence of Microstructure on the Physical Properties of Concrete Prepared by Substituting Mineral Powder for Part of Fine Aggregate. *Cem. Concr. Res.* **1996**, *26*, 101–111. [CrossRef]
99. Sua-iam, G.; Makul, N. Utilization of Limestone Powder to Improve the Properties of Self-Compacting Concrete Incorporating High Volumes of Untreated Rice Husk Ash as Fine Aggregate. *Constr. Build. Mater.* **2013**, *38*, 455–464. [CrossRef]
100. Binici, H.; Aksogan, O. Durability of Concrete Made with Natural Granular Granite, Silica Sand and Powders of Waste Marble and Basalt as Fine Aggregate. *J. Build. Eng.* **2018**, *19*, 109–121. [CrossRef]
101. Singhal, V.; Nagar, R.; Agrawal, V. Sustainable Use of Fly Ash and Waste Marble Slurry Powder in Concrete. *Mater. Today Proc.* **2020**, *32*, 975–981. [CrossRef]
102. Ashish, D.K. Concrete Made with Waste Marble Powder and Supplementary Cementitious Material for Sustainable Development. *J. Clean. Prod.* **2019**, *211*, 716–729. [CrossRef]
103. Varadharajan, S. Determination of Mechanical Properties and Environmental Impact Due to Inclusion of Flyash and Marble Waste Powder in Concrete. *Structures* **2020**, *25*, 613–630. [CrossRef]
104. Dobiszewska, M.; Beycioğlu, A. Physical Properties and Microstructure of Concrete with Waste Basalt Powder Addition. *Materials* **2020**, *13*, 3503. [CrossRef]
105. Binici, H.; Aksogan, O.; Görür, E.B.; Kaplan, H.; Bodur, M.N. Performance of Ground Blast Furnace Slag and Ground Basaltic Pumice Concrete against Seawater Attack. *Constr. Build. Mater.* **2008**, *22*, 1515–1526. [CrossRef]
106. Singh, S.; Khan, S.; Khandelwal, R.; Chugh, A.; Nagar, R. Performance of Sustainable Concrete Containing Granite Cutting Waste. *J. Clean. Prod.* **2016**, *119*, 86–98. [CrossRef]
107. Singh, S.; Nagar, R.; Agrawal, V.; Rana, A.; Tiwari, A. Sustainable Utilization of Granite Cutting Waste in High Strength Concrete. *J. Clean. Prod.* **2016**, *116*, 223–235. [CrossRef]
108. Joel, M. Use of Crushed Granite Fine as Replacement to River Sand in Concrete Production. *Leonardo Electron. J. Pract. Technol.* **2010**, *17*, 85–96.
109. Felixkala, T.; Partheeban, P. Granite Powder Concrete. *Indian J. Sci. Technol.* **2010**, *3*, 311–317. [CrossRef]
110. Raghavendra, R.; Sharada, S.A.; Ravindra, M.V. Compressive Strength of High Performance Concrete Using Granite Powder as Fine Aggregate. *Int. J. Res. Eng. Technol.* **2015**, *04*, 47–49. [CrossRef]
111. Turk, K.; Nehdi, M.L. Coupled Effects of Limestone Powder and High-Volume Fly Ash on Mechanical Properties of ECC. *Constr. Build. Mater.* **2018**, *164*, 185–192. [CrossRef]
112. Felekoglu, B. Utilisation of High Volumes of Limestone Quarry Wastes in Concrete Industry (Self-Compacting Concrete Case). *Resour. Conserv. Recycl.* **2007**, *51*, 770–791. [CrossRef]
113. Singh, M.; Srivastava, A.; Bhunia, D. An Investigation on Effect of Partial Replacement of Cement by Waste Marble Slurry. *Constr. Build. Mater.* **2017**, *134*, 471–488. [CrossRef]
114. Galan, I.; Briendl, L.; Thumann, M.; Steindl, F.; Röck, R.; Kusterle, W.; Mittermayr, F. Filler Effect in Shotcrete. *Materials* **2019**, *12*, 3221. [CrossRef]
115. Kępniak, M.; Woyciechowski, P.; Franus, W. Transition Zone Enhancement with Waste Limestone Powder as a Reason for Concrete Compressive Strength Increase. *Materials* **2021**, *14*, 7254. [CrossRef]
116. Lawrence, P.; Cyr, M.; Ringot, E. Mineral Admixtures in Mortars Effect of Type, Amount and Fineness of Fine Constituents on Compressive Strength. *Cem. Concr. Res.* **2005**, *35*, 1092–1105. [CrossRef]

117. Knop, Y.; Peled, A.; Cohen, R. Influences of Limestone Particle Size Distributions and Contents on Blended Cement Properties. *Constr. Build. Mater.* **2014**, *71*, 26–34. [CrossRef]
118. Kurdowski, W. *Cement and Concrete Chemistry*; Springer Netherlands: Dordrecht, The Netherlands, 2014; ISBN 9789400779440.
119. Roy, D.M.; Scheetz, B.E.; Silsbee, M.R. Processing of Optimized Cements and Concretes Via Particle Packing. *MRS Bull.* **1993**, *18*, 45–49. [CrossRef]
120. Brandt, A.M. *Cement Based Composites: Materials, Mechanical Properties and Performance*; Taylor & Francis Group: London, UK; New York, NY, USA, 2009.
121. Saraya, M.E.-S.I. Study Physico-Chemical Properties of Blended Cements Containing Fixed Amount of Silica Fume, Blast Furnace Slag, Basalt and Limestone, a Comparative Study. *Constr. Build. Mater.* **2014**, *72*, 104–112. [CrossRef]
122. Heikal, M.; El-Didamony, H.; Morsy, M.S. Limestone-Filled Pozzolanic Cement. *Cem. Concr. Res.* **2000**, *30*, 1827–1834. [CrossRef]
123. El-Didamony, H.; Salem, T.; Gabr, N.; Mohamed, T. Limestone as a Retarder and Filler in Limestone Blended Cement. *Ceram. -Silik.* **1995**, *39*, 15–19.
124. Liu, S.; Yan, P. Effect of Limestone Powder on Microstructure of Concrete. *Mater. Sci. Ed.* **2010**, *25*, 328–333. [CrossRef]
125. Bonavetti, V.; Donza, H.; Menéndez, G.; Cabrera, O.; Irassar, E.F. Limestone Filler Cement in Low w/c Concrete: A Rational Use of Energy. *Cem. Concr. Res.* **2003**, *33*, 865–871. [CrossRef]
126. Menéndez, G.; Bonavetti, V.; Irassar, E.F. Strength Development of Ternary Blended Cement with Limestone Filler and Blast-Furnace Slag. *Cem. Concr. Compos.* **2003**, *25*, 61–67. [CrossRef]
127. Benabed, B.; Kadri, E.-H.; Azzouz, L.; Kenai, S. Properties of Self-Compacting Mortar Made with Various Types of Sand. *Cem. Concr. Compos.* **2012**, *34*, 1167–1173. [CrossRef]
128. Kou, S.-C.; Poon, C.-S. Properties of Concrete Prepared with Crushed Fine Stone, Furnace Bottom Ash and Fine Recycled Aggregate as Fine Aggregates. *Constr. Build. Mater.* **2009**, *8*, 2877–2886. [CrossRef]
129. Hebhoub, H.; Aoun, H.; Belachia, M.; Houari, H.; Ghorbel, E. Use of Waste Marble Aggregates in Concrete. *Constr. Build. Mater.* **2011**, *25*, 1167–1171. [CrossRef]
130. Sakalkale, A.D.; Dhawale, G.D.; Kedar, R. Experimental Study on Use of Waste Marble Dust in Concrete. *Int. J. Eng. Res. Appl.* **2014**, *4*, 44–50.
131. Rai, B.; Naushad, K.; Kr, A.; Rushad, T. Assistant Influence of Marble Powder/Granules in Concrete Mix. *Int. J. Civ. Struct. Eng.* **2011**, *1*, 827–834.
132. Choudhary, R.; Gupta, R.; Nagar, R. Impact on Fresh, Mechanical, and Microstructural Properties of High Strength Self-Compacting Concrete by Marble Cutting Slurry Waste, Fly Ash, and Silica Fume. *Constr. Build. Mater.* **2019**, *239*, 117888. [CrossRef]
133. Struble, L.; Skalny, J.; Mindess, S. A Review of the Cement-Aggregate Bond. *Cem. Concr. Res.* **1980**, *10*, 277–286. [CrossRef]
134. Zhang, S.; Cao, K.; Wang, C.; Wang, X.; Wang, J.; Sun, B. Effect of Silica Fume and Waste Marble Powder on the Mechanical and Durability Properties of Cellular Concrete. *Constr. Build. Mater.* **2020**, *241*, 117980. [CrossRef]
135. Naaman, A.E.; Reinhardt, H.W. Proposed Classification of HPFRC Composites Based on Their Tensile Response. *Mater. Struct.* **2006**, *39*, 547–555. [CrossRef]
136. Singhal, V.; Nagar, R.; Agrawal, V. Use of Marble Slurry Powder and Fly Ash to Obtain Sustainable Concrete. *Mater. Today Proc.* **2021**, *44*, 4387–4392. [CrossRef]
137. Hameed, M.S.; Sekar, A.S.S.; Balamurugan, L.; Saraswathy, V. Self-Compacting Concrete Using Marble Sludge Powder and Crushed Rock Dust. *KSCE J. Civ. Eng.* **2012**, *16*, 980–988. [CrossRef]
138. Kępniak, M.; Woyciechowski, P.; Łukowski, P.; Kuziak, J.; Kobyłka, R. The Durability of Concrete Modified by Waste Limestone Powder in the Chemically Aggressive Environment. *Materials* **2019**, *12*, 1693. [CrossRef]
139. Ulubeyli, G.C.; Bilir, T.; Artir, R. Durability Properties of Concrete Produced by Marble Waste as Aggregate or Mineral Additives. *Procedia Eng.* **2016**, *161*, 543–548. [CrossRef]
140. Topçu, İ.B.; Bilir, T.; Uygunoğlu, T. Effect of Waste Marble Dust Content as Filler on Properties of Self-Compacting Concrete. *Constr. Build. Mater.* **2009**, *23*, 1947–1953. [CrossRef]
141. Tasdemir, C. Combined Effects of Mineral Admixtures and Curing Conditions on the Sorptivity Coefficient of Concrete. *Cem. Concr. Res.* **2003**, *33*, 1637–1642. [CrossRef]
142. Ilangovana, R.; Mahendrana, N.; Nagamanib, K. Strength and Durability Properties of Concrete Containing Quarry Rock Dust as Fine Aggregate. *ARPN J. Eng. Appl. Sci.* **2008**, *3*, 20–26.
143. Alli, O.O.; Alli, J.A.; Odewumi, T.O.; Yussuff, O.N. Strength and Durability Properties of Concrete Containing Quarry Rock Dust as Fine Aggregate. *Int. J. Sci. Res. IJSR* **2016**, *7*, 418–421.
144. Nilsson, L.-O. Durability Concept; Pore Structure and Transport Processes. In *Advanced Concrete Technology*; Newman, J., Choo, B.S., Eds.; Butterworth-Heinemann: Oxford, UK, 2003; Volume 1, pp. 3–29, ISBN 9780750656863.
145. Li, B.; Wang, J.; Zhou, M. Effect of limestone fines content in manufactured sand on durability of low- and high-strength concretes. *Constr. Build. Mater.* **2009**, *23*, 2846–2850. [CrossRef]

MDPI
St. Alban-Anlage 66
4052 Basel
Switzerland
www.mdpi.com

Materials Editorial Office
E-mail: materials@mdpi.com
www.mdpi.com/journal/materials

Disclaimer/Publisher's Note: The statements, opinions and data contained in all publications are solely those of the individual author(s) and contributor(s) and not of MDPI and/or the editor(s). MDPI and/or the editor(s) disclaim responsibility for any injury to people or property resulting from any ideas, methods, instructions or products referred to in the content.